HISTORICAL STUDIES
IN THE PHYSICAL SCIENCES
1

Historical Studies
in the
Physical Sciences

Russell McCormmach, *Editor*

Volume

I

UNIVERSITY OF PENNSYLVANIA PRESS · PHILADELPHIA · 1969

NOTICE TO CONTRIBUTORS

Historical Studies in the Physical Sciences, to be issued annually by the University of Pennsylvania Press, is devoted to articles in the history of the physical sciences from the eighteenth century on. Consideration is given both to the internal development of the physical sciences and to the relation of these sciences to intellectual, cultural, and social themes. Historiographic articles and review articles on the current state of scholarship are welcome.

To facilitate the refereeing process, all manuscripts should be accompanied by an additional carbon– or photocopy. Manuscripts must be typewritten and double-spaced; wide margins should be allowed. Since modern topics frequently demand length, no limitations have been set on the number of manuscript pages per article. Articles may include illustrations; these may be either glossy prints or directly reproducible line drawings. Articles may be submitted in foreign languages; if accepted, they will be published in English translation. Footnotes, also to be typed double-spaced, are to be numbered sequentially and collected at the end of the article. Contributors are referred to the *MLA Style Sheet* for detailed instructions on documentation and other stylistic matters.

An author receives twenty-five free reprints of his article. Additional reprints may be ordered at his expense.

Historical Studies in the Physical Sciences incorporates *Chymia,* the history of chemistry annual.

SBN:8122–7600–0

Printed in the United States of America

Contents

NIELS BOHR AROUND 1920

The publishers are grateful to the Bohr family for permission to reproduce this photograph and to Professor L. Rosenfeld for supplying the print.

Editor's Foreword

Historical Studies in the Physical Sciences is an annual journal devoted to the history of the physical sciences in the post-Scientific Revolution period. The rapidly growing interest in the history of the modern physical sciences and the increasing sophistication of writing in the field insures that a specialist journal of this nature can aspire to wide readership and high scholarly standards. I am convinced that specialization by period promises to be a more fruitful approach for a history of science journal than specialization by individual scientific discipline. There are facets of scientific activity that can only be understood when related sciences are treated as a group of interacting disciplines rather than as isolated specialties. This journal will encourage discipline-oriented studies of the individual physical sciences, of course. But I hope that it will also stimulate the study of the seldom-explored interdependence of the several physical sciences, and of their relation to the life sciences too. I hope as well that it will promote the study of the interdependence of the physical-science group with other, non-scientific aspects of modern civilization. I have focused on the modern period, since it holds especially challenging and timely problems, problems that so far have been little researched.

Though there is vigorous activity in the history of scientific thought, the outstanding work remains largely centered on the period prior to the eighteenth century. I hope that this journal will encourage the application of the techniques of intellectual history to problems of the modern period; one would like to understand the thought of Niels Bohr as well as that of Galileo, of Linus Pauling as well as that of Robert Boyle. Having said this, I want to stress that this journal will by no means be exclusively concerned with the physical scientists' intellectual heritage and equipment. It will be concerned as well with the context— cultural, political, moral, socio-economic, technological— of the physical sciences, a concern that is not yet strongly reflected in publications in the history of science. It is one of the purposes of this journal to stimulate the study of the social function of the physical sciences and the professional role of their practitioners. I will especially welcome studies of the institutions and organizations of modern physics and chemistry. The methods appropriate to the investigation of new problems may be drawn in part from philosophy and the social sciences, and this journal will accordingly be concerned with innovative methods of historical research.

I would like this journal to do more than reflect the usual genres of writing in the history of science. My goal is not merely to publish in one volume studies dealing with scientific ideas and others dealing with the context of science. The vision of the history of science that relegates the historiographic traditions of internal and external history to mutually exclusive roles is sterile, obstructing the synthesis of the intellectual and social history of science that must come. I will be strongly sympathetic to studies that implicitly challenge the cogency of the commonly held view of the internalist-externalist distinction. I would like, for example, to encourage articles exposing the intimate relation of institutions and ideas.

There are other highly useful types of articles that historians of science tend to overlook. I would like to see more good articles on historiographic matters and more articles that probe the current state of scholarship in well-defined topics in the history of the physical sciences. In this connection, I shall welcome review essays that are concerned with re-assessments of major issues and historiographic alignments. In general I would like to see articles that broaden the scope of the history of science, establishing needed connections with other branches of history.

I do not intend to publish articles that fall into categories that are no longer viable, e.g., heroic biography, the history of chains of discovery, the identification of precursors. It should be unnecessary to declare policy on such long-dead issues, but there is a staggering number of methodologically bad articles still making the rounds. As another point of policy, I do not plan to publish notes, short book reviews, antiquarian finds, or essays on slight themes, even though they may be narrowly illuminating and otherwise publishable. I want primarily to elicit substantial articles that illuminate major issues; an annual book is a fitting place for the publication of writings of this nature. As complex analyses demand a good deal of space, the articles of this series will frequently be lengthly. The book format of an annual journal offers advantages in this regard.

When possible I wish to bring together articles under one cover that bear thematic unity. For this and other reasons connected with planning future issues, I warmly invite authors to correspond about ideas for articles.

I am much concerned with the literary quality of the articles I publish. I urge authors to submit writing that is clear, direct, and free from jargon and pedantry. The literary models for the history of science should be drawn from historical, not scientific, practice.

I have indicated to authors and readers some of the perspectives I would like this journal to be associated with. I have undertaken this editorship in the belief that to promote and communicate an historical understanding of the modern physical sciences is an important task, and I am confident that this publication will contribute significantly to that understanding.

Sir John Herschel's Philosophy of Success

BY JOSEPH AGASSI*

This is an extended critical book-report on Sir John Herschel's *Preliminary Discourse on the Study of Natural Philosophy* (1831).[1] The book is not such a masterpiece as to deserve detailed study, but it contains one important discovery and a few interesting passages which I shall comment on. It expounds a popular philosophy of success which will become amply clear from the present summary of its presentation and ideas and which will be discussed in the concluding section. It expresses the atmosphere of the time in which it was written, as I shall endeavor to illustrate chiefly in my introductory and concluding sections. It influenced the literature considerably, if for no other reason than that the writings of both William Whewell and John Stuart Mill follow in its wake. Its very conception as an updated version of Bacon's *Novum Organum* is a forerunner to Whewell's *Novum Organum Renovatum*. But I shall not discuss Herschel's influence on posterity here, since this is a topic for a separate essay.

Herschel's *Preliminary Discourse* echoes, of course, d'Alembert's *Preliminary Discourse to the Encyclopedia;* indeed, Herschel's work, too, is a preliminary—to the Cabinet Encyclopedia. However, that encyclopedia falls far short, even in ambition, of its French predecessor, this quite in contrast with the fact that its preliminary (as we shall see)

* Department of Philosophy, Boston University, Boston, Mass. 02215.

[1] Sir John F. W. Herschel, *Preliminary Discourse, etc.* (London and Philadelphia, 1831; title page to English edition says 1830, but the page with portrait contains correct date). Facsimile by Johnson Reprint Corporation, with a new introduction by M. Partridge (London, 1967). Page numbers refer to the American edition. For the English edition page numbers read thus: 6→9, 12→15, 39→52, 60→79, 61→80, 63→83, 64→84, 73→96, 78→104, 79→105, 85→113, 108→149, 141→188, 143→190, 145→194, 147→196, 149→198, 153→204, 156→208, 162→215, 188→250, 190→253.

I

was very much more ambitious and pompous. Whereas d'Alembert had his encyclopedia as his main reason for writing his *Preliminary Discourse*, for Herschel the encyclopedia was only an excuse for writing his. Herschel's main reason for writing it lay elsewhere.

Herschel's work was much more successful than d'Alembert's— partly from the fact that d'Alembert's preliminary stood in the shadow of his monumental *Encyclopedia*, and partly because d'Alembert merely reaffirmed an accepted philosophy, whereas Herschel reaffirmed a philosophy whose foundations were shaking and whose tenets were thrown into serious doubt and confusion. Not only did philosophers like Whewell and Mill take it very seriously, but an original scientist like Faraday could write Herschel a fan letter telling him that it had encouraged him to pursue his scientific work. No doubt Faraday was a good friend, but no one even superficially familiar with him can doubt the sincerity of his gratitude.

It is therefore particularly intriguing to find out the core problem which Herschel and his contemporaries were facing. The choice of *prima-facie* hypotheses is rather obvious. In Herschel's day the great events were: (a) the rise of electrochemistry, Davy's overthrow of Lavoisier's doctrines, and Dalton's atomism; (b) the rise of electromagnetism; (c) the overthrow of Newton's optics by Young and Fresnel. Now chemistry did not particularly engage Herschel. Since his book is rather general, he refers in it only to atomism, and rather as to a success than as to a problem. Electromagnetism was a serious problem, but conviction was growing that Ampère's electrodynamic theory was an adequate and satisfactory solution. Herschel's book was written under the conviction that Ampère's success was unshakeable; this conviction, incidentally, was shaken very soon after the publication—by Faraday's discovery of magnetoelectricity of 1831 (and it took the scientific world many years to recognize this fact). This leaves us with the revolution in optics. I shall argue throughout this essay that the optical revolution proved Newton to have been not infallible and thus threw serious doubt on the widespread metaphysical belief in the existence of a foolproof scientific method. Herschel, as Whewell after him, tried to show that there exists such a method and that therefore there is no ground for the fear that Newton's mechanics will ever be superseded as Newton's optics had been. Herschel's book, in brief, comes to reinforce faith in science, in induction, and in Newtonian mechanics.

N.B. In all ensuing quotations, italics are mine, except in the obvious cases where they were required in the original.

I. THE PUBLIC SITUATION

Sir John Frederick William Herschel was the son of a poor Hanoverian musician, William Herschel, who settled in England as a court organist to King George III, and who became famous for his astronomical discoveries and cosmological speculations (not to mention his discovery of radiant heat). Sir John himself was a famous scientist and a fellow of a college in Cambridge where he had studied. He later became President of the Royal Society. During his first years in Cambridge he was one of the three heads of the undergraduate revolutionary club which ventured to overthrow Newton's symbolism and replace it with Leibnitz'.

This dispute concerned not merely symbolism but an idea too. Newton introduced calculus as a branch of geometry. Now today we know that even Lagrange's analytical mechanics can have a geometrical representation, but in the early nineteenth century this was not known, and Lagrange's novelties could not penetrate the wall of British conservatism. It may sound strange that a club of undergraduates can make such an easy and successful revolution. The need for the *new* mathematics was very strongly felt; in *Thomson's Annals* (1815), for instance, there was a long discussion over the British backwardness in mathematics, but no one before Babbage, Herschel, and Peacock dared to create the precedent.

The most revolutionary of these three was Herschel. Babbage is now remembered chiefly for his mathematical calculation-machine, and Peacock for his life of Thomas Young. Herschel is known as the one who introduced the wave theory of light (of Young and Fresnel) into England. The new wave theory of light had already been advocated in Britain by Brewster since 1814. In 1819 Herschel followed him, being the first who brought respectability to it in England.

One might have expected Herschel to be above the vulgar identification of all error with prejudice, as Brewster was; but this was not the case. On the contrary, Herschel, the revolutionary son of the great speculative thinker, tried to return to the old view that all errors are culpable and hence that all speculation is dangerous and to be avoided.

Nowadays, one can hardly imagine what a shock the refutation of

3

Newton's theory of light produced. Newton, the infallible, was found to be mistaken, and hence, allegedly, prejudiced; his theory of light, which was much more widely known than his theory of gravity, and which had been viewed in 1810 as doubly demonstrated beyond any shadow of doubt, was entirely deserted in 1820!

I cannot discuss all the repercussions of this shock wave. But it surely was a unique chance to dislodge the old Baconian myth and reject once and for all the identification of all error with prejudice. I quote only one passage from Captain Forman RN, who was neither a Newtonian nor a Huyghenian. Let me contrast the calm and far-seeing character of this passage from this elderly gentleman with the reaction of a young contemporary of his, Michael Faraday, a reaction dominated by a sense of disaster and by the urge to avoid wild speculations. Let me first quote the remarkable introduction to Forman's paper:

> Truth is the only legitimate object of philosophical research; and whoever believes his own opinion to be true, and fancies that he can add to the general stock of knowledge by imparting some new discovery, *has not only the right but is bound in duty to make his opinion known.* If among a number of erroneous opinions, he has afforded but one hint which in the hands of a wiser man may lead to important results, he has conferred a real benefit on society; while *his errors,* though they may outlive his own time, *will finally be dispelled* by the light of true philosophy, to which his own hint has so materially contributed.
>
> My design in writing this essay is to show the true cause of reflection and refraction of light; and as *Sir Isaac Newton* has already accounted for these phenomena by his hypothesis of alternate fits of easy refraction and reflection in the medium, it follows of course that *I intend to oppose his opinion. . . .* For this opposition however I shall offer no apology, because I am only excercising the right which he excercised before me. . . .
>
> For a great many ages the history of natural philosophy was little more than a record of errors. . . . Here and there we meet with a transient gleam of philosophy . . . which serves a traveller a few steps further on his way, and then left him to grope in the dark, perhaps for another generation. . . . *It is chiefly to Lord Bacon that we are indebted for the principle of making experiments the basis of philosophy. . . .* Before his time the *ipse dixit* of an acknowledged philosopher was sufficient to establish an opinion however absurd: but with this test, like another Hercules, he has cleansed the Augean stable of all its impurities, *and no hypothesis now can long hold its ground whose foundation is not laid*

on experiment.... There can now be no danger of my propagating error, even if my opinions should be wrong, because, in that case, they will not stand the test of experiment.[2]

As this passage indicates, the refutation of Newton's theory of light provided an opportunity for passing to a more explicit mode of arguing. Forman's arguments in favor of his proposal to endorse the explicit argumentative method of presentation are very strong. In his view scientific progress is slower than was generally believed; progress is made by presenting hints and rudimentary theories, and errors need not be feared since, being refutable, they will sooner or later be eliminated.

But Forman missed his target, and, I think, for an obvious reason. He refused to declare boldly that Newton's theory of light had been refuted. And he expresses his indebtedness to Bacon in the very passage in which he only implicitly rejects and severely criticizes Bacon's philosophy. Although he declared that refutations are not to be feared, but should, on the contrary, be considered as the means of improvement, he was not ready to declare both Newton's theory of light and Bacon's doctrine of prejudice as simply refuted errors.

That the idea is hard to concede is not hard to imagine, though documentation may not be easily available. There is a remarkable record of a shocked reaction in Faraday's lecture relating to the nature of light, in a series of lectures to the City Philosophical Society, delivered in 1819 (when he was twenty-seven years old), as quoted in his *Life and Letters* by Bence-Jones:

> In the constant investigation of nature pursued by curious and inquisitive man, some causes which retard his progress in no mean degree arise from the habits incurred by his exertions; and it not unfrequently happens, that the man who is the most successful in his pursuit of one branch of philosophy thereby raises up difficulties to his advancement in another....
>
> The evil of method in philosophical pursuits is indeed only apparent, and has no real existence but in the abuse. But the system-maker is unwilling to believe that his explanations are not perfect, the theorist to allow that incertitude hovers about him. Each condemns what does not agree with *his method,* and consequently each departs

[2] *Phil. Mag., 55* (1820), 417 ff. Forman, thus, is the discoverer of the chief error of Baconian philosophy, its fear of error. Note also that the reference to Bacon is echoed in Herschel's often quoted eulogy on Bacon, quoted here at the opening of Section 5.

from nature. And unfortunately, though no one can conceive why another should presume to bound the universe and its laws by his wild and fantastic imaginations, yet each has a reason for retaining and cherishing his own. . . .

As it regards natural philosophy, these bad, but more or less inevitable, effects are perhaps best opposed by cautious but frequent generalisations. . . .

Ultimately, however, facts are the only things which we are *sure* are worthy of trust. All our theories and explanations of the laws which govern them, whether particular or general, are necessarily deduced from insufficient data. They are probably most correct when they agree with the greatest number of phenomena, and when they do not appear incompatible with each other. The test of an opinion is its agreement in association with others, and we associate most when we generalise.

Hence I should recommend the practice of generalising as a sort of parsing in philosophy. It occasions a review of single opinions, requires a distinct impression of each, and ascertains their connection and government. . . .

Matter classed into four states—solid, liquid, gaseous, and radiant—which depend upon differences in the essential properties.

. . . . . .

Radiant state.—Purely hypothetical   Distinctions.

Reasons for belief in its existence. Experimental evidence. Kinds of radiant matter admitted. . . .

Nothing is more difficult and requires more care than philosophical deduction, nor is there anything more adverse to its accuracy than fixidity of opinion. The man who is certain he is right is almost sure to be wrong, and he has the additional misfortune of inevitably remaining so. All our theories are fixed upon uncertain data, and all of them want alteration and support. Ever since the world began, opinion has changed with the progress of things; and it is something more than absurd to suppose that we have a sure claim to perfection, or that we are in possession of the highest stretch of intellect which has or can result from human thought. Why our successors should not displace us in our opinions, as well as in our persons, it is difficult to say; it ever has been so, and from analogy would be supposed to continue so; and yet, with all this practical evidence of the fallibility of our opinions, all, and none more than philosophers, are ready to assert the real truth of their opinions. . . .

The history of the opinions on the general nature of matter would afford remarkable illustrations in support of what I have said, but it does not belong to my subject to *extend upon it*. All I wish to point out is, by a reference to light, heat, electricity, &c., and the opinions formed on them, the necessity of cautious and slow decision on philosophical points, the care with which evidence ought to be admitted, and the continual guard against philosophical prejudices which should be preserved in the mind. The man who wishes to advance in knowledge should never of himself fix obstacles in the way.[3]

There is a general pattern which the responses of Forman and Faraday illustrate. When a radical measure proves to be ineffective, one may become less of a radicalist and more of a reformist; alternatively one may become more radicalist. I have argued elsewhere[4] that this general pattern is illustrated both in social and political history and in the history of ideas. Here, however, I must add that each of these two responses is intellectual in its own way; and one need not be an intellectual. Instead one can respond with a mixture of a cock-sure manner, pretending that nothing has ever gone wrong as far as the principles are concerned, and a shocked manner as far as the deterioration of people's application of them is exposed.

One of those cock-sure responses to the crisis in optics can be found in the works of Sir John Herschel, who introduced this low standard of discussion even to the *Philosophical Transactions of the Royal Society* (though, it is true, only in a footnote):

> The author of the article on Polarization, in the 63rd Number of the Edinburgh Review, just published [1919], *is guilty of a most unpardonable mistake*, in asserting, (p. 188), as deducible from Dr. Brewster's experiments, that the Huygenian law is *incorrect*, for carbonate of lime. Dr. Brewster's general formulae for crystals with two axes resolve themselves into the Huygenian law when the axes coincide, of which case it is only an extention. That excellent philosopher, if I understand English, in the paragraph which gave rise to this strange assertion, only means to declare his opinion that it remains undemonstrated.[5]

---

[3] H. Bence-Jones, *The Life and Letters of Faraday,* in two volumes, *1* (London, 1870), 303–311, extracts from a lecture "On the Forms of Matter," from which the above is an extraction. The full lecture is extant in the Royal Institution. I hope someone will soon publish Faraday's early works.
[4] See my "Methodological Individualism," *Brit. J. Sociology, 11* (1960), and my *Towards an Historiography of Science, History and Theory,* Beiheft 2 (The Hague, 1963; facsimile reprint, Wesleyan Univ. Press, 1967). See also J. W. N. Watkins, "Epistemology and Politics," *Proc. Arist. Soc.* (London, 1957).
[5] *Phil. Trans.* (1820), 45 n.

For the sake of completeness, or for the curious, the passage which enraged him is: "We shall show, from the experiments of Dr. Brewster, that the Hughenian law *is not general;* that it *is not even correct* for the phenomena of calcareous spar [= carbonate of lime]; and that the explanation which Laplace attempted . . . falls to the ground." Both Herschel and the writer whom Herschel declares unpardonably mistaken agree that Huygens' law explains certain phenomena of double refraction but *not* all of them (as it was originally intended), and that Brewster generalized it into a law which explains more phenomena. The author, then, is blamed for an "unpardonable mistake" even though he supports the right views by the right arguments. The unpardonable mistake is purely stylistic—the right argument for the right theory was put in the wrong language, not in the usual cliches. What exactly these cliches were, how exactly one had to apply them, and what one had to do when they were inapplicable or applicable but not to one's taste—these were difficult and vexing questions. Consequently, sensitive people found it ever harder to find the right words before submitting their works for publication. (Even in our own relatively very permissive age the difficulties still exist and often cause much tension between author and editor.) Postponing publication, however, could incur a different kind of censure—the impatient demand for publication (publication pressure). Let me quote Herschel again, this time a footnote published in 1845: "This memoir [of Fresnel] was read in the Institute, Oct. 7, 1816. A supplement was received Jan. 19, 1818. M. Arago's report on it was read June 4, 1821. And while every optical philosopher in Europe has been *impatiently* expecting its appearance for seven years, it lies as yet unpublished, and is known to us only by meagre notices in a periodical journal."[6]

I do not wonder that people preferred to communicate their views verbally rather than in writing. The method of writing was standardized in the inductive style which was out of date. In England two people were in the authoritative position to suggest changes of standards, the two friends Herschel and Faraday. I have discussed Faraday's activities elsewhere.[7] Herschel was a conservative *par excellence;*

---

[6] *Encyclopedia Metropolitana* (London, 1845), *4,* 533. As to Fresnel's sensitivity, see, for example, his correspondence with Young concerning priority in Peacock's life of Young.

[7] See my "An Unpublished Paper by the Young Faraday," *Isis, 52* (1961); also "The Confusion between Physics and Metaphysics in Standard Histories of Science," in *Ithaca, 1962* (Paris, 1964); also *Towards an Historiography of Science, op. cit.* (note 4), Section 5 and notes; also my *Faraday as a Natural Philosopher,* forthcoming.

he demanded strict adherence to the inductive style of presentation.

In another essay[8] I have discussed the standards of publication of scientific essays as proposed by Boyle, instituted by the Royal Society, and rubber-stamped by Newton. With the refutation of Newton's optics all became fluid. Doubts about Newton, about science, and about modes of research and modes of publication all found their way to public attention. Herschel tried to handle them and restore an order of sorts in the scientific community. He did so, I propose, by publishing his views in a definitive essay about science in general.

2. A GENERAL VIEW OF SCIENCE

Herschel's *Preliminary Discourse* is composed of three parts. The first is on the nature and advantages of physics; the second is on prejudices and induction, illustrated by historical examples; the third is about the classification of physics, illustrated by further historical examples. Here is a brief summary of the work, starting with part 1, "The Nature and Advantages of Physical Science."

In the struggle for survival man has proved the fittest. This is due to his rationality. Man is driven to learning both by utilitarian motives and because of intellectual needs. Herschel's story of man's development of his intellectual powers sounds to modern readers a bit naive: it seems that God created man as a nineteenth-century thinker, but without nineteenth-century knowledge. I do not know how historical Herschel considered this picture; its basis is surely in the classical theory of the intellect, whose *locus classicus* is in Buffon's work and whose origins, as Paul Hazard has observed, are traced to Bacon's philosophy. Yet, it may strike one as odd that Herschel is more of a Darwinian-before-Darwin than the eminent naturalist Buffon; this may be related to the fact that the struggle for survival first occurred in the economic literature, especially in Malthus, to whom, of course, Darwin acknowledges his great debt.

Having established the value of science, Herschel now launches an attack on those who oppose science as anti-religious and on those who support science from purely utilitarian considerations. Science is opposed to the religion of Galileo's persecutors, not to religion as such. Science can be of great utility, but to say that it is merely useful is

---

[8] "Who discovered Boyle's Law?," forthcoming; also my doctoral dissertation, *The Function of Interpretations in Physics* (University of London, 1956).

humiliating to the rationalist and the speculative philosopher. Besides, the utility value of a theory cannot be predicted. The theory is remote from everyday life at first (though it will sooner or later be connected with simple facts no less than with uncommon phenomena).

In Chapter II Herschel tells us that science is composed of abstract and empirical components. The abstract component seems, at least casually, to be nothing less than a Kantian *apriori* scheme of language, logic, and mathematics—the scheme of *possible* science. Man is rational, and he would not accept the authority of science unless he understood what is behind scientists' seemingly mad assertions. Only when one understands science, one sees the certainty of these assertions. The certainty is achieved by verification, by the occurrence of the results predicted by the theory, whether it be Newton's theory of gravity or Fresnel's wave theory of light.

This, however, was *apropos* of *abstract* science, the *a priori* framework. Next, in Chapter III, comes the law of causality, which, Herschel says, is "the first thing [which] impresses us from our earliest infancy." This statement seems puzzling not so much because one may doubt it as because Herschel does not need it at all: he deliberately admits the law of causality to be *a priori* valid, and so he need not claim it to be provable by observations, from infancy or otherwise. The law, or any specific causal law, is not written by God in a way in which human legislators write laws; it is not isolated but a part of a system, a part of the whole truth about matter. Thus God comes *ex machina* to help Herschel in his traditional Cartesian role of the defender of scientific truth.

Atomism, for example, incorporates, in its modern form, various laws. But "the ascent to the origin of things," the finding of such laws, must not be through the act of speculating about the nature of things, but the empirical discovery of these particular laws (p. 29). Herschel has by now forgotten that he had previously (p. 12) spoken favorably of the philosopher as the person who speculates on everything. He now claims that the business of the natural philosopher is decidedly not to speculate but humbly to search for particular causes. The ancients, who lived in the infancy of mankind, had good reasons to study the problem of whether or not there exist fixed laws of the universe; "but to us, who have the experience of some additional thousands of years, the question of permanence is already, in great measure, decided in the affirmative. The refined speculations of modern astronomy . . . have proved to demonstration, that . . . the force of gravitation . . . has undergone no change in intensity from a high antiquity" (p. 30).

This dual allusion to Laplace's inductive investigations and to his own father's speculations makes it very hard to decide whether to Herschel causality is *a priori* or *inductive,* and whether speculations are permissible or forbidden. Be it as it may, Herschel expresses explicitly the view that *prior* to science there exists a twofold framework. The one side of it is "abstract science" and it is composed of language, logic, and mathematics. Herschel's abstract science corresponds roughly to Kant's *analytic a priori* knowledge. The other part, quite different from abstract science, includes causality, the immutability of natural laws, and the indestructibility of matter. This corresponds to a very substantial part of Kant's *synthetic a priori,* provided that the seeming inconsistencies I have alluded to be removed in one way or another. (Even if causality and such were considered *a posteriori* in the *epistemological* sense, no interpreter would deny that for Herschel they have *a priori* status, at least from the *methodological* point of view: they are prerequisites for induction. Indeed, this very distinction may offer the clue to the resolution of the seeming inconsistencies.) I do not wish, however, to go further into this topic, or even to suggest that Herschel, who had but little epistemological interest, had ever read Kant, whose works after all, were rather unknown in England before the efforts of De Quincy to popularize them. On the whole, Herschel's view is much too empiricist to be considered even mildly Kantian; yet *a priorist* arguments are contained in it all the same. This seems to me to represent a general trend which historians of philosophy unjustly overlook, and which I have discussed at length elsewhere.[9] Let me only add here that Herschel studied Dr. Isaac Watts's once very popular *Logic* (1724), where the distinction is drawn between essences which are grasped *a priori* (Platonic ideas) and ones which are arrived at by induction (Aristotelian essences proper).

From the *a priori* we come to the *a posteriori;* Part II deals with the empirical aspects of science.

3. THE DOCTRINE OF PREJUDICE

Herschel is one of the last exponents of the orthodox Baconian doctrine of prejudice. In line with Bacon, Watts, and many other

---

[9] See my "Unity and Diversity in Science" in R. S. Cohen and M. Wartofsky, eds., *Boston Studies in the Philosophy of Science, 4* (New York, 1969).

thinkers, he sees the doctrine of induction and the doctrine of prejudice as merely two sides of the same coin. After Whewell's defense of the role of the imagination in science, a change took place in the inductive tradition. But for Herschel, still, the theory of induction is the same as, or the contrapositive of, the theory of prejudice. The ultimate source of knowledge, he says, is experience: observation and experiments. From this, he concludes, the doctrine of prejudice follows:

> Experience once recognized as the fountain of all our knowledge of nature, *it follows* that, in the study of nature and its laws, we ought *at once* to make up our minds to *dismiss* as *idle prejudices,* or at least *suspend as premature, any preconceived notion* of what might or what ought to be . . . the case, and content ourselves with observing, as plain matter of fact what *is.* (p. 60)

But what should we observe first? Of course preconceived notions are premature and might be refuted by fact; but why dismiss them? Merely because they may be false! The science which Herschel advocates is infallible! It is infallible, however, only if we make "one preliminary step," which is "the absolute dismissal and clearing the mind of all prejudices, and the determination to stand and fall by the results of a direct appeal to facts in the first instance, and of strict logical deduction from them afterwards" (pp. 60–61).

This is absurd. If we dismiss all prejudices what remains "to stand and fall by the results of a direct appeal to facts in the first instance"? And how, for example, can atomism, to which Herschel had already committed himself, be obtained by "strict logical deduction" from facts? But Herschel proceeds undisturbed. In accord with Bacon and Watts, he now classifies prejudices into two groups: prejudices of opinion and prejudices of the senses. He also defines them echoing Watts. At least five of his six examples were endorsed by Bacon as truths. I do not know whether he did this on purpose; he may have merely tried to bring examples of unambiguously refuted theories, for which it was safer to go to the distant past. And of course, Herschel stresses, a refuted theory never was a part of science; it was always a prejudice.

Herschel discusses the method of avoiding prejudices of opinion, which shows again the Wattsian origin of his view: "To combat and destroy such prejudices we may proceed in two ways, either by demonstrating the falsehood of the facts alleged in their support, or by showing how the appearances, which seem to countenance them, are more satisfactorily accounted for without their admission" (p. 61). Again we see that we may have to refer to a theory in order to show that it is false and thus a prejudice. But then those who had held it prior to its refutation may have been in simple and honest error; why call it, then, a prejudice? Herschel continues the above quoted passage, as if to answer my question:

> But it is unfortunately the nature of prejudice of opinion to adhere, in certain degree, in every mind, and to some . . . [even] after all grounds for their reasonable entertainment is destroyed. Against such a disposition the student of natural science must contend with all his power. Not that we are so unreasonable as to demand from him an instant and peremptory dismissal of all his former opinion and judgment; all we require is, that he will hold it without bigotry, retain till he sees reason to question them, and be ready to resign them when fairly proved untenable, and to doubt them when the weight of probability is shown to lie against them. If he refuses this, he is incapable of science. (pp. 61–62)

It is strange that Herschel suggests that one should try to refute a prejudice, and then he admits that of necessity the refutation "to a certain degree" fails to dissuade: either the "certain degree" is marginal and one may ignore it, or it is substantial and then perhaps investing efforts in refutations is not productive enough to be commendable. It seems obvious to me, but I shall not insist on the point, that the ambiguity here reflects a real ambivalence: on the one hand Herschel adumbrates the Baconian doctrine according to which it is futile to try to make people change their minds by criticizing them, and, on the other hand, Herschel, only one decade before, when he was young and without authority, had succeeded in convincing the whole of his English public that the infallible Newton, the greatest scientist on earth, had been wrong. He never expressed a word of appreciation for such a "candid and sincere" public (to use Boyle's idiom). Nowhere did he mention that the success of science of which he is so rightly proud is at least partly due to the good will and curiosity of the general public. Instead, one decade after his victory, he tells his public *ex cathedra* that no matter how strongly and sincerely they discard their past views, they are still prejudiced "to a certain extent" somewhere in the corners of their minds.

What, then, can we do if our natural tendency is to be prejudiced?

13

The student of natural science must fight this tendency with all his power, Herschel says, echoing the puritanic Watts again. "Not that we are so unreasonable to demand of him an instant and peremptory dismission of all his former opinions and judgments," Herschel adds by way of qualifying Watts's doctrine of empty-mindedness. "All we require" from our student, Herschel says in all modesty, is non-dogmatism; let him by all means "retain" his past judgments "till he sees reason to question them and be ready to resign them when fairly proved untenable and to doubt them" in proportion, as Watts had demanded.

By now we have before us an admixture of Bacon's, Boyle's, and Watts's views together within some new adjustments—which is not easy to put together cogently. Now we come to the method of eliminating prejudices of the senses. This elimination in no way advocates any degree of mistrust of the senses; rather, it aims at destroying unconscious judgments by appealing to the strict use of the senses themselves. A prejudice of the senses can only be destroyed by the highest court of appeal, namely, the senses themselves, in all their purity. The senses may, it is true, be prejudiced, or influenced by faulty judgment, by an admixture of opinion and sensation; we may, therefore, receive from the senses some contradictory evidence; and we must in such cases discard a part of the evidence. But once we realize that the error is originally that of judgment, we shall not fail to pin the fault on the sensation which is less direct and which is less purely sensational.

I need not discuss this view at length: I have discussed it and explained its historical significance elsewhere.[10] Rather, let me mention that Herschel had in mind specific instances of prejudices of the senses which intervened with scientific research; I will illustrate with an example. In his *The Decline of Science In England* (1830) Charles Babbage concludes with a comparative sketch of Davy and Wollaston. He refers to Wollaston's refined senses and the allegation that these made him a good scientist, and dismisses this allegation with disdain. To reinforce his view he tells the story of how Herschel had showed him solar absorption spectra: I shall put the instrument before you so that you will be able to see them, but, Herschel predicted, you will not see them from not knowing how to see them; then I shall tell you how to look

---

[10] See my *Towards an Historiography of Science, op. cit.* (note 4), Sections 2–5 and my "Sensationalism," *Mind*, 75 (1966), 1 ff.

at them and you will be unable to comprehend how they had eluded you before. And, Babbage adds with satisfaction, Herschel's prediction came true.

Having described some sense illusions as examples of prejudices of the senses, Herschel concludes (p. 63, italics in the original) that the *"sensible impressions* made by external objects on us" depend on the circumstances and therefore we must be careful; we must lay confidence in our observations only in a limited degree at first, correcting them appropriately before being certain of them. He does not tell us how to do all this, but in this respect he is no worse than Bacon and Watts. Next comes his theory of observation. As we do not sense with our mind, he says, all sensations are in fact signals, the mechanism of which is inexplicable to us. This, it seems to me, is, in a humble way, a statement of the problem of observation: all signals, observations included, need *interpretation* before they convey a message; how do we interpret them? Herschel's solution is stated as briefly and psychologistically as the problem: the interpretation is made by *association,* which means, I understand, induction by simple enumeration: "we can only regard sensible impressions as signals conveyed . . . to our minds . . . which receives and reviews them, and by *habit and association,* connects them with corresponding qualities . . . of objects; just as a person writing down and comparing the signals of a telegraph might interpret their meaning" (p. 64).

I shall not discuss the psychology involved. It is hypothetical and thus in need of testing; it was actually tested, and it was promptly refuted. But as Herschel had no access to this refutation we should overlook it here; Herschel's linguistic simile, however, is rather unfortunate, since in his own view language is *a priori* valid, and so should be, then, the language of the signals, i.e., the rule of association. But, the signals we receive from our senses, he also says, should convey messages which do not in any way depend on our theories!

Immediately following the simile, Herschel brings in an example: "As, for instance, if he had *constantly observed* that the exhibition of a certain signal was *sure to be* followed next day by the announcement of the arrival of a ship at Portsmouth, he would connect the two facts by a link of the very same nature with that which connects the notion of a large wooden building, filled with sailors, with the impression of her outline on the retina of a spectator on the beach" (p. 64). This is more puzzling than helpful: can one *observe* that a signal is *"sure to be* followed" by another signal? Why is it so certain that if two signals are always consecutive one would connect them in the same way as an impression of an outline of a ship is connected by an observer with the notion of a ship? Is the notion of a ship identical with "the notion of a large wooden building, filled with sailors"?

I have an answer only to the last question. It is perhaps not very satisfactory, but it is the only one I found: in Watts's *Logic* there is an arbitrary example of a definition: "A ship may be defined as a large hollow building made to pass over the sea with sails" (Pt. I, Ch. VI, Sec. 6). I admit that Watts's "hollow building" is different from Herschel's "wooden building" just as his "sails" differ from Herschel's "sailors"; but the latter is probably a misprint and the former is an insignificant variation. Moreover, as Watts is not dogmatic about his definition, and presents it only as a possible example, Herschel may have taken the liberty to deviate a little.

What matters, however, is not Herschel's problem but rather his claim that we can actually solve it with *certainty*. Herschel ends the chapter with the beautiful image of a guide who knew, by the sight of the flight of condors soaring in circles miles away, that below them lay a carcass over which stood a lion.

4. INDUCTION

From the doctrine of prejudices of thought and of sensation Herschel goes over to discuss "the analysis of phenomena," the creation of experimental or observational set-ups in which complicated phenomena appear to have a simplicity, with fewer ingredients than in everyday life, so that *causes* of fact can be found with their aid:

> But, it will be asked, how we are to proceed to analyse a composite phenomenon into simpler ones, and whether any general rule can be given for this important process? We answer, None. . . . Such rules, could they be discovered, would include the whole of natural science. . . . However, . . . the analysis of phenomena . . . is useful . . . as it enables us to recognize, and mark for special investigation, those which appear to us simple; to set *methodically* about determining their laws, and thus to facilitate the work of *raising up general axioms* . . . which shall, as it were, transplant them out of the external into the intellectual world, and enable us to reason them out *a priori.* (p. 73)

This, he continues, enables us, by "reasoning back from generals

16

to particulars," to devise more simple set-ups, and come to more dis-coveries. While the preparation of the original simple experiment is unaided methodologically, the *a priori* reasoning is; the method is Ba-con's "true ladder of axioms." Any abstract theory can have a *class in-terpretation,* and when we find a similarity between two laws *we search for the class to which this similarity applies,* thus arriving at a higher level law.

This is the whole of Herschel's methodology, a version of refined empiricism; it can be briefly summed up thus. We give up all preju-dices and start afresh. We see a complicated world. Somehow, unaided by any method, we isolate a set-up A which leads to a result B invari-ably, i.e., repeatedly. This we generalize into a law: all A is B. We meditate over many such laws and by a method of similarity we de-duce new laws from these, of a higher level of abstraction. To put it more sharply, Herschel has three stages of induction: (1) fact finding (and temporary classification), (2) finding immediate causes, and (3) finding the higher causes until we find the very first cause. Which of these stages is induction or the "analysis of the phenomena" in Her-schel's terms? Induction, it seems, is alleged to be the *finding of causes;* hence, it belongs to, if it is not identical with, part (2), which raises the problem: why does Herschel present induction before (1)? The an-swer is imposed on us by the logic of the situation: only those facts the causes of which we already know (or at least suspect) come under (1) [if (1) will include all observed phenomena it will be a mad col-lection]; so (1) is in fact only a Baconian retrospective myth. Herschel's claim is in fact this: we start with (2), with causes, and *then* pretend to have started with (1), with facts. And, when discussing (2), the causes, Herschel does offer, to use Bacon's term, some "aids to the in-tellect": he recommends the employment of what he calls "proximate causes" and others customarily call "hypotheses." But before coming to this, we should stop to notice Herschel's explanation to the reader of why he must accept induction. Induction, Herschel says, is the ba-sis of science, as we can see from the "fact" that before Bacon there was no science.

5. A EULOGY ON BACON

It is to our immortal countryman Bacon that we owe the broad prin-ciple and the fertile principle; and the development of the idea, that the whole of natural philosophy consists entirely of a series of inductive generalizations, commencing with the most circumstantially stated particulars, and carried up to universal laws, or axioms, etc. etc. (p. 78)

So opens Chapter III of Herschel's *Preliminary Discourse*. It is the only passage in the book that was quoted repeatedly and highly approvingly by a variety of authors in the nineteenth, and even twentieth, century. It was even quoted in Whewell's *Novum Organum Renovatum*, Whewell's sharp criticism of Bacon's notwithstanding. This shows us that one only has to make a broad and unqualified statement in favor of a thinker in order to be quoted by all the adherents of this thinker. Of course, the merits of Bacon are not those which Herschel enumerates: the ladder of axioms is credited by Bacon to Plato, and the theory of circumstantial descriptions is Boyle's not Bacon's, and inductive generalizations are Aristotle's and Newton's (Bacon viewed them as puerile). Not one of those who quote this passage says where in his writings did Bacon ever mention, or allude to, circumstantial description. It is more a matter of rhetoric than of serious history of ideas.

Herschel's proof of the fertility of Bacon's teaching, which is the thesis of Chapter III with which the above quoted passage opens, is very surprising indeed: "Previous to the publication of the *Novum Organum* of Bacon [1620], natural philosophy, in any legitimate and extensive sense of the word, could hardly be said to exist" (p. 79). The Greek philosophers were great abstract thinkers and excellent arguers, (see *Novum Organum*, I, Aph. 71 and 79), but as empirical scientists they were defective in the manner described by Bacon: they were prejudiced. Yet, even Herschel had to admit that at least Aristotle had displayed some knowledge of nature: he had a system of physics, and he happened to make some observations. But his physics was prejudiced by his biological approach and led to dogmatism (see *Novum Organum*, I, Aph. 77). Archimedes is mentioned only in other parts of Herschel's book; in the discussion about Greek science he is conspicuously absent. In another passage (p. 72) he states that Archimedes was "too late," which is an echo of another ploy of Bacon's. Next comes the routine mockery of scholasticism and alchemy, with Arabic science not even mentioned (see *Novum Organum*, I, Aph., 71). True, prior to Bacon there were Roger Bacon, Paracelsus, Agricola, and Gilbert; but these were only rudiments of the beginning. The real beginning was Copernicus, Kepler, and Galileo. But even they can be belittled, and Herschel does belittle them—not to spite them, I am sure, but in order to show that science begins with induction by showing that it begins with Bacon:

> By the discoveries of Copernicus, Kepler, and Galileo, the errors of the Aristotelian philosophy were effectually overturned on a *plain appeal to the facts of nature;* but it *remained to show on broad and general principles,* how and why Aristotle was wrong; to set in evidence the particular weakness of his method of philosophizing, and to substitute in its place a stronger and better [one]. . . . (p. 85)

If we turn back to Herschel's discussion of Aristotle (p. 110) we find in a footnote that it was Galileo who "exposes unsparingly the Aristotelian style of reasoning"; he gives there an example of an Aristotelian "string of nonsense" and, ironically enough, this same example of Aristotelianism can be found in Bacon's *Novum Organum* (Book II, Aph. 48).

There is little doubt that Galileo did more than expose details of Aristotle's errors. But we should not go into needling particulars. Herschel sticks to generality and ignores Galileo's methodology as most of his contemporaries do. He wants to praise Bacon for his invention of modern inductivism and here he is on safe ground, except that his thesis (science = induction) is not proven by historical evidence, but rather is imposed upon history; accepting induction, he is determined to praise its inventor:

> This important task was executed by Francis Bacon, Lord Verulam, who will, therefore, justly be looked upon in all future ages as the great reformer of philosophy, though his own actual contributions to the stock of physical truths were small, which were the fault rather of the general want of physical information of the age than of any narrowness of view on his own part; and of this he was fully aware. (p. 86)

Here Herschel speculates about Bacon's place "in all future ages" and about Bacon's own state of mind. This is sheer hero-worship. Was Bacon so unsuccessful because of the lack of information? Robert Leslie Ellis states in his general introduction to Bacon's *Works* that it was the insincerity of Bacon which stood in the way of his scientific achievement. Charles Singer argues in the *Encyclopedia Britannica* (Art.

"Bacon") that empirical refutations of many of Bacon's views as expressed in the *Novum Organum* were known prior to the publication of that work.

An immense impulse was *now* given to science, Herschel's eulogy continues, and everybody "rushed eagerly" to observe "matters of fact"; art and even nature herself helped by kindly supplying the telescope and microscope, and by displaying two rare astronomical phenomena (the two novas). The word "now" makes sense only if it means "after 1620"; but then the claim is to be interpreted as implying that the novas appeared after 1620. To repudiate this silly implication, perhaps, Herschel gives the dates of the novas in a footnote. It is not that he is ignorant of the history of astronomy, of course; he was merely misusing the word "now." As I have discussed this misuse twice already, I shall not go into that further here. Let me simply state that Herschel did not explicitly state that the "immense impulse" to observe and experiment was *caused by* Bacon; he only mentions it, opening with "now," and ending with the phrase "lifetime of Galileo." But all this eulogy to the Renaissance of science follows the eulogy to Bacon which follows the abuse of the people who made this Renaissance. To mislead the reader still further, Herschel goes on to judge "the immediate followers of Bacon and Galileo," namely, Boyle and his associates, who observed facts, but were still under the spell of "natural magic"; they were not sufficiently theoretically minded.

The "immense impulse" peters out into "natural magic"; both, however, relate to observation: the impulse was to observe, and the magic kept within observation and away from theory. The general impression and thrust is rather clear, though nothing else is: Bacon's star shines in a dark period; his faults are few and due to others, while others' stars are dim in comparison with his and only reflect his light. Copernicus, Kepler, Galileo, and Boyle all found some facts, partly due to his own inspiration, but he, Bacon, was the reformer of philosophy, the herald of Newton's success!

Herschel wrote his book at the time when the claim that Bacon had greatly contributed to philosophy was highly contested in some quarters. This, again, is the worst thing about Herschel's mode of arguing: he does not mention the contrary opinion which he was exorcizing. He did not mention, in particular, that both Bacon's alleged greatness and his methodology were equally challenged.

Herschel seems merely to reiterate naively Bacon's view that no real science existed before Bacon, and to present the rest of the history of science as the history of inductive reasoning.

Herschel's *Preliminary Discourse* seems to be a modernized version of Bacon's *Novum Organum*—the same thesis with more modern illustration, and perhaps with some additional new ideas. In order to know what Herschel accepted from Bacon and what he did not one has to compare these two works; but I cannot say how to find out Herschel's reasons for having selected what he did. In the end of the *Novum Organum* there are the twenty-seven so-called "Prerogative instances." Herschel selected only six of them, though without telling his selection rule. Besides the *crucial instance,* his instances share with Bacon's only the labels. I shall choose one instance in order to show how apologetic Herschel could be.

Take the "travelling instances" (*Novum Organum,* II, Aph., 23), which are instances of generation and corruption, or of increase and decrease of certain qualities. I do not think that it is possible to understand Bacon's muddle unless we see that he speaks of Telesio's theory in an Aristotelian idiom. Telesio tried to explain the variety in the phenomena by reducing them to series of dualities. His hot-cold duality was perhaps the better known; but as Bacon's discussion of induction is mainly illustrated by Telesio's theory of heat, he had to use for an example of his "travelling instance" something else. He chose Telesio's theory of light, and his examples, like most of his factual optical examples, from Telesio's *Consentini De Colorum Generatione Opusclum* (Naples, 1570). In Telesio's theory of light, whiteness and light are opposites. Transparent things like glass and water can be turned white, by chopping and foaming them respectively. Telesio, and likewise Bacon, explained these two facts by making air responsible for whiteness. This explanation Herschel knew to be false. The explanation is confirmed by the fact that white paper is rendered less opaque when wetted; the assumption is that water replaces air in its pores. This confirmation is true, of course. We are now ready for Herschel's quotation from Bacon; it is one of six which he chose because he wished to praise Bacon and which I choose in order to poke fun at his uncritical apologia:

> Bacon's "travelling instances" are those in which the *nature* or quality under investigation travels in degree; and thus affords an indication

of cause by a gradation of intensiy in effect. One of his instances is very happy, being that of "paper which is white when dry, but proves less when wet, and comes nearer to the state of transparency upon the exclusion of air and the admission of water." In reading this and many other instances in the *Novum Organum,* one would almost suppose (had it been written) that the author had taken them from Newton's *Opticks.* (p. 141)

6. AIDS FOR THE INTELLECT

After the eulogy on Bacon and the unfair attack on all the rest of Newton's predecessors there remains little for Herschel to do but to expand his remarks on induction and to give some examples. Chapter IV is on collection of facts. Scientists make many observations, measure what they can, and make, by the way, some accidental discoveries. Chapter V is on classification of this multitude of facts. The classification should not be held as a theory but merely as a temporary nomenclature. Chapter VI is on the first stage of induction, on lowest level generalizations and their verifications. Here the concept "force" is introduced (p. 108): Herschel identifies "force" with "cause" and alleges that we can have a "direct perception" of forces! But forces are only sometimes visible, not always; and so, when the cause of a phenomenon is not obvious, a crucial experiment should be made to decide between the possible causes. Moreover, observability is not an essential matter, but a matter of external conditions: usually forces are observable only in simple experimental set-ups.

How do we arrive at the idea of preparing an experiment which makes forces manifest? First of all we postulate "proximate causes," in fact hypotheses, to the effect that under certain circumstances certain results occur. The name "proximate causes" shows that Herschel considered the possibility that the result would not be quite the same as what was conceived. Yet he ignores the fact that most experiments lead to no new results.

So we suppose that by the method of "proximate causes" we have an experiment with a cause captured in it. How do we observe or find it? There are four or five necessary and sufficient clues. A cause is found (1) by the presence of the effect whenever it is present, (2) by the absence of the effect in its absence, (3) by the increase and the decrease of the effect with the increase and decrease of the cause (even in strict proportionality!), and (4) by the reversal of cause and effect. This is an improved Baconian theory of causation, which is in origin medieval and draws on Aristotle. It is most questionable.

For one thing, it prohibits all cases of delayed effect, such as phosphorescence. One may argue that cause and effect are contiguous, and delayed effects are linked by causal chains; this, however, will be a more Renaissance and less medieval theory than Herschel presents, where cause and effect are not contiguous but coincidental. Similar arguments may be easily found against strict proportionality, or even monotony, of cause and effect: already Hume used the case that increase in heat increases comfort up to an optimum and then increases discomfort. There are many equally incontestable instances in which an optimum effect is reached while cause continues to increase. Herschel even takes it for granted that all causes are strict, quite contrary to commonsense; who ever doubted that a gun is caused to fire by the finger's pressure on the trigger?—least of all the soldier who has experienced hundreds of cases in which the cause was present and the effect absent. And, need one add, all too often the effect is unfortunately present in spite of the absence of the cause; the gun fires by itself, as they say.

The above rules are followed by ten "observations" or "rules of philosophizing" (the phrase is due to Newton, who opens Book III of the *Principia* with his four famous rules of philosophizing), which, unlike the previous rules, are not general—indeed they are mere rules of thumb.

Next comes Chapter VII, on the method of verification (p. 142). It involves checking that all cases are observed, and that all empirical exceptions to the conclusion to be verified are carefully re-examined: if the exception is explicable by finding a different cause, the induction will be saved, but if there are too many exceptions "our faith in the conclusion will be proportionally shaken" (a splendid Pickwickian sense of "verification"!).

No wonder that inductivists need the doctrine of prejudice. Since they teach that all scientific theory is verified, they do indeed need extra admonition to accept refutations of allegedly verified theories. To add confusion Herschel uses—in places but not systematically—"verification" and "confirmation" as synonymous. There also exist as yet "unverified inductions," which are none other than testable but as yet untested hypotheses. We should not place confidence in them, we are admonished, until they are verified. Kepler's laws are

Herschel's example. He does not mention, even in this context, a testable but refuted hypothesis: all refutations are dealt with along with the doctrine of prejudice: all refuted theories are, and always were, sheer prejudices.

7. HIGHER LEVEL GENERALIZATIONS

We have thus far observed facts, analysed them, based causal laws on them, and now we must proceed since we should not rest content with mere facts. We are now willing and ready to develop theories. How? Having been warned up to this point to be on guard against the method of speculating, we now learn the secret of theorizing—speculate. There is no other method of theorizing than that of inventing a hypothesis, and Herschel has a reason for not telling us until page 143: "*The liberty of speculation* which we possess in the domain of theory is not like the wild licence of the slave broke loose from his fetters, but rather like that of the freeman who has learned the lesson of self-restraint in the school of just subordination." Though I much disagree with the sentiment of this passage, I should only draw attention to the following. At least *prima facie* this passage contradicts most of what precedes it. It is, of course, possible to reconcile the seeming contradiction, but how to do so satisfactorily I, for one, do not know. And even if this were no serious problem, the fact remains that Herschel conceals the idea of this passage from his reader for a long time, thus treating him as a bad schoolmaster treats an unruly schoolboy.

Let us, however, observe the methods of self-restraint of the "freeman," the scientist in action. We have to know the mechanisms of the universe; regrettably, however, these are often either too big or too small. "Yet we are not to despair since we see regular and beautiful results brought about in human works," which are incredible, like printing and steam engines—even though *why* a steam engine works we shall not know for a long time. In the meantime there is a ray of light: we may be able to make *hypotheses* (p. 145).

This may invite trouble: we may have too many hypotheses on our hands; there may be two, or even more, hypotheses with which to explain the same phenomena. Now, Herschel says, "are we to be deterred from framing hypotheses and constructing theories, because we meet with such dilemmas, and find ourselves frequently beyond

24

our depth? Undoubtedly not" (p. 147). Hypotheses, he strongly claims, are not theories. This is clearly so, since theories are certain, while hypotheses are uncertain forever. The hypothesis can be refuted in the course of research, and theories are arrived at by generalization. What Herschel means, but does not say, seems to be this: after having *tested* the hypothesis, if it is not refuted, we give it a *class interpretation* which then becomes the (causal) generalization and thus the irrefutable theory. "A well imagined hypothesis," Herschel continues, is very helpful, though he does not tell us how a hypothesis is "well imagined." Anyway, it must lead to a theory, and if it is verified, it is of the highest importance (p. 148). Therefore, Herschel says, hypotheses must concern the *agents* of the phenomena: "These agents are *not to be arbitrarily assumed;* they must be such as we ALL have good inductive grounds to believe they do exist in nature" (p. 148).

This stress on unanimity may be legitimate, even appropriate. But one cannot help asking what is the place of a rebel like Thomas Young; and the answer seems inescapable. Science, it is true, is like the Polish Sejm; accepting a law is like voting there: every member has the right to veto. But, to be a member you must be unprejudiced. The unprejudiced bases his judgment on "good inductive grounds." The unanimity, then, is quite spurious; what matters are the good inductive grounds.

Here the whole of Chapter VII stands or falls. For, it is here that Herschel permits us to speculate for the first time, provided that we speculate with restraint, and restraint is a matter of having good inductive grounds recognizable by all who are unprejudiced. It is somewhat distressing, I confess, that such strong claims are made within a rather obscure passage concerning a very obscure entity—the agent—whose entry into the scene was unheralded.

Herschel continues: we made a hypothesis concerning the agents, and "we have next to consider the laws which regulate the actions of these primary agents" (p. 149). So now the agents have become primary. This rule, it seems to me, is quite unscientific. If we first make a hypothesis about the agent and then about its mechanism, laws, or properties, then the first hypothesis is untestable. I have no objection to such a procedure, having repeatedly claimed that the roots of scientific hypotheses are in metaphysics; but Herschel should object. It seems rather obvious to me that Herschel implicitly suggested accepting Newtonian dynamistic metaphysics as prior to any hypothe-

25

sis concerning the particular mechanism or force acting under certain circumstances. Herschel demanded first that we assume the existence of an agent, a force or a fluid, and *then* assume its mode of action, its mechanism, its specific laws of force and motion; this is because the science of his day partly proceeded on such Newtonian lines, and he did not notice that far from having been verified, the first assumption concerning the agent cannot be tested as long as it is not supplemented by a hypothesis concerning the specific mode of action of that agent. If one first assumes the existence of the luminous ether or, to take Herschel's example (in slight modification), the electric fluid, and only generations later find an adequate theory of the mode of action of the ether (Young) or of the electric force (Coulomb), then only the second step is testable. But in Herschel's view not only can the mode of action be verified, but the existence of the medium or substance is usually verified first. He knew very well, but simply did not consider relevant here, that although the assumption concerning the existence of the ether as the agent of light was pre-Newtonian, only Young's assumption concerning its mode of action as being *transversal* waves convinced people to try out the ether theory again.

Be that as it may, by now we have verified the theory concerning the existence of the agent and we wish next to find its mode of action. Three ways are open to us: (1) "inductive reasoning," (2) "a *bold* hypothesis," and (3) a combination of these two: a bold hypothesis supported by much previous knowledge (p. 149). This is a novelty in the inductive literature. Until then the prevailing idea was that a mild hypothesis, or, to use Laplace's terminology, a "natural" hypothesis, is preferable to a bold one. This novelty, I think, is thanks to Young and Fresnel. Young was bold enough to present his hypothesis as a hypothesis, and Fresnel was too. It seems that these facts were too remarkable for Herschel to ignore; he had to give bold hypotheses a prominent place in science, even though only as tentative tools until verified and converted into theories proper. Here is how Herschel views the relation between theory and hypothesis:

> In estimating, however, the value of a theory, we are not to look, *in the first instance,* to the question, whether it establishes satisfactorily, or not, a particular process or mechanism; for of this after all, we can never obtain more than that indirect evidence which consists in its leading to the same result. What . . . is far more important for us to know, is whether our theory truly represents *all* the facts, and in-

cludes *all* the laws, to which observation and induction lead. A theory which did this would, no doubt, go a great way to establish any hypothesis of mechanism or structure, which might form an essential part of it: but this is very far from being the case, except in a few limited instances; and till it is so, to lay any great stress on hypotheses of the kind, except in as much as they serve a scaffold for the erection of general laws, is to "quite mistake the scaffold for the pile." Regarded in this light, hypotheses have often eminent USE: and a facility in framing them, if attended with an equal facility in laying them aside when they have *served their turn,* is one of the most valuable qualities a philosopher can possess; while, on the other hand, a bigoted adherence to them, or indeed to peculiar views of any kind, in opposition to the tenor of facts as they arise, is the bane of all philosophy. (p. 153)

The quotation comparing hypotheses to scaffoldings is very common and, I think, belongs to Goethe.[11] Notice that the chief preoccupation of the whole passage is typically classical inductive: it is that of order of procedure: which comes first, the hen or the egg?

Herschel had earlier suggested that we should make a hypothesis first about the agent and then about its mechanism: namely, from the more general to the more particular aspects of the specific explanation. Now, without any indication of a change of mind, he suggests we start with the more particular aspect, the better testable part of the theory (the more directly observable, to use the inductive idiom). The mechanical model is only the scaffold which may serve to find a theory. A theory is verifiable; a hypothesis is a refutable instrument.

By viewing a mechanical hypothesis as an instrument Herschel shows that we may arrive at a theory by a "hypothesis." The "theory" is nearer to facts (more testable) than the hypothesis or model; but it may be the hypothesis or model which helped us to build the theory: the hypothesis is only the scaffold. The real thing is the verified theory which should be viewed, and even presented, as *prior* to the hypothesis. The idea of a scaffold is already used by Bacon, who believed that Thales had arrived at his speculation by induction and afterwards removed the scaffold—the *observed facts* (*Novum Organum,* I, Aph. 125).

Herschel never says that the use of hypotheses is essential to the method of science; he does not say if it is ever part of the theory, or

---

[11] See Goethe, *Gedenkausgabe der Werke, Briefe, und Gesprache* (Zurich, 1949), *9,* 653, para. 1222. See also my "Unity and Diversity in Science," *op. cit.* (note 9), 376.

if, as a part of the theory, it is verifiable. Herschel is simply not clear; he gives two examples from the theories of caloric and of light, but the examples do not help much. True, one can distinguish between Fourier's facts and his theory without much difficulty, but I, for one, do not know how to distinguish between Fourier's theory and his model, especially since he persistently stood above the controversy between the caloric and the heat-as-motion schools. The caloric hypothesis was a model, to be sure; it is not clear whether or not Herschel accepted it as verified, nor how he would distinguish the caloric theory from the caloric hypothesis or model. In the passage above he says that the theory even supports the model, so that it may have been verified first. But the problem is how to present such a theory, how to relate it to the model. Lavoisier had spoken of caloric without committing himself to what it meant exactly. Only *after* having presented the theory did Lavoisier explain that caloric is the matter of heat. Perhaps Herschel approved of this.

After discussing the use of hypotheses, Herschel moves on to make an important contribution to the theory of testing hypotheses—the theory of *independent* tests, which, I think, is almost entirely his own invention—and a very important one. True, he claimed that Kepler's theory is in accordance with Newton's. But he also claimed that planets deviate from Kepler's orbits, a fact which enabled one to find independent tests of Newton's theory. His new and important idea, then, is this: a theory must be tested by other facts than those by which previous theories were tested; otherwise it is not a new theory. Now, to avoid confusion, let me note that we have nowadays two very closely connected criteria for the novelty of a theory: one is that the new theory must contradict existing theories (Popper), and the other is that a new test for it can be devised (Herschel). If a theory is empirical, i.e., testable, and if it contradicts previous theories, then it is certainly testable *independently* by *new tests* (the crucial tests). Possibly, however, we may find a theory which is testable by new facts but does not contradict any previously accepted view. If this possibility is denied, then the two criteria (Popper's and Herschel's) are coextensive. When one accepts this possibility (as Herschel does), one must conclude that Herschel's criterion is the broader of the two. The interesting fact, however, is that Herschel began a trend of postulating stringency of tests.

In this vein Herschel requires that we test a theory by "a great mass of observed facts" (p. 156). But this does not mean many repetitions. We repeat one test *many times*, making the observations more accurate (using statistical laws, p. 162) in order to compare them with more exact results of the theory. The moment we are satisfied that fact and theory accord we look for *other* facts, and this raises the problem of the criterion for *otherness* or novelty of facts. Herschel does not cope with this problem, but he does suggest that we somehow use background knowledge, or common sense, to solve it. In this he surely is right, and his claim that we use deviations from old theories as new tests (Keplerian irregularities to test Newtonianism) is strikingly new. The idea was later used by Herschel in his paper attacking Hegel's accusation of Newton as a plagiarist. It was later still used by Duhem to disprove the classical inductive theory of the ladder of axioms: Newton's theory does not rest on Kepler's, but rather modifies it. It is here, I think, that Herschel's ideas led to serious and important developments; in any case, most of the methodological literature is still too glib concerning the problem of independent evidence and the difference between new evidence and a variant of old evidence.

8. THE HISTORY OF SCIENCE

Part III of Herschel's book is a history of the physical sciences. First comes a classification of the sciences. This includes a passage on light, where Newton's corpuscular hypothesis is described (p. 188). In defense of Newton's conduct Herschel says that the corpuscular hypothesis had explained all the then known phenomena, including Newton's own discoveries. He was confident that "had the properties of light remained confined to these, there would have been no occasion to have resorted to any other mode of conceiving it." This has been meanwhile refuted: Whewell has convincingly argued that Newton's theory of light had never been properly tested and confirmed; Mach has shown that Newton had overlooked the most important part of Grimaldi's discovery.

Huygens, Herschel says, had a rival hypothesis, but it seemed to be less capable of explaining diffraction. Other phenomena were discovered which could serve as a further trial of the explanatory power of these hypotheses; e.g., Grimaldi's diffraction. Moreover, Newton's rings were explained by Newton and not by Huygens. Here the history of science is presented for a while as the history of competing hypotheses: Herschel even implies that *there is a criterion of choice of a hypothesis which is not inductive:* the theory with a *higher explanatory power* is to be preferred. Although Newton's theory is false while Huygens' theory is true, our predecessors were right in adhering to the false theory because it had a higher explanatory power, where explanatory power is assumed (as usual) to be monotonic with the paucity of assumptions and the multitude of explained facts.

Herschel was historically mistaken. The explanatory power of Newton's theory of light was much poorer than it appears, because, as Whewell showed, for each new fact explained by Newton a new hypothesis was made. Herschel could have found this out had he asked himself if Newton's theory had ever been tested by *independent* tests. Instead, he merely based his argument that the explanatory power of Newton's theory was *greater* on the claim that it explained refraction, diffraction, and Newton's rings. Instead of examining whether Newton's explanations of these phenomena included new assumptions or not, Herschel was engaged in defending Newton against the charge that he had presented theories as facts. Herschel approached the problem in a roundabout manner, using Biot's hypothesis to illustrate Newton's hypothesis of fits of easy transmission and easy reflection:

> The simplest way in which the reader may conceive this hypothesis, is to regard every particle of light as a sort of little magnet revolving rapidly about its centre while it advances in its course, and thus alternatively presenting its attractive and repulsive pole, so that when it arrives at the surface of the body with its repulsive pole foremost it is repelled and reflected; and when the contrary, attracted and so enters the surface.[12] Newton, however, very cautiously avoided announcing his theory in this or any similar form, confining himself entirely to *general language.* In consequence, it has been confidently asserted by all his followers, that the doctrine . . . as laid by him, is substantially nothing more than a statement of facts. (p. 190)

So, first of all we have cleared Newton, although at the cost of an unjust smearing of all his followers and a distortion of Newton's text. And this is the lesson to draw from the mistaken belief, allegedly

[12] This, by the way, was Biot's attempt to reconcile Newton's theory with the facts of polarization and double refraction, which, he showed, could be explained by assuming that the spin of the photon (to use the modern idiom) assumes only definite discrete values. But this assumption, now employed in quantum theory, was then rightly rejected as too arbitrary.

of Newton's followers, that the theory of fits is only a statement of fact:

> Were it so, it is clear that any other theory which should offer a just account of the same phenomena must ultimately involve and coincide with that of Newton. But this . . . is not the case, and this instance ought to serve to make us extremely cautious how to employ, in stating physical laws derived from experiment, language which involves any thing in *the slightest degree theoretical,* if we would present the laws themselves in a form which *no future research shall modify or subvert.* (p. 190)

This is Bacon's great discovery: if we want our factual reports to be unassailable, we must employ a language which involves no theory whatsoever. At the period of transition from the corpuscular to the wave theory of light physicists were engaged in an interesting and difficult exercise: translating statements of general facts from the language of one theory to that of another. Duhem used the existence of this procedure (his example is, indeed, Newton's rings) as an argument against induction, claiming the general impossibility of divorcing any so-called factual statement from some theoretical language, though it could be divorced from any one theoretical language by translating it into another.

In concluding my examination of Herschel's historical examples, I should say, in fairness, that I chose the worst of his examples. Not that I think that his history of galvanism or magneto-electricity is sufficiently candid, but at least there he does not distort history so obviously as when he presents his own interpretations as hard historical facts. The major subject of his history was, I think, the revolution in optics, no matter how well he concealed this behind observations on the history of other fields of science.

9. THE PHILOSOPHY OF SUCCESS

There is a fundamental difference between eighteenth- and nineteenth-century views on induction. Eighteenth-century philosophers were enthusiasts. They collected information and made theories with the hope that soon something grand would come about, something similar to the creation of Newton's mechanics, only much more universal and significant. It was felt that the smallest contribution was welcome and important because it hastened the coming of the kingdom of Reason on earth. This was the philosophical, intellectual, moral, and political atmosphere of the age of Enlightenment.

By the 1820's the picture had changed. Lavoisier's revolution had been overthrown by Davy, and Newton's optics by Young and Fresnel. Confusion reigned. The difficulties inherent in Ampère's theory caused even despair in some quarters. There was a feeling that Rationalism was going a little too far, slightly beyond its own natural limitation. It is not true, of course, that everybody was confused, nor that there was no confusion before. But a glance at contemporary works dealing with the problems of atomism or electrodynamics or science and religion or the social organization of science (see especially Babbage's *The Decline of Science in England* [1830]) shows how much the atmosphere had changed.

But then there came a sharp turn. The nineteenth century will be known as the century of the philosophy of *success*. Science rapidly progressed, even though it had more problems, and more formidable ones. The prevailing feeling was that everything was improving fast, and that theoretical physics was approaching its ultimate stage. The eighteenth-century ideals shrank in a number of respects; in particular, due to the failures of the French Revolution, scientists qua scientists shied away from public affairs. But the old ideals seemed no longer so much ideal as reality, or the reality of the next day. The philosophy of induction was utopian since its beginning, but its features had altered. The seventeenth century was one of hope, the eighteenth, one of progressive work, of the advancement of learning and the improvement of the mind; the nineteenth century was the reaping of the dazzling harvest of success.

The philosophy of success is, of course, as old as the success of Newton's theory. It shows its first mark in the preface to the second edition of Newton's *Principia,* and it plays a very significant role in Laplace's philosophy. Still, Laplace's philosophy is more of a philosophy of improvement than of success. The shift of emphasis, one may argue, from improvement to success, is insignificant and gradual, and this indeed may be the case. Yet, in my view, the change was not so smooth.

I do not know how much Herschel's work was a part of this trend and how much he was one of its causes. I have too little evidence to support my own view that he was more of an originator than reflector of the philosophy of success. In either case, Herschel's philosophy of success deserves mention. Herschel's emphasis on success permeates his book. The idea that science is identical with scientific success, intellectual as well as material, is implicit throughout, and explicit in quite a number of places. Failure is mentioned only once: the success of science is predicated on the avoidance of possible failures, the results of working against the laws of nature. One failure which science could have helped to avoid is the disaster of an inventor of a submarine who sank with it at the very first trial (p. 45).

It is a measure of the difference between the atmosphere in which Herschel wrote his book, and the one in which these lines are written, that I find it hard to explain to my reader that Herschel knew that failure in scientific inquiry is unavoidable without sounding as if I am calling him a liar: it seems so strange to us today that a man of Herschel's stature should have felt obliged to mention only failure which science can prevent and pass silently over failure which is unavoidable. He felt obliged, I suppose, to fuse philosophy with some degree of propaganda. This is the highway to self-deception.

The propaganda is at places rather thick. Herschel's philosopher is "accustomed" to create science (p. 12); he is entirely disinterested and free from authority (p. 13); he is "deeply imbued with the best principles of sound philosophy" (p. 39); and he goes on verifying his theories and sagaciously generalizing them to the moral and physical benefit of all mankind. As a portrait this is not new in the least: it is Bacon's philosopher, with a small difference. The ideal is now the real.

In Herschel's methodology there exist no more problems than in Bacon's. He names only tasks—to observe, to find causes, to generalize—but no problems; and, like Bacon, he even offers arguments from success. The task of Herschel's philosophy of success is, however, not really to eliminate problems; rather it is to put them in their place—within the workshop—and leave no trace of them in the shop-window, where success alone is to be displayed. The greater the problems in the workshop, the greater was the display of success in the shop-window. This display was the method of concealing real issues, perhaps also of burying them, but not intentionally.

Herschel's attitude has one important consequence, probably unintended: it widens the cleavage between researcher and general public, between insider and outsider. Insiders, professional scientists or dedicated amateurs, existed throughout the history of science; in the eighteenth century, however, the pretense was that everyone was an amateur and even somewhat of a dilettante; there were allegedly no insiders, no mysteries, no esoteric teachings. The inductive style was instituted, as I have discussed elsewhere,[13] in order to encourage the amateur, and in a variety of ways. The inductive style of writing encouraged self-training in experiments, as well as the publication of even very minor discoveries; it also pushed aside the method of explicit argument and criticism as too frightening, not to say offensive, to the amateur. With the advent of nineteenth-century science, style changed with pace (Sir Humphry Davy says: I shall not start with a description of an electrochemical pile, since those who do not know it will not understand me anyhow). Thus, the chief purpose of employing the inductive style was being ignored even before Herschel arrived on the scene. More and more educated people were losing contact with natural science and were told instead that the scientist almighty is going to set everything right. The argumentative method was pushed even further out of sight and *the inductive style was now functioning not only as a method of concealing arguments, but also as a method of concealing problems.*

True, Faraday was just starting to violate the taboos of the inductive style, and towards the end of the century his openly critical style was gaining ground. For the greatest majority, however, the method of avoiding argument remained that of presenting the history of science in a doctored version as the history of inductive success, and of concealing all current problems by presenting present theories as if they were utterly unproblematic.

It was in 1860 that a paper by Newland concerning his hypothesis of the periodic table of the elements was rejected because the Royal Chemical Society *"made it a rule not to publish* papers of purely theoretical nature, since it was likely to lead to correspondence of a *controversial* character."[14]

Not only the unpublished controversial material of the period suffers from the philosophy of success, but the published material does so too. Distortion of this kind still makes us present most of nineteenth-century science as if it involved no controversy. The controversial character of Faraday's ideas has still not found adequate expression in the historical literature. Disagreements galore concerning theories of force, energy, and heat are still ignored almost regularly. Kirchhoff's claim that Balfour Stewart had no priority over himself concerning the ra-

[13] See note 8 above.
[14] J. A. R. Newland, *The Periodic Law* (London, 1884), 23; quoted by E. T. Whittaker, *A History of the Theories of Aether and Electricity*, 2 (New York, 1960), 11.

diation law is based on the fact that Stewart had made quite a few errors and offered an unsatisfactory proof. Rayleigh's famous counter-claim is based on the fact that Stewart was successful in his experiments. Historians accept one claim or the other and the matter is still waiting to be clarified. I had myself accepted a version of the story from a nineteenth-century source and found later how mistaken I had been.[15]

In Herschel's work I have found one interesting departure from the philosophy of success, or at least a statement in a different mood. When attacking irrationalism he mentions (pp. 6–7) that science is based on the *honesty* of the witness and on universal scepticism; truth is capable of standing up to *all* tests "and coming unchanged out of every possible form of *fair* discussion." If I am not mistaken this is the only reference Herschel makes to discussion. It seems to me quite traditional that when arguing with irrationalists, rationalists are more humble and critical—more rational—than when talking to other rationalists. This is so because critical rationalism suffices for the rejection of irrationalism. Most rationalists, however, at least feel that criticism must be supplemented with something more positive, something more than what is necessary to combat irrationalism. The negation of other views is not enough; we must show the strength of our views to those whom we wish to recruit. Bacon expressed this mood: "the greatest obstacle to the progress of the sciences, and the undertaking new tasks and provinces in the same, is found in the *despair of men and the supposition of impossibility*" (*Novum Organum*, I, Aph. 92). And therefore we must teach our pupils that science is demonstrable. Unfortunately or otherwise, the claim for demonstrability is baseless, and so it leads to despair, or at least to much dissatisfaction. Nevertheless, it is clear that Bacon had hit on a significant problem. Permanent ill-success *may* lead to despair, to barren skepticism, and to cynicism. But great success may even be worse, since it *may* spoil. The only solid hope there is is that people work because they are *curious*, because they are ready to *try*, no matter what the outcome may be. We must base rationalism on other foundations than past success or even the hope of entering the promised land of success.

In conclusion, I may state that, in view of the great fame and wide influence of Herschel, I was somewhat puzzled, upon reading his *Pre-*

---

[15] See my "The Kirchhoff-Planck Radiation Law," *Science* (April 17, 1967). The part on Balfour Stewart (p. 32) is based on A. Cotton's excellent but historically inaccurate paper in *Astrophys. J., 9* (1899), 237. See my note 20 there.

liminary Discourse, to find so much in it which is so very uncritical. The explanation of this puzzle is possibly that Herschel was arguing against the surge of skepticism—from a position of strength[16]—claiming that the new skepticism may spoil all past achievement. His philosophy of success was to put down the threat from skepticism, accounting for its presentation in a mood rather than a statement.

Herschel was certainly not the first philosopher of success. Bacon's philosophy of hope is in a way its early predecessor, and thinkers from Mersenne to Laplace spoke increasingly about the importance of success. With Herschel failures became entirely uninteresting. There is one positive element in this transition from hope and feigned success to real success which may merit mention—at least for historians of science who care about the spirit of one age as contrasted with the spirit of another age. During the period of hope popular error was attacked with the hope that it would be soon eliminated; during the period of success error was tolerated as a permanent feature of public ignorance to be overlooked as much as possible.[17] A fervent quarrel, like that between the phlogistonists and the anti-phlogistonists or between the anti-phlogistonists and Davy, could hardly take place in the mid- or late-nineteenth century. Success to a certain measure may make us less apprehensive and thus less intolerant. Though Herschel did not excel in toleration, his followers did. Some knight-errants of the enlightenment tried to raise indignation concerning prejudices even during the nineteenth century; most scientists, however, felt that positive popular lecturing and positive research would do. And so, perhaps social reformists who prefer the fervent battle against prejudice to token popular lectures may point out, not without some measure of justice, that success also makes us less sensitive. It is always a mixed blessing.

---

[16] See W. W. Bartley, "Approaches to Science and Skepticism," *The Philosophical Forum, 1* (1969).
[17] See L. Pearce Williams, "The Politics of Science in the French Revolution" in Marshall Clagett, ed., *Critical Problems in the History of Science, Proceedings of the Institute for the History of Science at the University of Wisconsin, September 1–11, 1957* (Madison, 1959).

Of course, the nineteenth century had more success in popular adult education than its predecessors. Boyle's *Seraphick Love,* and Spinoza's and Locke's equivalents, were all meant for the intellectual. Isaac Watts's *Logic* and *Improvement of the Mind* were much wider in influence, but nothing like Sam Smile's *Self-Help.* Yet the combination of the industrial revolution and the rise of socialism and allied radicalist movements was the crucial factor. Professors' lectures and the likes were but the trimmings. Doubtless, the rise of the British Association was an expression of the rise of the new class of technologists and experts, but it too was not as important as the movement of literacy classes for workers run by radicalists—as described, e.g., by C. P. Snow in his *The Two Cultures and The Scientific Revolution.* There is no doubt that here lies the greatest of Snow's historical errors; nineteenth-century scientists mostly kept out of the movement and worked only on its fringe.

Wollaston and the Atomic Theory of Dalton

BY D. C. GOODMAN*

The atomic theory of John Dalton, announced at the beginning of the nineteenth century, was generally opposed into the middle of the century, and was still unacceptable to some at the close. Certain of the main issues raised by Dalton's work are revealed through a study of the papers of the English scientist William Hyde Wollaston,[1] a well-known contemporary of Dalton. Wollaston's views on the atomic theory show a surprising inconsistency, varying remarkably as his mood alternated between bold speculation and utter scepticism.

According to Dalton the chemical elements consisted of different kinds of indivisible atoms, distinguished by their weight. The atoms of hydrogen were identical, but differed in weight from those of oxygen or carbon. Chemical combination occurred through the union of the various atoms in the simplest multiple ratios. If two elements A and B formed several compounds, Dalton assumed that one atom of A combined with one atom of B in one compound, with two atoms of B in another, and that two atoms of B combined with one of A, and so on. For him this represented an order of preferred combinations from which atomic weights could be calculated. Since he knew of only one compound of oxygen and hydrogen, he supposed wrongly that water was a compound of one atom of each of these elements. From the known

---

* 4 Park Way, Temple Fortune, London, N.W. 11.

[1] W. H. Wollaston (1766–1828) studied medicine at Caius College, Cambridge and later practiced in London. He became interested in chemistry while at Cambridge, and when he left medicine in 1800 this was the science to which he devoted most of his time. In particular he solved the difficult problem of finding a satisfactory process for making platinum malleable, discovering palladium and rhodium in the course of this research. He also invented the reflective goniometer, which, in an improved form, is still a standard measuring instrument in crystallography. He was Secretary of the Royal Society in 1804–16 and President in 1820. In 1823 he was elected one of the eight foreign associates of the Académie des Sciences.

combining weights he calculated that the atomic weight of oxygen was 7, relative to a hydrogen standard of unity. He later admitted that such calculations occasionally brought him doubts, since he could not be certain that his assumed atomic constitutions were always correct.

In his excellent recent study,[2] D. M. Knight has discussed the reasons for the widespread opposition to this theory. For Davy and Faraday the theory was too complicated to be true. Dalton's apparent belief in numerous types of atoms, as many as the elements and irreducible to one another, conflicted with their vision of a simple world in which only one kind of atom existed. In addition the theory was commonly opposed on methodological grounds; to chemists, who regarded their science as an empirical discipline, speculations on unobservables, such as Dalton's atoms, seemed remote and out of place. A contemporary journal expressed this attitude in the following way: "Let chemists make more use of their crucibles and furnace, than of their pen and ink. The man who, by one accurate analysis, makes us acquainted, experimentally, with the constituents of a single compound, is entitled to more thanks than he who, by calculation, forces the whole catalogue of chemistry into the trammels of a theory, and does more towards establishing the theory, if it be worth preserving."[3]

This indeed was the sort of advice which Wollaston gave in what was to be the most influential of his papers on the atomic theory. But this was by no means his only opinion, for his earliest discussion of the subject recommended an entirely different approach.

I. WOLLASTON'S ATOMIC SPECULATIONS IN 1808 AND 1813

In 1808 Wollaston reported the results of his investigations on the composition of various salts[4] in a paper which directed increased attention to Dalton's theory; the accurate experiments described in it proved that chemical combination really did occur in simple multiple proportions as predicted.[5] Some examples had been pointed out

---

[2] David M. Knight, *Atoms and Elements* (London, 1967).

[3] "Review of a System of Chemistry," *Journal of Science and the Arts, 4* (1818), 316–317.

[4] Wollaston, "On Super-acid and Sub-acid Salts," *Phil. Trans., 98* (1808), 96–102.

[5] For example, Berzelius said that Wollaston's study first showed him how to express experimental results in a way that would reveal this law. (*Jöns Jacob Berzelius,* trans. O. Larsell [Baltimore, 1934], p. 172.) Subsequent expositions of the laws of chemical combination frequently referred to Wollaston's experiments. The numerical prefixes "bi-" and "quadro-", which he invented to express the composition of the oxalates, came to be the accepted nomenclature for salts, in preference to the earlier terms "super-" and "sub-". (Berzelius, *Essai sur la Théorie des Proportions chimiques* [Paris, 1819], p. 169; E. Turner, *Elements of Chemistry* [Edinburgh, 1827], p. 98.)

only a fortnight before by Thomas Thomson in an analysis of oxalates,[6] and Wollaston, commenting on this work, said that he had noticed several instances of this mode of combination and had hoped to discover the cause. But Wollaston added that further inquiry was now unnecessary, since Dalton's recently published theory at once explained these observations.

Wollaston's experiments on carbonates, sulphates, and oxalates were simple and decisive. For example, he measured the volumes of "carbonic acid," or carbon dioxide, liberated by the action of acid on known weights of carbonate and bicarbonate of potash, and found that they were in the ratio of one to two respectively. Therefore, he concluded, potash combined with a certain weight of carbonic acid to form the neutral carbonate, and with twice this weight to form the acid bicarbonate. His analysis of the oxalates of potash showed that a fixed weight of potash would combine with three different weights of oxalic acid to form distinct salts. The weights of acid were in the proportions 1:2:4, and the absence of a compound with three portions of acid seemed anomalous; but he explained:

> To account for this want of disposition to unite in the proportion of three to one by Mr. Dalton's theory, I apprehend he might consider the neutral salt as consisting of
>
> 2 particles potash with 1 acid
> The binoxalate   as 1 and 1, or 2   with 2,
> The quadroxalate as 1 and 2, or 2   with 4,
> in which cases the ratios which I have observed of the acids to each other in these salts would respectively obtain.[7]

Such attempts to discern the atomic constitution of particular substances were later to cause Wollaston doubts. Here they inspired him to a bold development of the atomic theory:

> But an explanation, which admits the supposition of a double share of potash in the neutral salt, is not altogether satisfactory; and I am further inclined to think, that when our views are sufficiently extended, to enable us to reason with precision concerning the proportions of elementary atoms, we shall find the arithmetical relation alone will not be sufficient to explain their mutual action, and that we shall be obliged to acquire a geometrical conception of their relative arrangement in all the three dimensions of solid extension.

[6] T. Thomson, "On Oxalic Acid," *Phil. Trans., 98* (1808), 63–95.
[7] Wollaston, *op. cit.* (note 4), 101.

For instance, if we suppose the limit to the approach of particles to be the same in all directions, and hence their virtual extent to be spherical (which is the most simple hypothesis); in this case, when different sorts combine singly there is but one mode of union. If they unite in the proportion of two to one, the two particles will naturally arrange themselves at opposite poles of that to which they unite. If there be three, they might be arranged with regularity, at the angles of an equilateral triangle in a great circle surrounding the single spherule; but in this arrangement, for want of similar matter at the poles of this circle, the equilibrium would be unstable, and would be liable to be deranged by the slightest force of adjacent combinations; but when the number of one set of particles exceeds in the proportion of four to one, then, on the contrary, a stable equilibrium may again take place, if the four particles are situated at the angles of the four equilateral triangles composing a regular tetrahedron.[8]

Wollaston's admitted delay in publishing his results was later criticized by Whewell as "scrupulous timidity."[9] There was, however, anything but timidity in his speculations on the spatial arrangement of atoms and the relative stabilities of the various configurations. Although in the second half of the nineteenth century, the development of stereochemistry would extend the atomic theory in the way he had envisaged and confirm his intuition of the stable tetrahedral array, at the time his conjecture passed almost unnoticed. However, it impressed Andrew Ure, professor of chemistry at the Andersonian Institution of Glasgow, who introduced it in his new *Dictionary of Chemistry:* "We cannot withhold from our readers the following masterly observations, which must make everyone regret, that the first development of the atomic theory had not fallen into such philosophical hands."[10] Ure was convinced that he had found experimental evidence supporting these speculations in the variation in the physical properties of acids at different dilutions, which he thought due to the formation of different compounds of the dry acid with various numbers of water atoms:

I have, therefore, much pleasure in referring to my researches on the constitution of liquid nitric acid, as unfolding a striking confir-

[8] *Ibid.,* 101–102. It should be remarked that Dalton had considered geometrical arrangements of atoms and had constructed models using balls and pins. J. Dalton, "On a New and Easy Method of Analysing Sugar," pp. 3–4. Dalton published this with some other essays, all paginated separately, in *On the Phosphates and Arseniates* (Manchester, 1840).
[9] W. Whewell, *History of the Inductive Sciences* (London, 1837), *3,* 150.
[10] A. Ure, *A Dictionary of Chemistry* (London, 1821), Article "Equivalents (Chemical)."

mation of Dr. Wollaston's true philosophy of atomical combination. When I wrote the following sentence, I had no recollection whatever of Dr. Wollaston's profound speculations on tetrahedral arrangement: "We perceive that the liquid acid of [s.g.] 1.420, composed of 4 primes of water + 1 of dry acid, possesses the greatest power of resisting the influence of temperature to change its state. It requires the *maximum* heat to boil it, when it distils unchanged; and the maximum cold to effect its congelation. . . .

Here we have a fine example of the stability of equilibrium, introduced by the combination of four atoms with one.[11]

Similarly, a solution of dilute sulphuric acid, which Ure assumed to be a compound of one atom of dry sulphuric acid with three water atoms, exhibited the instability predicted by Wollaston for such combinations. For the addition of a small quantity of water produced a great change in the specific gravity of the solution. But Ure's confidence was not justified, and his explanations passed without comment, for he had proved nothing: apart from the question of whether he had dealt with mixtures or compounds, he did not know the relative numbers of atoms involved in combination.

In his first discussion of the atomic theory, in 1808, Wollaston had felt it necessary to extend Dalton's treatment into three dimensions, and in 1813 he further developed this spatial theory in a theoretical discussion of crystal structure.[12] He supposed that crystals consisted of spherical units, which could be arranged in various ways to produce different crystal shapes. He described the structure of boracite in terms of a regular arrangement of black and white balls, and added that this model fitted the atomic theory. However, here and in his earlier paper, he mentioned an alternative to Dalton's extended corpuscles. In 1808, he considered the possibility that atoms were of "virtual extent," and he elaborated on this in 1813: "And though the existence of ultimate physical atoms absolutely indivisible may require demonstration, their existence is by no means necessary to any hypothesis here advanced, which requires merely mathematical points endued with powers of attraction and repulsion on all sides, so that their extent is *virtually* spherical, for from the union of such particles the same solids will result as from the combination of spheres

---

[11] *Ibid.*
[12] Wollaston, "On the elementary Particles of certain Crystals," *Phil. Trans., 103* (1813), 51–63. Dalton claimed that Wollaston had borrowed his own ideas on crystal structure. See his letter in C. Daubeny, *An Introduction to the Atomic Theory* (Oxford, 1831), p. 137.

impenetrably hard." [13] This amounted to an hypothesis of atoms as point-centers of force, the theory put forward in the eighteenth century by Boscovich, who considered the stabilities of atomic groups to be a function of inter-atomic distances. This theory might well have been the inspiration for Wollaston's geometrical speculations, though the particular structures which he describes do not come from Boscovich. After 1813, Wollaston abandoned the theory of point atoms.[14]

II. WOLLASTON'S EQUIVALENTS: HIS REJECTION
OF ATOMISM IN 1814

Wollaston's next paper[15] on the subject was totally different from his two earlier ones. Abandoning atomic explanations, he urged chemists to keep closely to the facts and avoid theory. His discussion of the laws of chemical combination made it quite clear that he no longer supported Dalton:

> According to Mr. Dalton's theory, by which these facts are best explained, chemical union in the state of neutralization takes place between single atoms of the substances combined; and in cases where there is a redundance of either ingredient, then two or more atoms of this kind are united to only one of the other.
>
> According to this view, when we estimate the relative weights of equivalents, Mr. Dalton conceives that we are estimating the aggregate weights of a given number of atoms, and consequently the proportion which the ultimate single atoms bear to each other. But since it is impossible in several instances, where only two combinations of the same ingredients are known, to discover which of the compounds is to be regarded as consisting of a pair of single atoms, and since the decision of these questions is purely theoretical, and by no means necessary to the formation of a table adapted to most practical purposes, I have not been desirous of warping my numbers according to an atomic theory, but have endeavoured to make practical convenience my sole guide. . . .[16]

This positivistic approach had already been put forward by

---

[13] Wollaston, *op. cit.* (note 12), 61.
[14] The theory of Boscovich interested others at this time; Humphry Davy commented on it in his *Elements of Chemical Philosophy* (London, 1812), p. 56.
[15] Wollaston, "A Synoptic Scale of Chemical Equivalents," *Phil. Trans., 104* (1814), 1–22.
[16] *Ibid.,* 7.

Humphry Davy.[17] He too had said that numerical statements in chemistry should refer only to the facts of combination in definite and multiple proportions. It was imprudent to estimate the numbers of atoms in combination, their weights or shapes, about which nothing was known. Further it was unnecessary to introduce the idea of indivisible particles at all in a factual account of combining weights. This interpretation, particularly as developed by Wollaston, was to have considerable sway over chemists, who took it to embody a sound philosophy of their science.

Wollaston collected the best analyses available and proceeded to assign characteristic numbers to elements and compounds to represent their combining weights. These he called "equivalents," a term invented by Cavendish in the previous century to express the weights of acids and bases which neutralized each other. The German chemist Richter had calculated the equivalents of various acids and bases. Wollaston was to extend the use of this term in an unfortunate way. He said that his numbers would be based on an oxygen standard of 10,[18] partly because of a belief in the central importance of this element in chemistry, earlier emphasized by Lavoisier. His numbers however were calculated on the basis of tacit assumptions, just as arbitrary as those he had objected to in Dalton. Worse, by calling them chemical equivalents he gave a false meaning to this term which misled later chemists. For example, he dealt with the oxides of carbon as follows:

> The first question, consequently, to be resolved is, by what number are we to express the relative weight of carbonic acid, if oxygen be fixed at 10. It seems to be very well ascertained, that a given quantity of oxygen yields exactly an equal measure of carbonic acid by union with carbon; and since the specific gravities of these gases are as 10 to 13.77, or as 20 to 27.54, the weight of carbon may be justly represented by 7.54, which in this instance, is combined with 2[o] of oxygen forming the deutoxide, and carbonic oxide being the protoxide will be duly represented by 17.54.[19]

---

[17] Davy, *op. cit.* (note 14), 114. In fact Davy did not follow this program, which appears to exclude all atomic considerations in chemistry. He joined the critics of Dalton, but his real reason for doing so was his belief in the alternative theory of Boscovich, within which he speculated freely on atoms.

[18] Berzelius chose this standard for the same reason. He and Wollaston had agreed on this during his visit to London in 1812. Afterwards a mild dispute arose over who had thought of this first. *Jac. Berzelius Bref,* ed. H. G. Söderbaum (Uppsala, 1912–1941), *1,* Part II, 32 and Part III, 51, 65, and 69.

[19] Wollaston, *op. cit.* (note 15), 8.

This calculation went beyond the facts in assuming constitutions for the oxides of carbon, which we would now represent as CO_2 and CO. Although this later proved correct Wollaston had only guessed the truth. He might have represented the oxides another way, corresponding to CO and C_2O. This would equally have explained the fact that a fixed weight of carbon combined with two weights of oxygen, one double the other. His numbers were after all as "warped" as Dalton's. It was clear also that he was not referring to a combination with a fixed weight of 10 of oxygen, which would have correctly given two values for the equivalent of carbon. He wanted one number only for each substance. His treatment of the metallic oxides involved the same maneuvers. His table of equivalents had the entries:

$$
\begin{aligned}
\text{Iron} \quad 34.5 + 10 \text{ Oxyg.} &= 44.5 \text{ Green Oxid of Iron} \\
+ 15 \text{ Oxyg.} &= 49.5 \text{ Red Oxid} \\
\text{Mercury } 125.5 + 10 \text{ Oxyg.} &= 135.5 \text{ Red Oxid Mercury} \\
+ 125.5 \text{ [Mercury]} & \\
&= 261 \quad \text{Protoxid [Mercury]}^{20}
\end{aligned}
$$

Here again he guessed the correct constitutions for the oxides of mercury, corresponding to HgO and Hg_2O. But for all he knew he might have expressed the multiple proportions in another way, which others preferred, agreeing with HgO_2 and HgO.

With practical considerations uppermost in mind, he used his numbers in a way which proved very attractive. He distributed them on a slide-rule.[21] He inscribed the names of commonly used chemicals along a line, logarithmically divided from 10 to 320, positioning them according to the values of their so-called equivalents. When the slider was moved until the number 10 was next to oxygen, the numbers alongside all the other substances were at once read off as equivalents. When the number 100 was moved to common salt, the quantities of other reagents needed to decompose this weight of salt were immediately given, avoiding the usual multiplications and divisions. Here was an instrument, handy and compact, which gave quick and reliable answers to the routine problems of the laboratory.

Wollaston had tried to strip chemistry of all but the bare bones of

[20] *Ibid.,* 18 and 19.
[21]In the Whipple Museum, Cambridge there are slide-rules which once belonged to Wollaston. They show his attempts to apply them to gauging, foreign exchange, and the timber business. They are described in R. Gunther, *Early Science in Cambridge* (Oxford, 1937), pp. 34-35.

experimental data. There remained, he believed, a purely descriptive body of knowledge which no one could object to. Dismissing all theory he took pains to point out that his value for chlorine would remain, whatever the outcome of the controversy over its elementary status.[22] Above all he had recommended a chemistry without atoms. He could hardly have moved further from his earlier thought.

Wollaston's 1814 paper was later criticized by Auguste Comte, who regretted that it fell short of his cherished positivistic ideals. He claimed that the switch from atoms to equivalents amounted to no more than "a mere artifice of language" and said that atomic ideas were still discernible.[23] This shrewd analysis escaped the chemists, who succumbed in general to the linguistic substitution, unaware that any atomic theory was involved.

Wollaston's term "equivalent" was one of several invented to express the facts of chemical combination without the hypothetical dress of "atom." Davy spoke of "proportional numbers" and Berzelius used "volumes"[24] in this sense; others referred to "doses,"[25] "combining quantities,"[26] and "stoichiometrical numbers."[27] This new terminology seemed to describe mere empirical quantities. It was an illusion, yet it proved to be compelling and generally went undetected. For although this nomenclature avoided images of physical corpuscles, it did not express the bare facts as was often claimed. In each case the preferred vocabulary referred to numerical values calculated on the basis of arbitrary assumptions about constitutions, that is to say by the use of theoretical formulas. But this was something which chemists simply passed over, apparently satisfied, once the obvious atomic ideas had been cleared away, that an unobjectionable practical science remained. A dilemma is clearly present in Davy's proposal:

> As in all well known compounds, the proportions of the elements are in certain definite ratios to each other; it is evident, that these ratios may be expressed by numbers; and if one number be employed to denote the smallest quantity in which a body combines, all other quantities of the same body will be multiples of this number; and the

[22] Wollaston inclined to the older view that this substance was a compound of muriatic acid. *Jac. Berzelius Bref, 3,* Part VII, 291–292.
[23] A. Comte, *Cours de Philosophie Positive* (Paris, 1830–42), *3,* 149.
[24] Berzelius, "Experiments on the Nature of Azote . . . ," *Annals of Philosophy, 2* (1813), 359.
[25] M. Donovan, *Treatise on Chemistry,* 4th ed. (London, 1839), p. 368.
[26] W. Brande, *Manual of Chemistry,* 3rd American ed. (New York, 1829), p. 638.
[27] L. Gmelin, *Handbook of Chemistry,* trans. H. Watts (London, 1848–1872), *1,* 42.

smallest proportions in which the undecomposed bodies enter into union being known, the constitution of the compounds they form may be learnt, and the element which unites chemically in the smallest quantity being expressed by unity, all the other elements may be represented by the relations of their quantities to unity.[28]

The problem here was to know what value to give to the smallest combining quantity of an element. This was not as straightforward as both Wollaston and Davy had made it appear. For this was to ask for the atomic weights of elements. In the disguised language of equivalents and proportions one property of Dalton's atoms survived: the chemical indivisibility of the smallest combining weights. The calculation of these weights required a knowledge of the molecular formulas of the various compounds of the elements. This was only made possible by appealing to Avogadro's hypothesis, which was not accepted until the second half of the nineteenth century. Until then such calculations rested on unproved assumptions.

However there was general praise for Davy and Wollaston. It was felt that they had extracted what was of importance in Dalton's theory, namely, the combination of the elements in fixed and multiple proportions, leaving aside all unnecessary atomic speculations. A typical reaction came from William Brande, who had succeeded Davy as professor of chemistry at the Royal Institution. In his introductory lecture on the atomic theory he confined his discussion to the laws of combination:

> Happy illustrations of these laws, divested of all hypothetical aspect, and immediately brought to bear upon the minute and accurate details of practical chemistry, are to be found in some of Dr. Wollaston's papers. . . .
>
> These facts have been explained upon the idea that certain indivisible atoms of matter, some simple and some compound, unite to each other, and that their relative weights are represented by those in which they combine—an assumption altogether gratuitous and hypothetical, but which has led to the adoption of the term *Atomic* theory; we shall, however, rather prefer the term, *theory of definite proportionals,* or of *chemical equivalents,* in our discussions of these matters.[29]

[28] Davy, *op. cit.* (note 14), 112.
[29] "Proceedings of the Royal Institution," *Quarterly Journal of Science, Literature and the Arts,* 21 (1826), 109–110. Like other chemical writers of the period, Brande misused the word "theory" to signify a law.

Wollaston's influence was nowhere more obvious than in Ure's dictionary entry "Atomic Theory. See Equivalents (Chemical)." Ure commented: "In the article Equivalents (Chemical), as well as under the individual substances, the reader will find the primitive combining ratios, or atoms as they are hypothetically called, fully, and I trust fairly, investigated from experiment. This is the sheet anchor of scientific research, which we must never part with, or we shall drift into interminable intricacies." [30] The article contained a long description of Wollaston's paper, and his table of equivalents was reproduced in the appendix. Curiously it combined approval of Wollaston's scepticism with an admiration for his earlier atomic speculations. But no inconsistency was mentioned in the review of Ure's article by the *Quarterly Journal of Science*, which described its account of the atomic theory as "the most complete and most philosophical hitherto offered to the public." [31]

The strength of the sceptical reaction was also apparent in Davy's well-known presidential address to the Royal Society in 1826, when a Royal Medal was awarded to Dalton. Davy emphasized that the prize was for the discovery of the law of multiple proportions, not for a theory of atoms. He also pointed out with approval that Wollaston's equivalents separated the practical part of Dalton's work from the theoretical atomic part.[32] One Fellow of the Royal Society went further. Claiming to represent a large body of opinion, he did not think that Dalton deserved any reward at all:

> I next ask, whether Mr. Dalton has materially contributed to its development and extended application? Whether we owe to his suggestions those "Tables of Equivalents," which are so useful in the Laboratory, and so important to the manufacturer of chemical products?—No: all this is due to Dr. Wollaston, whose *logometric scale of chemical equivalents* brought the theory into practice, and rendered that, which was a mere abstract subject of chemical inquiry, little understood and less investigated—an instrument of the utmost usefulness and value. He did that for the theory of definite proportions which Mr. Watt effected for the steam-engine: it was a comparatively useless and cumbersome machine; he rendered it generally applicable and divested it of its clogs and incongruities. Had Mr. Dalton stood in the place

---

[30] Ure, *op. cit.* (note 10), xxiii.
[31] "Analysis of Scientific Books," *Quarterly Journal of Science, 11* (1821), 348.
[32] Davy, *Six Discourses delivered before the Royal Society* ... (London, 1827), p. 128.

of Dr. Wollaston, I should have gladly witnessed the honour which has fallen upon him; for the merit of those who fertilize barren inventions, and bring them home "to the business and bosoms of men," I am inclined to rate very highly, and even sometimes to consider it as superior to that of the original inventor.[33]

On the continent also there were similar expressions of appreciation for Wollaston's cautious approach. Applause came from Schweigger, the editor of a leading German scientific journal: "I was delighted to see Dalton's theory understood by one of his most important countrymen in the way it appeared to me from the start. I mean Wollaston, whose paper expresses Dalton's atomic theory in terms barely different from Richter's stoichiometry. . . . The paper is distinguished not only by its freedom from all atomic subtlety, but generally by its frankness and clarity, characteristic of all the work of this brilliant philosopher." [34] This was echoed by Carl Bischof, professor of chemistry and technology at the University of Berlin: "It contains an exposition, so simple and purified from all hypothesis, that it can in fact be regarded as a model for all future stoichiometrical investigations." [35] Döbereiner, professor of chemistry at Jena, who later put forward his triadic law relating the atomic weights of similar elements, dedicated an early book to Wollaston.[36] The atomic weights which he later used were those of Berzelius, who, from 1818, published tables of values. These were calculated from unproved formulas, which Berzelius occasionally changed. The resulting fluctuations in his atomic weights emphasized their uncertain basis. By contrast Wollaston's scepticism seemed to offer a safe course. This was what Liebig turned to when he gave up his support for Berzelius: "The equivalents will never alter, but I doubt very much whether there will ever be accord on the expression of atomic weights. The study of chemistry will be infinitely facilitated if all chemists decided to return to equivalents." [37] This opinion became popular on the continent in the 1840's through the textbook of Gmelin.

[33] "On the recent Adjudgment of the Royal Medals by the President and Council of the Royal Society," *Quarterly Journal of Science*, *1* (1827), 15.

[34] J. Schweigger, "Ueber die festen chemischen Mischungsverhältnisse nebst stöchiometrischen Tafeln," *Journal für Chemie und Physik*, *14* (1815), 500.

[35] C. G. Bischof, *Lehrbuch der Stöchiometrie* (Erlangen, 1819), p. 82.

[36] J. W. Döbereiner, *Beytrage zur chemischen Proportions—Lehre* (Jena, 1816).

[37] J. v. Liebig, "Bemerkungen zu Berzelius' Abhandlung über einige Tages-Fragen der organischen Chemie," *Annalen der Pharmacie, 31* (1839), 36. For a similar statement see E. Turner, *Elements of Chemistry*, 3rd ed. (London, 1831), p. 178. Liebig's statements on atoms are not consistent; see note 44.

In 1856 David Low, professor of agriculture at Edinburgh, was still able to praise Wollaston's equivalents in words that might have been written forty years before.[38] Even the international congress of chemists which met in Carlsruhe in 1860 to settle disputes on formulas decided that equivalents were empirical and independent of atoms. Cannizzaro's pamphlet, distributed at the end of that conference, removed a source of objection to Dalton's theory which Wollaston, Davy, and many others had referred to. It showed how a set of atomic weights could be established, without a series of arbitrary assumptions, from molecular formulas determined in accordance with Avogadro's law.

The remarkable success of Wollaston's positivistic memoir must be attributed in part to the outstanding problem finally solved by Cannizzaro. (It is ironical that Wollaston's equivalents were open to the same charge of arbitrariness.) But more than this was involved, since lively opposition to Dalton continued after Cannizzaro until the end of the century. As Knight[39] points out, it was not until the second half of the century that Dalton's theory was conspicuously successful, in structural organic chemistry. Until then chemists could maintain with some justice that it was unnecessary. So, in the meantime, Wollaston's plan to eliminate theoretical entities altogether from chemistry was endorsed by attempts to construct a non-atomic chemistry, such as those of Brodie and Ostwald.

It was regrettable that Wollaston should have called his numbers "equivalents," since it led to the erroneous use of a term, which, until then, had been correctly employed for the weights of acids and bases involved in neutralization. A glance over his table of values was sufficient to show that they could not represent quantities that were chemically equivalent,[40] for they were referred to a weight of oxygen that varied. That there was no fixed standard for comparison concealed the fact that a substance can have more than one equivalent, depending on the way it reacts. The false idea of a constant chemical equivalent misled Dumas,[41] who wondered what single equivalent to assign to copper, which combined with different weights of oxygen to form more than one oxide. Today reference would be made to a fixed weight of oxygen and two values would be given for the equivalents of copper.

[38] D. Low, *An Inquiry into the Nature of the Simple Bodies of Chemistry*, 3rd ed. (Edinburgh, 1856), p. 18.

[39] Knight, *op. cit.* (note 2).

[40] This was clearly pointed out by Bischof, *op. cit.* (note 35), 82.

[41] J. B. Dumas, *Traité de Chimie appliquée aux Arts* (Paris, 1828–1846), *1*, xxxi.

49

The mistake seems to have been spotted first by Laurent, who insisted that a substance could have several equivalents depending on the reaction.[42]

Further confusion arose from the use of equivalents and atomic weights as synonyms. Wollaston's values were commonly referred to by each of these terms. This was due, on the one hand, to the retention of "atom" in the weak sense of a combining weight; it was a conveniently short word.[43] On the other hand, admitting indivisible corpuscles, some believed that these combined in such numbers that their relative weights really were expressed by equivalents.[44] This erroneous identification, not entirely due to Wollaston, continued until Cannizzaro's use of Avogadro's hypothesis.

III. CALCULATIONS WITH WOLLASTON'S CHEMICAL SLIDE-RULE

Wollaston's chemical slide-rule, with or without his numbers, was widely adopted by chemists as a calculating device, irrespective of their views on the atomic theory. For over twenty years it was a standard piece of laboratory equipment. The instrument could be bought cheaply in the Strand with printed instructions.[45]

It was soon used by William Prout[46] to discover what would now be called the empirical formulas of organic compounds. He pasted on the rule slips of paper marked with ascending multiples of Wollaston's values for carbon, hydrogen, and oxygen. He took these to be numbers of atoms. Referring to the percentage composition of various organic compounds, he moved the slider, looking for marked multiples which agreed with the experimental data. These were then read off as the atomic constitutions of the compounds, a practice which was hardly Wollaston's intention. He soon gave up these numbers and adopted others on the basis of his famous unitary theory of matter.

---

[42] A. Laurent, *Chemical Method,* trans. W. Odling (London, 1855), p. 11.

[43] E. Turner, *An Introduction to the Study of the Laws of Chemical Combination and the Atomic Theory* (London, 1825), p. 33; T. Thomson, *An Attempt to Establish the First Principles of Chemistry by Experiment* (London, 1825), *1,* 36.

[44] Liebig, *Familiar Letters on Chemistry,* ed. J. Gardner, 2nd ser. (London, 1844), p. 85. This confusion was carefully avoided by L. J. Thenard, *Traité de Chimie élémentaire, théorique et pratique,* 4th ed. (Paris, 1824), *1,* 285.

[45] For a detailed description of the instrument that was available see L. J. Thenard, *A Treatise on the general Principles of Chemical Analysis,* trans. A. Merrick (London, 1818), p. 298.

[46] W. Prout, "Some Observations on the Analysis of Organic Substances," *Annals of Philosophy, 6* (1815), 269–271.

Brande of course welcomed the new practical aid, recommending it as part of a portable laboratory.[47] He published several tables of equivalents for use with it,[48] expressed as whole numbers relative to a standard hydrogen unit. He thought that Wollaston's invention, more than anything else, had popularized the laws of combination, and claimed that it was essential to the student, and to industry as a check for yields. His successful textbook continued to describe it as late as 1848, adding to it weights, measures, and temperatures.[49]

This enthusiasm was shared by David Boswell Reid, who conducted classes of practical chemistry at Edinburgh University. He was delighted with the mass of information condensed in the instrument and urged his students to employ it in their analyses. He enlarged the number of substances to be put on the rule, though like Brande he used values other than Wollaston's.[50]

William West, later secretary of the Philosophical Society of Leeds, lectured on the sliding scale, which he had enlarged to contain data for over two hundred substances.[51] But the most elaborate modification was proposed by Prideaux, a member of the Plymouth Institution. The scale was doubled and opened on hinges like a book. Symbols for about five hundred substances were packed into thirteen columns.[52] From this it is easy to appreciate Thomson's remark that, with so many substances on the slide-rule, more time would be lost in looking for them than in carrying out the usual arithmetical calculations it was designed to avoid.[53]

In his manual of laboratory technique, Faraday included a detailed description of Wollaston's instrument, which he said was constantly employed.[54] He warned students that the sliding scales from the instrument maker could never be accepted as accurate. For they were made of paper printed with the required divisions and pasted on wood, and moisture caused the paper to stretch, distorting the scale. He ad-

---

[47] W. Brande, *Manual of Chemistry,* 2nd ed. (London, 1821), 2, 486 and Plate 1.

[48] Brande, *Tables in Illustration of the Theory of Definite Proportionals* (London, 1828).

[49] Brande, *Manual of Chemistry,* 6th ed. (London, 1848), *1,* 142.

[50] D. B. Reid, *Directions for using the Improved Sliding Scale of Chemical Equivalents* (Edinburgh, 1826).

[51] W. West and E. S. George, *Prospectus of a Course of Lectures on the Elements of Experimental Chemistry* (Leeds, 1821), p. 8. Wollaston wrote to West, encouraging any attempts to improve his invention. The letter, dated 26 Nov. 1819, is at the Wellcome Historical Medical Library, London.

[52] J. Prideaux, "Continuation of the Table of Atomic Weights, and Notice of a new Scale of Equivalents," *Philosophical Magazine,* 8 (1830), 423–433.

[53] T. Thomson, *op. cit.* (note 43), 2, 519.

[54] M. Faraday, *Chemical Manipulation* (London, 1827), pp. 551 f.

vised students to check the divisions with a pair of compasses, and insisted that whenever accurate calculations were needed, these should be obtained arithmetically from tabulated values and not from the chemical slide-rule. This practice was later to supersede Wollaston's mechanical method.

In Sweden the industrious Berzelius used it a great deal, particularly to calculate the quantities of reagents where it was not necessary to weigh with greater accuracy than two places of decimals. He remarked: "All these results will be found within two minutes, without fear of error, whilst in calculating in the ordinary way it would take at least a quarter of an hour for a man used to it and he would still be open to mistakes. I do not speak of calculations by logarithms, which practical chemists probably use only rarely, and which besides need as much time as the scale." [55] French chemists could read a full account of the instrument in the standard textbook of Thenard. The edition of 1824 stated that the scale was widely employed in commerce.[56] In Germany Schweigger inserted paper scales in his journal, to be cut out and pasted on backings.[57] He travelled to workshops in Augsburg to discuss the manufacture of this invention, which he felt was a boon to the practical chemist.[58] He also lengthened the rule in the hope of achieving greater accuracy.[59] Other modifications were suggested by Grotthuss[60] and Scholz,[61] the director of a porcelain works in Vienna. Wollaston's design was converted into moveable concentric discs in Italy.[62] In New York his invention was on sale in book-stores.[63]

This active interest gradually came to an end, as the device which had often been referred to as indispensable was given up in the search for greater accuracy. Signs of its decline appeared in 1837, when a

[55] Berzelius, *Essai sur la Théorie des Proportions chimiques*, p. 189. See also *Jac. Berzelius Bref, 1,* Part III, 84 and 102.

[56] L. J. Thenard, *Traité de Chimie,* 3rd ed. (Paris, 1824), *5,* 264. The second edition adopted many of Wollaston's values: *Traité de Chimie,* 2nd ed. (Paris, 1817–18), *4,* 214.

[57] "Synoptische Scale der chemischen Aequivalente," *Journal für Chemie und Physik, 12* (1814), 105.

[58] Schweigger, "Anzeigen," *Journal für Chemie und Physik, 12* (1814), 357–358.

[59] Schweigger, "Ueber die Verfertigung und Benutzung der logarithmischen Rechenstäbe . . . ," *Journal für Chemie und Physik, 14* (1815), 115–129.

[60] T. v. Grotthuss, "Raumverhältnisse der gasformigen Substanzen . . . ," *Journal für Chemie und Physik, 33* (1821), 154–162.

[61] P. Meissner, *Handbuch der allgemeinen und technischen Chemie* (Vienna, 1819), *1,* 59.

[62] "Tavola Circolare degli Equivalenti Chimici," *Giornale di Fisica, Chimica, e Storia naturale, 10* (1817), 28–39.

[63] W. Macneven, "Exposition of the Atomic Theory of Chemistry; and the Doctrine of Definite Proportions," *Annals of Philosophy, 16* (1820), 348.

German textbook[64] could no longer admit the rough calculations that it had been designed for. A few years later Thomas Graham wrote that the instrument was "not itself of much practical value" and substituted logarithmic tables.[65] Kopp, writing his history of chemistry at this time, could already look back on Wollaston's invention as a thing of the past.[66] In 1850 seven-figure logarithms were published for chemical calculations,[67] indicating the demand for precision which Wollaston's slide-rule could not satisfy.

IV. WOLLASTON'S RETURN TO ATOMISM IN 1822:
AN ASTRONOMICAL PROOF

Wollaston's sceptical paper on chemical equivalents represented an abrupt reversal of his earlier position. But this was not his final word on the subject; in 1822, without any explanation, he made a second *volte-face* and returned to Dalton's theory by a most unusual route. This was nothing less than an attempt to prove the existence of atoms from the height of the earth's atmosphere.[68]

He argued that the atmosphere would have a sharp limiting height if, and only if, there were a limit to the divisibility of matter. For then the expansion of the atmosphere, caused by repulsive forces between its particles, would be balanced by a definite gravitational attraction acting on finite, indivisible masses. But if atmospheric particles were infinitely divisible, the gravitational attraction would be infinitely reduced. Its expansion no longer checked, the earth's atmosphere would circulate throughout space and collect about the other planets. Therefore the classical philosophical problem of the divisibility of matter could be settled once and for all by a crucial astronomical test. If observations failed to show the presence of atmospheres on other planets then atoms must exist; if there were planetary atmospheres other than our own then atoms were fictions. Further he pointed out that the result of this test on the gaseous atmosphere would apply to all substances and states of matter: "For, since the law of definite proportions discovered by chemists is the same for all kinds of matter, whether solid,

---

[64] B. Scholz, *Anfangsgründe der Physik* (Vienna, 1837), p. 171.
[65] T. Graham, *Elements of Chemistry* (London, 1842), pp. 117 and 1071.
[66] H. Kopp, *Geschichte der Chemie* (Braunschweig, 1843–1847), 2, 375.
[67] W. Dexter, *Tabulae Atomicae* (Boston, 1850).
[68] Wollaston, "On the Finite Extent of the Atmosphere," *Phil. Trans., 112* (1822), 89–98.

or fluid, or elastic, if it can be ascertained that any one body consists of particles no longer divisible, we then can scarcely doubt that all other bodies are similarly constituted; and we may without hesitation conclude that those equivalent quantities, which we have learned to appreciate by proportionate numbers, do really express the relative weights of elementary atoms." [69] There was no indication here of his former concern to know the relative numbers of combining atoms, though of course the problem still existed. He seems to have forgotten this in his concentration on the existential question.

In May 1821 Wollaston saw an opportune moment for the test, for Venus was passing very close to the sun. If there were a solar atmosphere, it would be dense on account of the huge attracting mass of the sun. An observer on earth should perceive effects of refraction in the motion of Venus through such an atmosphere. With the help of his friend Captain Kater, who had worked on trigonometrical surveys, he followed the path of Venus. They detected no irregularities, concluding that the sun was without an atmosphere. In case the great solar heat rarefied any existing atmosphere to such a degree that it was imperceptible, Wollaston turned to Jupiter, a cold planet. The occultation of its satellites occurred regularly, which would not be the case if this event were viewed through a dense atmosphere. He concluded that the earth's atmosphere was of finite height, not reaching other planets; therefore atoms existed.

This paper was hardly likely to cause a general conversion to Dalton's theory, which Wollaston seemed to expect, since it took no account of the objections to the theory. There was no need to convince scientists that atoms existed; this was already a general Newtonian belief.[70] The real point was whether or not theoretical discussions on the ultimate nature of matter had a place in chemistry. The common answer throughout the nineteenth century was that this was a metaphysical concern, not a chemical one.

The new argument attracted a mixed reception. Turner, soon to become professor of chemistry at University College, London, accepted the proof as decisive.[71] Dalton used the same argument, though with-

---

[69] *Ibid.,* 90.
[70] This was explicitly stated by Thomson in his article "Atomic Theory," *Encyclopaedia Britannica,* Supplement to the 4th, 5th and 6th eds. (Edinburgh, 1824), p. 605.
[71] E. Turner, *Elements of Chemistry* (Edinburgh, 1827), pp. 130 and 176.

out reference to Wollaston.[72] Faraday admired Wollaston's reasoning and proceeded to apply it to the process of evaporation.[73] He argued from the equilibrium between the forces of repulsion and of gravity in vapours that there were limits to the production of vapours. For example as silver cooled from a white heat, the repulsion between the particles of vapour surrounding the metal gradually decreased until it was less than the force of gravity. At this point he falsely supposed that silver vapour would no longer exist, its particles fixed to the solid metal by the superior attraction. Faraday did not apply these thoughts further to the question of the existence of atoms, but Daubeny, professor of chemistry at Oxford, did do so and believed they confirmed Wollaston's conclusions.[74]

There next appeared some attempts to deny the necessity of Wollaston's conclusion. Thomas Graham, soon to publish his important studies on gaseous diffusion, pointed out that while Wollaston had suggested a possible cause for limited atmospheric expansion, another explanation seemed likely.[75] He said that the temperature of the atmosphere decreased with increasing height, and that this cooling alone would condense the atmosphere at a height of about 27.27 miles. He believed that the effects of such a condensation were visible at the poles where latent heat was supposedly liberated as the aurora borealis, thus explaining the phenomenon without mentioning atoms.

The same objection was made by Dumas, whose faith in Avogadro's theory had recently been shaken by anomalous results in vapour density studies. In a lecture[76] at the Collège de France he told students that they would find no proof of the existence of atoms in Dalton's work, since he had simply assumed them. Furthermore, he did not think it was necessary for chemistry to consider the ultimate divisibility of matter:

> Suppose chemical reactions only take place between *masses of a certain order*, divisible, if you like, by forces of another nature: all the phenomena of chemistry would be explained just as easily as if indi-

---

[72] Dalton, "On the Constitution of the Atmosphere," *Phil. Trans.*, *116* (1826), 174–187.

[73] Faraday, "On the Existence of a Limit to Vaporization," *Phil. Trans.*, *116* (1826), 484–493.

[74] C. Daubeny, *op. cit.* (note 12), 106.

[75] T. Graham, "On the Finite Extent of the Atmosphere," *Philosophical Magazine, 1* (1827), 107–109.

[76] J. B. Dumas, *Leçons sur la Philosophie Chimique* (Paris, 1837), pp. 231 f.

visibility was accepted as an essential property of these masses. Even if such masses were infinitely divisible with respect to non-chemical forces, this would not be relevant to explanations of chemical phenomena. . . . Do not all the conceptions of chemists exist independently of this ulterior divisibility? [77]

However Dumas was prepared to consider the evidence for atoms, since if their existence were established it might lead to a mathematical treatment of molecular phenomena as exact as Newtonian astronomy. His lecture continued with a detailed account of Wollaston's argument, which he said deserved close attention. But he was forced to reject it: "the whole scaffolding" of the reasoning collapsed once it was admitted that air could liquefy or solidify in the upper regions of the atmosphere. Low temperatures would condense the oxygen and nitrogen into layers, and this would suffice to limit the atmosphere. He concluded: "Therefore we will say to Wollaston: you have well established the absence of an atmosphere around the sun and Jupiter, but you have found nothing applicable to the question of atoms. It is unimportant whether matter is infinitely divisible or whether its division stops at a certain term: your observations will be explained without serious difficulty on either system." [78] Dumas found no evidence for atoms here or anywhere else. He wished the very word "atom" would be taken out of chemistry, since it went beyond experience.

This criticism did not convince Daubeny, who did not believe that temperatures in the upper atmosphere were low enough to liquefy oxygen and nitrogen. He saw Dumas' condensed enveloping layers as revived Greek crystalline spheres, interfering with the transmission of solar light and heat. [79] But his enthusiasm for Wollaston's argument was dampened by Whewell's remarks at the ninth meeting of the British Association. [80] Whewell, who was unusual for his vigorous opposition

---

[77] *Ibid.*, pp. 233–234. Thomson also wrote that the question of the divisibility of matter was best avoided in chemistry (*op. cit.* [note 43], *1*, 31). Dumas' conception of relative indivisibility, determined by the available powers of chemical analysis, was common. Further statements can be found in J. v. Liebig, *op. cit.* (note 44), 77–78, and W. Prout, *Chemistry, Meteorology and the Function of Digestion* (London, 1834), p. 22.

[78] Dumas, *op. cit.* (note 76), 242.

[79] "Ninth Meeting of the British Association for the Advancement of Science," *The Athenaeum Journal* (1839), p. 727.

[80] W. Whewell, "Remarks on Dr. Wollaston's Argument . . . ," *Reports of the British Association for the Advancement of Science* (1839), p. 26. Also his *The Philosophy of the Inductive Sciences* (London, 1840), *1*, 419–420.

to atomism in general, attacked Wollaston for inventing a baseless argument. He argued that the phenomenon of a finite atmosphere had nothing to do with atoms; it could be explained by supposing that the density of the air decreased with height in such a way that it became zero at a certain finite height.

It seems odd that none of Wollaston's critics brought forward the conflicting astronomical evidence for the existence of atmospheres on other planets, which was more cogent than speculations on the temperature and density of the upper atmosphere. They could have referred to Schroeter's observations on Venus or those of William Herschel on Saturn. In each case the irregular brightness of the surface was attributed to a planetary atmosphere. The belts of Jupiter were also well known. Yet, unaccountably, Wollaston's denial went unchallenged.

Odder still, the logical weakness of the argument was missed by Graham, Dumas, and Whewell. For there was a concealed assumption; it was that the repulsive particles of the finite atmosphere were atoms. But they might be particles of higher order, such as undivided molecules. This conclusive refutation of Wollaston's discussion was first made by George Wilson,[81] a lecturer in chemistry who had worked under Graham. He was anxious to expose a fallacy which he thought might mislead beginners. He pointed out that our atmosphere contained water and carbon dioxide. These were chemical compounds, so their particles could hardly be the ultimate indivisibles. As for oxygen and nitrogen, these might be compounds also. But even if they were elements, there was no reason to assume with Wollaston that they existed as monatomic particles. He continued:

Does it follow as a necessary inference, that because a body is simple, its gaseous repelling molecule must consist of but one atom? The answer is assuredly in the negative. The molecule might, on the other hand, be made up of a pair of atoms, like a binary star, with a centre of repulsion common to the two; or of 10, or of 100, or 1000 atoms (if such bodies there be), grouped together into a compound whole. . . . To prove that the atmosphere consisted of finite molecules, was only to reach the threshold of the difficulty; for each molecule supplied as good a text whereon to dispute the question of infinite

[81] G. Wilson, "On Wollaston's Argument . . . ," *Transactions of the Royal Society of Edinburgh, 16* (1849), 79–86.

divisibility, as the whole atmosphere out of which it was taken. The point which most of all demanded proof, namely, that the molecule was an atom, was the very one which he took for granted.[82]

Wollaston's last paper on the atomic theory had, he said, left the classical philosophical debate in its former state of indecision. The argument ceased to be used in atomic discussions, though it continued to be referred to in another context as opposing the theory of an ethereal medium filling the whole of space.[83]

Chemists did not mourn the loss of Wollaston's proof. They continued to avoid the philosophical question he had discussed. As the Irish chemist Robert Kane put it: "With such abstract speculations, however, chemistry had no connexion; its fundamental condition, that there exist many kinds of elementary matter, of which the quantity is measured by their weight, being totally independent of any abstract idea of what matter is, or how its properties may have their source."[84]

This position was not reversed until the very end of the nineteenth century. The discovery of fundamental particles finally convinced chemists that the structure of matter played an essential part in their science. But by then the indivisible atom, which Dalton envisaged and Wollaston believed he had established, had become obsolete.

V. CONCLUSION

No explanation can be given of Wollaston's wavering attitude to Dalton's theory. Neither in his other published papers, nor in his notebooks at Cambridge, is there any further information on the reasons for his repeated changes of opinion. Nor can they be related to internal developments in chemistry. It is true that in the period between his paper on equivalents and his attempted astronomical proof, the law of isomorphism and Dulong and Petit's law of specific heats

---

[82] *Ibid.*, 84.

[83] A. v. Humboldt, *Kosmos* (Stuttgart and Tübingen, 1845–62), *3*, 52; E. E. Schmid, *Lehrbuch der Meteorologie* (Leipzig, 1860), p. 51. On modern views, the atmosphere held by a planet depends on the gravitational mass and its temperature, though molecules can escape if they have sufficient energy, and a few do dissociate into atoms and ions. It is interesting to note that a recent attempt to determine the mean molecular weight of substances in Jupiter's atmosphere involved a photometric study of the light of a star, which faded as it sank behind that atmosphere. (W. A. Baum and A. D. Cole, "A Photometric Observation of the Occultation of σ Arietis by Jupiter," *The Astronomical Journal, 58* [1953], 108–112.)

[84] R. Kane, *Elements of Chemistry,* 2nd ed. (Dublin, 1849), p. 7.

were discovered which supported atomic chemistry. But the numerous exceptions to these, as well as the disagreement over chemical formulas, inhibited their application and eventually disappointed those, especially Berzelius, whose hopes had been raised. However there is no indication that these discoveries had anything to do with Wollaston's final return to atomism.

An attitude of enthusiasm for atomic chemistry, tempered by doubts, can be found in some of the leading chemists of the first half of the nineteenth century. Although Berzelius never accepted the details of Dalton's theory, he believed that some type of corpuscular theory was an essential part of chemistry. This remained for him a distant goal; his optimistic visions were checked by his admitted failure to discover how atomic weights could be calculated with certainty. Similarly Dumas began by considering atoms and polyatomic particles in chemistry. But his plan to determine atomic weights through vapour density studies encountered difficulties, later to be explained by thermal association or dissociation of particles. Consequently he wanted to eliminate atoms from chemistry, since, he said, there was no empirical evidence for them.

Wollaston's discussions differ from these in their extreme alternations, ending, through a most unusual argument, in a bold statement of belief in atoms. Yet it was his sceptical essay on equivalents rather than his other atomic speculations, or those of Berzelius, Dumas, or Davy, which expressed the most common feeling toward Dalton's theory. The frequent demand in the chemical literature of the period to draw a clear dividing line between fact and theory was reasonable enough, though as we have seen this was not achieved by adopting Wollaston's equivalents. The further tendency to insist that useful knowledge consisted of facts alone and to regard theory as an unnecessary encumberance resulted in a weak conception of chemistry.

Dalton's atomic theory did nothing to reverse this position, since it remained weak in explanation and prediction until the later developments by Kekulé, van't Hoff, and Lebel in the theory of valence and in stereochemistry. It is significant that Wollaston's paper on equivalents should have described Dalton's ideas as "purely theoretical" and of no practical use. For this was the actual state of the theory at the time. The favorable reception of Wollaston's equivalents is a valuable indication of the continuing weakness of Dalton's theory in the first half of the nineteenth century.

The Electric Current in Early Nineteenth-Century French Physics[1]

BY THEODORE M. BROWN*

The principal events in the early history of electric-circuit theory are well known. Luigi Galvani, an Italian, discovered the basic current effects in the 1780's, which Alessandro Volta, another Italian, first interpreted as physical phenomena in the 1790's; Humphry Davy, an Englishman, explored the electrochemical activities of Volta's invention, the "pile," during the first two decades of the nineteenth century; and H. C. Oersted, a Dane, with the help of the Frenchman, A. M. Ampère, opened the way to direct galvanometric measurement by noting current-magnet interaction in 1820. As the story is usually told, these advances, and intervening ones, were international in scope, rapid in pace, and cumulative in effect.[2] Cutting across national lines, nineteenth-century scientists marched quickly and without significant interruption to a circuit theory in which current was a continuous flow of electricity generated by chemical reactions in the pile and maintained in directly measurable amounts in the "connecting wire."

The real story, however, was not quite so simple. For one thing, national styles characterized European science at this time, and differing national styles were reflected in electrical research.[3] French physicists, with whom this paper is concerned, were by far the best

* Program in History and Philosophy of Science, Princeton University, Princeton, New Jersey 08540.
[1] An earlier version of this paper was written for Professor Thomas S. Kuhn's seminar in the history of physics at Princeton University. Since that time I have greatly benefitted from the advice of several friends and colleagues, particularly Dr. R. M. Young of Cambridge University, but my intellectual debt to Professor Kuhn remains foremost.
[2] See Sir Edmund Whittaker, *A History of the Theories of Aether and Electricity* (New York, 1960), *1*, Chapter III, for a detailed and fairly sophisticated version of the "usual story."
[3] For the "national styles" of nineteenth-century physics, see John T. Merz, *A History of European Thought in the Nineteenth Century* (New York, 1965), *1*, Chapters I–III. A recent account of French science during the Napoleonic era is Maurice Crosland's *The Society of Arcueil* (Cambridge, Mass., 1967).

organized and most sophisticated physicists in Europe. Grouped in the *Institut* (the *Académie* of the Napoleonic era), they maintained standards of instrumental precision best represented by the work of Coulomb, and they employed techniques of Newtonian analysis epitomized in the treatises of Laplace. What distinguished French physics in general also distinguished French electric circuit theory in particular; the electric circuit was analyzed differently and measured more precisely in France than it was in England, Italy, or Germany. Moreover (and perhaps most important), French theories of the electric circuit did not develop in a continuous or cumulative fashion, and certainly not according to the simple pattern outlined above. An examination of the printed sources for this period reveals, in fact, that from the turn of the nineteenth century until 1820, when only electrostatic measuring instruments were available, France's physicists made the pile, not the circuit of which it was a part, the principal object of research. After some initial hesitancy, they regularly used instrumental techniques borrowed from Coulomb to study the electrostatic configuration of the pile, and, in the face of chemical evidence vigorously advanced by the followers of Humphry Davy, they built and defended a theory of the piles operation based on deliberate and detailed electrostatic analogies. Only after 1820, when Ampère's interpretation of Oersted's discovery created, simultaneously, new instrumental and conceptual possibilities, did non-electrostatic phenomena in the pile and connecting wire become the main experimental concerns, while in theory the electric current began to look like a current for the first time.

It is the purpose of this paper to examine in detail the history of French electric-circuit theory just outlined. In the process new information will emerge about the practice of physics in early nineteenth-century France and, more generally, about the nature of conceptual transformation within a highly-organized and sophisticated community of scientists.

I. GALVANI, VOLTA, AND THE FIRST FRENCH REACTIONS TO THE NEW ELECTRICAL PHENOMENA

Electric-current phenomena first came to the attention of French scientists as the result of a convenient accident. In 1780, the anatomist Galvani observed unusual twitches in the legs of a laboratory frog;[4]

---

[4] Whittaker, *op. cit.* (note 2), 67 ff.

Galvani, who with his associates had been studying a wide range of irritable responses, noticed, quite by accident, that muscular convulsions were stimulated when two pieces of dissimilar metal were put in contact with the moist tissues of a frog.[5] He offered a tentative conjecture about these strange and confusing effects: the frog's tissues contain a special sort of "animal electricity," which produces convulsions when it passes from one part of the frog's body to another.[6] Two pieces of metal, zinc and copper for example, complete a circuit containing the frog's muscles and nerves; when the "animal electricity" discharges from the muscles, it stimulates the tissue to convulsive contraction.

Galvani's discovery was studied with considerable enthusiasm in many parts of Europe, not least so in France. Although there were few, if any, relevant papers in the French journals in the 1780's, by the 1790's members of the *Institut* had started to consider the multiplying and perplexing phenomena of "le Galvanisme." [7] Reports on new aspects of "l'influence de l'application des metaux sur l'irritabilité et la sensibilité" were frequent in 1796 and 1797, and during 1798 a regularly functioning "Commission du Galvanisme" consistently reported new discoveries and experiments.[8] By 1800, the variations and complexities of galvanic irritability were still pursued as popular and respectable research topics.

But in 1800 the discussion of "animal electricity" entered a new phase in France as in the rest of Europe. Volta, a professor of physics who had been supporting a rival theory to Galvani's since 1792, announced his invention of a device, the "pile," with which he claimed to overthrow once and for all the theory of the "animal electricians." [9]

---

[5] Galvani's investigations seem to belong to a long series of studies relating irritability and electricity undertaken in the latter part of the eighteenth century. Some of these studies have been described by H. E. Hoff, "Galvani and the Pre-Galvani Electrophysiologists," *Annals of Science, 1* (1936), 157–172.

[6] Whittaker, *op. cit.* (note 2), 69.

[7] I have not been able to find any papers in the *Memoires* or *Annales de Chimie* for the 1780's and early 1790's which treat these topics. The *Journal de Physique*, however, contains several communications of the early 1790's; see, for example, *41* (1792), 57 ff, and 66 ff. Papers of the late 1790's will be discussed below.

[8] This claim is based on a reading of the *Procès-Verbaux* of the French *Académie, 1* and 2. See also the *Annales de Chimie* for several of the important papers: for example, "Extrait d'une lettre de M. Humboldt à M. Blumenbach, contenant de nouvelles expériences sur l'irritation causée par les metaux," read at the *Institut* in *Frimaire, An* 5 (November 1796), *Annales de Chimie, 22* (1797), 51 ff. The *Commission du Galvanisme* was appointed on 21 *Brumaire, An* 5 and was made up of Charles, Coulomb, Fourcroy, Halle, Pelletan, Sabatier, and Vauquelin.

[9] See Whittaker, *op. cit.* (note 2), 69 ff.

Volta's "Pile"

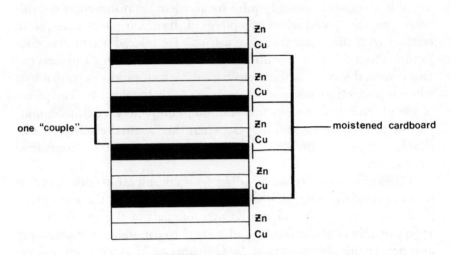

one "couple"

Zn
Cu
Zn
Cu
Zn
Cu   moistened cardboard
Zn
Cu
Zn
Cu

Volta constructed his pile as follows. Copper and zinc discs were placed in direct contact, and several of these copper-zinc "couples" were piled one on top of the other with pieces of moistened cardboard inserted between (see figure). The pile was a primitive wet-cell battery, but Volta did not explain his device's operation as we would. He saw statical electric "fluid" somehow set free and forced into motion at each zinc-copper surface of contact. This electric fluid was impelled from the first couple, freely traversed the cardboard layer, received an additional impulsion at the copper-zinc interface of the next higher couple, and so on.[10] Although Volta did not explain what the strange "contact force" was or how it operated, he asserted that its propulsive action on the electric fluid could be surmised from the strong and continuous *courant* of electricity one felt upon touching the terminals of the pile; in fact, galvanic spasms generally result when the electric fluid, generated and impelled by the contact force, stimulates ordinarily passive muscle tissue to contraction. All that the animal electricians really see in their so-called "galvanic" phenomena, he argued, are the physiological effects of contact-induced and pile-multiplied "voltaic" current.[11]

[10] See Volta, "On the electricity excited by the mere contact of conducting substances of different kinds," *Philosophical Transactions*, 90 (1800), 403 ff.
[11] Volta implies this throughout his article. His most explicit statement concerns the phenomena of taste, *ibid.*, 423–424.

Volta's theory was received with little enthusiasm, and the old objection of the animal electricians, that physiological effects depend rather on a special animal fluid than on ordinary electricity, was still raised, even against the pile. In fact, in what seems to have been the first serious public consideration of Volta's apparatus at a meeting of the *Institut,* French scientists, who generally seemed partial to animal electricity theories, ignored Volta's explanation of his own invention.[12] In *Fructidor, An* 8 (August 1800), M. Robertson read a memoir entitled "Expériences nouvelles sur le fluide galvanique," which tried to show that the sensations caused by touching the voltaic column were considerably different from those caused by the ordinary "electric commotion." [13] The column's effect depended on a special sort of irritation which varied according to the part of the body touched. Robertson proposed physical distinctions between galvanic and electric shocks; whereas the galvanic shock is obviously strong and continuous, an ordinary electric one is merely instantaneous. Moreover, during the galvanic commotion no sign of ordinary electricity can be detected by an electrometer. Robertson concluded with a vigorous rejection of Volta's opinion that the electric fluid was the principle of galvanism, and he argued, instead, for an acid *sui generis* which circulated in a continuous current as the "fluide galvanique." [14]

Volta soon had more to contend with than the continued vitality of animal electricity theories. In the first few months after announcing his invention, he was frequently informed of chemical changes observed during the column's operation. These chemical reactions attracted considerable attention in several parts of Europe and Britain, and they created fresh difficulties for Volta's simple contact theory.[15] Carlisle and Nicholson, and then Humphry Davy, noticed both gas-generating and oxidation reactions; the latter led Davy to suggest that "the oxydation of the zinc in the pile, and the chemical changes connected with it, are somehow the cause of the electrical effects it produces." [16] In France, too, scientists saw in the pile's chemical reactions

---

[12] Volta, however, had previously been known to the *Institut;* see *Procès-Verbaux, 1,* 162. See also, "Extrait d'une lettre de M. *Volta* à M. *Gren,* sur l'électricité dite animale," *Annales, 23* (1797), 276 ff.

[13] See *Procès-Verbaux 2,* 224. The paper was printed in the *Annales, 37* (1801), 132 ff.

[14] Robertson wrote, "J'ai rejetté le fluide electrique, comme principe du *galvanisme,* pour attribuer à un *acide, sui generis,*" *Annales, loc. cit.,* 145.

[15] Whittaker, *op. cit.* (note 2), 73 ff.

[16] *Ibid.,* 75.

convincing reasons to reject Volta's overly "simple" theory.[17] By 1801, some of France's scientists were so busy collecting new data on the pile's various actions that they had neither time nor desire to speculate on their cause. M. Gautherot, one of the most active contributors at this time to the *Institut's* consideration of voltaic electricity, read a memoir "Sur le Galvanisme" in *Ventose, An* 9 (February 1801), in which he claimed that the new phenomena were so strange and unique that it was hazardous to advance a theory. One ought to, he argued, merely accumulate new data: "j'ai pensé qu'il serait utile aux progrès de ces connaissances, de ne m'occuper d'abord que de la recherche des faits, et d'observer avec soin leur modification." [18]

During this period of intense experimental activity and continued controversy over theory, Volta had an opportunity to present his own case directly to the *Institut*. Late in 1801 he came before the *Classe de Physique et Mathematique* to read a paper "De l'électricité dite galvanique," in which he reasserted the identity of galvanic and electric fluids in spite of seemingly powerful counter-evidence.[19] Volta listed the principal objections to his own position. First, certain signs of ordinary electricity are absent during galvanic effects; second, substances transmit the galvanic fluid very poorly in comparison with the electric fluid; and third, water can be decomposed by the galvanic fluid with little or no sign of directly measurable electrical activity, whereas rapid and "high tension" static electricity is needed for the same effects. Volta replied to these objections by demonstrating that the force which gives the electric fluid its impulsion comes only from the contact of metal with metal. With combinations of collecting plates, condensers, and electroscopes (the standard electrostatic instruments then available), he obtained faint but clearly recognizable signs of static electricity from pairs of metals. By joining insulated pieces of zinc and silver and then discharging them individually onto a condensing electroscope he demonstrated that contact develops static electricity, positive in the first metal and negative in the second.[20] In the pile, sum-

---

[17] M. Désormes claims that his observations cannot be accounted for by the "simple théorie de Volta." Désormes, "Expériences et observations sur les phénomènes phisiques et chimiques que présente l'appareil électrique de Volta," read to the *Institut* in *Ventose, An* 9 (Feb. 1801), *Annales, 37* (1801), 284 ff.

[18] Gautherot, "Mémoire sur le galvanisme," *Annales, 39* (1801), 205.

[19] Volta's appearance is recorded in the *Procès-Verbaux, 2,* 425. His paper was published in the *Annales, 40* (1802), 225 ff.

[20] *Ibid.,* 236–248.

ming together electricity from many couples merely multiplies the small individual contribution of each.

The scientists of the *Institut* were now challenged to decide between theories or to give allegiance to no theory at all. After some hesitation and perhaps in response to certain political pressures from above, they chose to appoint a special committee, composed, among others, of Laplace, Coulomb, Fourcroy, Guyton, and Biot,[21] to investigate and report on the experiments and theories of Volta. In order to understand this committee's actions, it is necessary briefly to consider first another set of electrical investigations undertaken at the end of the eighteenth century.

2. COULOMB'S ELECTROSTATICS

Shortly before the discoveries of Galvani and Volta, Charles Augustin Coulomb completed a series of memoirs on well-known and generally understood electrostatic effects.[22] Coulomb was one of France's great engineer-physicists, combining in his work remarkable skill in experimental technique with ease and precision in mathematical analysis. He employed his talents to remodel completely the science of electrostatics in the 1780's.[23] A few samples will serve to give the flavor

[21] The activities and deliberations of the special commission are still obscure, but the final results were publicly announced in the official *Mémoires*, 5 (1804), "Histoire" section, 195 ff., in the form of a "Rapport" by Biot, which will be discussed further below. The commission for which Biot reported seems to have been originally constituted in *Brumaire, An* 10 (October 1801), and politics seems to have played a role in its appointment. The entry in the *Procès-Verbaux* for 21 *Brumaire, An* 10 reads: "Le C^n *Volta*, professeur de Pavie, lit la première partie d'un Mémoire sur la théorie du galvanisme et particulièrement sur la nature du fluide galvanique. Le C^n *Bonaparte* propose que la Classe, manifestant, dès les premiers moments de la paix générale, le désir de recueillir les lumières de tous ceux qui cultivent les sciences, donne une médaille d'or au C^n Volta, la premier savant étranger qui, depuis la paix, ait lu un Mémoire dans le sein de la Classe, comme une marque de son estime particulière pour ce professeur et de son empressement à accueillir les travaux de tous les savans étrangers. Il propose de plus qu'une Commission soit chargée par la Classe de faire en grand toutes les expériences propres à répandre un nouveau jour sur l'importante branche de la physique dont le C^n Volta vient d'entretenir la Classe, et il demande que ses propositions soient renvoyées à une Commission. La Classe renvoye les propositions du C^n Bonaparte à la Commission déjà nommée pour s'occuper du galvanisme." It is important to note that the special commission for which Biot began to report on 11 *Frimaire, An* 10 was *not* the same committee previously responsible for reporting on galvanic phenomena. Biot was not a member of the old commission; see note 8 above.

[22] Coulomb's research covered the period from 1785 to 1789. His memoirs are collected in *Collection de Mémoires Relatifs à la Physique*, 1 (Paris, 1884).

[23] This view is presented in a doctoral dissertation, *Charles Augustin Coulomb: Physics and Engineering in Eighteenth-century France*, written by Charles Stewart Gillmor for Princeton University (May 1968).

and style of his work. Using the torsion balance of his own invention he verified the inverse-square law of electrostatic attraction.[24] He also determined the laws of electrostatic leakage, employing an experimental procedure that was as simple as it was elegant. With the balance he measured the static force of a test body and, several minutes later, he measured that force again;[25] he found that the force sensibly decreased from the first to the second reading; this he attributed to the dissipation of supposedly isolated electricity. By altering the initial charge, length, and substance of the support, and by noting variations in atmospheric conditions, he correlated his findings as laws of electrostatic leakage. He found, generally, that, under given atmospheric conditions, the rate of dissipation varied in neat mathematical proportion: directly as the quantity of the original charge and inversely as the squared length of the dielectric support.[26]

Fully exploiting these dissipation studies, Coulomb moved on to a more significant and impressive examination of the electrostatic configuration of charged bodies. Here he found that he needed supplementary devices, for the balance, however precise, was too bulky for most experimental circumstances. He met this difficulty by designing his "proof plane," a small disc of gold leaf stuck to a thin wire of gum lac. This plane was applied to many points of a conductor, receiving, because of its small diameter, a charge proportional to the electric density at each point touched. The charge of the plane was then measured by the balance; in this way the complicated distribution of electricity on bodies of different figures could be determined as a straightforward laboratory exercise.[27]

As a result of this work, French physicists of the late eighteenth century possessed a highly refined science of electrostatics, consisting of Coulomb's recent experimental and mathematical results and techniques adequate for all known electrical phenomena. Almost immediately after Coulomb completed these investigations, however, Galvani, Volta, and their new discoveries intruded upon the scene, and

[24] Some account of Coulomb's experimental techniques can be found in W. James King, "The Quantification of the Concepts of Electric Charge and Electric Current," *The Natural Philosopher*, 2 (1963), 107–127.

[25] Coulomb's memoir on leakage, the third in the *Mémoires Relatifs*, is called "De la quantité d'électricité qu'un corps isolé perd un temps donné."

[26] *Ibid.*, 166.

[27] The "proof plane" is described and then employed in the fifth memoir, "Sur la manière dont le fluide électrique se partage entre deux corps conducteurs mis en contact," Sections XLIV and XLV.

French scientists, in their first reactions, seemed temporarily to forget Coulomb, his balance, and the new science of electrostatics. Only when Volta came before the *Institut* and the special committee was appointed to consider his results did Coulomb, who figured prominently on that committee, regain full control of French electrical science.

3. BIOT'S "RAPPORT" AND THE ESTABLISHMENT
OF THE ELECTROSTATIC INTERPRETATION OF THE PILE

The *Rapport* of the committee was written by Jean Baptiste Biot, a rising young physicist who seems to have been deeply influenced by Coulomb and Laplace, the two senior physicists on the commission.[28] Before his appointment as *rapporteur,* Biot had contributed a few papers on mathematics, and, like Laplace, he seems to have maintained his mathematician's sensibilities while studying the voltaic apparatus,[29] treating the pile as a device which was best understood in terms of rigorous, quantitative, Coulombian electrostatics.

Biot's *Rapport* begins with a brief history of galvanic and voltaic electricity and moves into a graceful apology for the *Institut*'s too long delay in supporting Volta's side in the dispute. Volta's experiments on the electrostatic effects of copper-zinc contact are reported next and commended. On the basis of these observations and careful electrostatic reasoning, Biot claims that the pile's operation can be easily explained: in a stack of copper-zinc-pasteboard elements, kept on an isolating support, each zinc element will, because of contact, be in an electric state different from that of the copper plate immediately beneath it. Regarding the excess electricity of the zinc relative to that of the copper as unity, Biot explains:

> Si la pile n'est composée de deux pieces, l'une inférieure de cuivre, l'autre supérieure de zinc, l'état électrique de la première sera représenté par $-\frac{1}{2}$, et celui de la seconde par $+\frac{1}{2}$.
>
> Si l'on ajoute une troisième pièce qui doit être de cuivre, il faudra, pour qu'il se fasse un déplacement de fluide, la séparer, par un carton mouillé, de la pièce de zinc inférieure; alors elle devra acquérir le même état électrique que cette dernière, du moins en négligeant

---

[28] For a brief biography of Biot, see *Dictionnaire de Biographie Française* (Paris, 1954), 6, 506.
[29] On 1 *Brumaire, An* 8 (October 1799), for example, Biot "lit un Mémoire intitulé *Considèrations sur les équations aux différences mêleès*"; see *Procès-Verbaux*, 2, 18. For the "maintenance" of Biot's mathematical sensibilities, see below.

l'action propre de l'eau qui paroît fort petite, et peut-être encore la très-foible résistance que ce liquide, comme conducteur imparfait de l'électricité, peut opposer à la communication. L'appareil étant isolé, l'excès de la pièce supérieure ne peut s'acquérir qu'aux dépens de la pièce de cuivre qui est au-dessous: alors les états respectifs de ces pièces ne seront plus les mêmes que dans l'expérience précédente, et deviendront:

Pour la pièce inférieure, qui est de cuivre − ⅔;
Pour la seconde, qui la touche et qui est de zinc, − ⅔ + 1 ou ⅓.

La troisième qui est de cuivre, et qui est séparée de la précédente par un carton mouillé, aura la même quantité d'électricité, c'est à dire + ⅓; et la somme des quantités d'électricité perdu par la première pièce, et acquise par les deux autres, sera encore égale à zéro, comme dans le cas de deux pièces.

Si nous ajoutons une quatrième pièce, qui sera de zinc, elle devra avoir une unité de plus que celle de cuivre, à laquelle elle est immédiatement superposée: cet excès ne pouvant s'acquérir qu'aux dépens des pièces inférieures, puisque la pile est isolée, on aura:

Pour la pièce inférieure, qui est de cuivre − 1;
Pour la seconde de pièce, qui la touche et qui est de zinc, 0, c'est-à-dire qu'elle sera dans l'état naturel;
Pour la troisième pièce, qui est de cuivre, et qui est séparée de la précédente par un carton mouillé, 0; elle sera aussi dans l'état naturel.
Enfin, pour la pièce supérieure, qui est de zinc, et qui est en contact avec la précédente, + 1.

En poursuivant le même raisonnement on trouveras les états électriques de chaque pièce de la pile, en la supposant isolée et formée d'un nombre quelconque d'élémens; les quantités d'électricité croîtront, pour chacun d'eux, de la base au sommet de la colonne suivant une progression arithmétique, dont la somme égale à zéro.[30]

This detailed account of the electrostatic configuration of the column is the principal theme of Biot's *Rapport*. There are also three appendices filled with supplementary algebraic details and instructions for calculating the electric state of the pile.[31] Volta had never mentioned anything like this in his papers. He had spoken of the "perpetual impul-

---

[30] "Rapport," *op. cit.* (note 21), 199–201.
[31] See the three "Notes," *ibid.*, 211–222.

sion" of the electric fluids and had simply (and vaguely) referred to the increasing intensity of the communicated shocks resulting from the addition of elements to the column.[32] Biot's concentration on electrostatics thus seems more a sign of Coulomb's influence than of Volta's.

By so rigorously examining the distribution of static electricity in the pile, the commission consciously extended the domain of Coulombian electrostatics. At the same time, and for the same reason, the committee, probably unconsciously, distorted some of Volta's most important ideas. The pile of Biot's report differs from Volta's pile in two subtle but crucial respects: in the nature of electromotive force, and in the meaning of electrical "circulation."

The perpetual impulsion of electricity due to the contact force was a point on which Volta had insisted from the first and had reaffirmed as recently as in his address to the *Institut.* In Biot's *Rapport,* the unexplained electromotive force of contact does *not* impel electricity; contact force merely holds already separated electricities apart. Each zinc element in a stack of couples is, by virtue of the contact force, in an electric state different from that of the matching copper plate, and the liquid or moist cardboard between each couple simply conducts the electric fluid separated at the copper-zinc junctions. Thus, while the zinc-cardboard interface allows electricity to pass unidirectionally, the contact force at the zinc-copper interface prevents the once separated fluids from recombining. Contact force clearly serves a dual function: first, it sorts electricity, and, second, it keeps it sorted. Both, obviously, are electrostatic functions, and very little seems left of Volta's *active impulsion.*

The second important change between Volta's and Biot's pile follows from the commission's shift of interest from the "commotion" administered externally to the electrostatic state maintained internally by the column. Volta's unqualifiedly continuous current becomes a complicated kind of static discharge. These are Biot's words:

> Alors, si l'on touche d'une main le sommet de la pile, et de l'autre sa base, ces excès d'électricité se déchargeront à travers les organes dans le réservoir commun, et exciteront une commotion d'autant plus sen-

[32] In his paper in the *Philosophical Transactions* for 1800, *op. cit.* (note 10), 405, Volta merely explains that adding more elements to the column intensifies the resulting "commotions"; he makes no more specific correlation than this. Similarly, in his memoir, "De l'électricité dite galvanique," *op. cit.* (note 19), he explained that multiplying couples multiplies observable tension, but he left his remarks at this.

sible, que cette perte se réparant aux dépens du sol, il doit en résulter un courant électrique dont la rapidité plus grande dans l'intérieur de la pile que dans les organes, qui sont des conducteurs imparfaits, permet à la partie intérieure de la pile de reprendre un degré de tension qui s'approche de celui qu'elle avoit dans l'état d'équilibre.[33]

Volta's *courant* of perpetually impelled electricity is here broken into a multitude of instantaneous discharge currents. And the pile now delivers only a *seemingly* continuous shock, whose appearance of continuity depends on the accidental circumstance that the pile repolarizes before the organs of the body fully discharge the electricity they have already received from it.

By these subtle alterations, Biot expressed the commission's preoccupation with the theoretical side of Coulomb's electrostatics. When he commented on unsolved difficulties, he proclaimed a similar dedication to the experimental half of the Coulombian science:

> ... il reste à la déterminer d'une manière rigoureuse, à chercher si elle est constante pour les mêmes metaux, ou si elle varie avec les quantitiés d'électricité qu'ils contiennent, et avec leur température. Il faut évaluer avec la même précision l'action propre que les liquides exercent les uns sur les autres et sur les métaux. C'est alors que l'on pourra établir le calcul sur des données exactes, s'élever ainsi à la véritable loi que suivent, dans l'appareil du citoyen Volta, la distribution et le mouvement de l'électricité, et compléter l'explication de tous les phénomènes que cet appareil présente. Mais ces recherches délicates exigent l'emploi des instrumens les plus précis qu'aient inventés les physiciens pour mesurer la force du fluide électrique.[34]

Biot seems to have had Coulomb's balance in mind when he wrote this, especially the last sentence; not content to leave his desire a thing of words and promises, he soon began new research on the pile in which he employed this "most precise instrument."

Biot's results were announced to the *Institut* in *Prairial, An* 11 (May 1803) and published in the *Annales de Chemie* as a forty-page memoir, "Recherches physiques sur cette question: quelle est l'influence de l'oxidation sur l'électricité developpée par la colonne de Volta." [35] The immediate stimulus for Biot's research probably was the con-

---

[33] "Rapport," *op. cit.* (note 21), 203. The term "courant électrique" here seems to have a fully electrostatic import.
[34] *Ibid.*, 208.
[35] See *Annales, 47* (1803), 5 ff.

tinued insistence by some writers that oxidation was the principal source of electrical action in the column.[36] He had already tried to refute this claim in 1801, when he collaborated with Frederic Cuvier on a brief study of the absorption of oxygen by the working column, and he probably wanted to strengthen his refutation.[37] But Biot, as *Rapporteur* for the special commission and representative of Coulomb, had, in 1803, a more general goal as well. He wanted to show the immense power of the electrostatic interpretation of the pile.

Biot states his underlying intention at the outset of the 1803 memoir: to collect "expériences incontestables" for contact force (in the commission's special sense) as the sole operating principle in the voltaic column.[38] Because Volta's experimental techniques were persuasive but *not precise enough,* Biot could not accept Volta's claim to have proved his hypothesis. He took the contrary opinions of Davy, Nicholson, Ritter, Pfaff, Van Marum, and Wollaston no more seriously, for they too lacked the precise sort of experimental data which such research requires. For Biot, only Coulomb's electric balance in the cabinet of the *Institut* yielded a sufficient degree of exactitude; he would be satisfied with no other measuring instrument.

It was not easy to use Coulomb's balance on the pile. Biot had, for example, to eliminate variable or accidental grounding, and for this he designed a special apparatus which ensured constant contact with the ground, "au moyen d'un parallélipipède de bois revêtu d'une feuille d'étain, et fixement attaché à la table où se faisoient les expériences; ce parallélipipède servoit de support à la pile, et portoit à une de ses extrémités une tige de cuivre verticale et mobile, terminée par un plateau métallique horisontal qui servoit de support au condensateur. On pouvoit ainsi amener cet instrument à la hauteur de la pile soumise à l'expérience, sans changer en rien les communications établies." [39] Biot next guaranteed a constant contact of the condenser with the top of the pile by placing on the latter "un petit vase de fer rempli de mercure; le bouton du condensateur à l'extrémité de sa

---

[36] See Whittaker, *op. cit.* (note 2), 75. Biot may also have been attacking specific opinions of Frenchmen, such as those of Hachette and Désormes. He reported on their memoir on 3 *Prairial, An* 11 and read his own "oxidation" memoir on 17 *Prairial, An* 11; see *Procès-Verbaux,* 2, 669 and 674.

[37] In 1801 Biot's aim was probably to refute the claims for the oxidative generation of voltaic electricity made at the *Institut* by Gautherot. Biot's joint paper with Cuvier was printed in the *Annales, 39* (1801), 242 ff.

[38] Biot, "oxidation" memoir, *op. cit.* (note 35), 6.

[39] *Ibid.,* 18.

tige flexible étoit aussi en fer. De cette manière, lorsqu'il étoit amené à la hauteur de la pile, il suffisoit d'abaisser son bouton dans le mercure, à l'aide d'un tube de verre verni, et en abandonnant ensuite la tige à sa propre élasticité, on étoit certain d'avoir l'effet d'un simple contact aussi exactement que possible." [40] Biot then measured the electricity on the condenser knob with the electric balance. He also varied the material moistening the inter-element pasteboards and thus determined directly and precisely the influence of the conductor. These were some of his results:[41]

Conducting Solution	Mean Torsion Angle of Balance
Soda	70
Pure Water	77
Sulfate of Aluminum	81 + ½

While Biot found that as he changed the conductor from soda to sulfate of aluminum the electric "tension" of the column, measured by the torsion balance, varied markedly, he still did not know whether the variations depended on the chemical activity of the conducting substances or merely on their physical conductivities. To answer this question, he varied the quantity of chemical substance on the pasteboards, noting if a significant shift took place in the column's measurable electrostatic properties. Biot's fundamental assumption was that if chemical actions are to be considered important in the operation of the pile, then they must produce more than negligible effects.

Biot constructed two voltaic columns, each of twenty zinc-copper elements, and generally similar, except that in one an aluminum sulfate conducting solution moistened the entire surface of the pasteboard discs, while in the other the liquid covered only one-ninth of the surface-area. Using condenser and balance as before, Biot again determined mean torsion angles. For the pile with aluminum sulfate on the whole of the pasteboards the angle was 83 + ½°; for the other it was 85°.[42] These variations were far too small to weaken Biot's faith in the electrostatic theory of the pile, and the experiment seemed to confirm, or at least not to falsify, the hypotheses of the special commission.

The arguments and techniques on which Biot relied were of considerable influence, for it soon became evident that he and the other

[40] *Ibid.*, 19.
[41] *Ibid.*, 20.
[42] *Ibid.*, 34–35.

members of the commission were not the only prominent French scientists publicly favoring the reasoning of the *Rapport* and of the oxidation memoir. In his *Traité Élémentaire de Physique* (1803), R.-J. Haüy dates the modern era of voltaic science from the commission's *Rapport.* Section V of this influential textbook contains an historical survey which passes quickly from Galvani's and Volta's experiments on animal and contact electricity to an exposition of Biot's new and up-to-date investigations. The most original aspects of Haüy's account, however, are the extensions of the electrostatic explanation of the pile's activity.[43] For one thing, he makes explicit the commission's implicit distortion of Volta's notion of contact electromotive force. Rejecting the Italian scientist's "impulsion qui agit pour chasser dans le zinc une partie du fluide électrique que possedait le cuivre," Haüy favors a "simple énoncé des faits, sans entrer dans la considération de la force motrice, qui ne semble pas être encore bien connue."[44] For Haüy all electrical "facts" are strictly electrostatic, and, therefore, contact force cannot be responsible for any sort of "impulsion." Secondly, Haüy makes the commission's probably unconscious alteration of Volta's meaning of circulation explicit in his *Traité.* He calls attention to the *successive* nature of the extremely rapid but nonetheless discrete shocks into which Biot's *Rapport* had implicity converted Volta's continuous current.[45] Finally, Haüy ends his discussion of voltaic electricity with a section entitled "Analogie entre l'électricité galvanique et électricité ordinaire"; the evidence cited here in support of the analogy, which thus becomes almost an identity, had been collected mainly by Van Marum, who had successfully duplicated several voltaic phenomena, such as chemical decomposition, with electricity from static machines.[46]

In another place, Haüy illustrates again the vast power of the electrostatic theory of the pile. Turning his attention to the curious phenomena which the chemists Thenard, Vauquelin, and Fourcroy had reported,[47] unexplained, to the *Institut* in 1801, he shows how even chemical "anomalies" can be accounted for in electrostatic terms.

---

[43] R.-J. Haüy, *Traité Élémentaire de Physique* (Paris, 1803). A second edition appeared in 1806, the one cited here. Haüy claims that Biot's "Rapport" is "le plus méthodique et plus lumineux qui ait paru sur cette matière," *ibid.*, 2, 15.

[44] *Ibid.*, 15.

[45] Haüy contrasts the single influence of the Leyden jar to that of the pile, which "attaque par une succession rapide de petites impulsions . . . ," *ibid.*, 56.

[46] *Ibid.*, 50 ff.

[47] One of their papers was printed in the *Annales, 39* (1801), 103–104.

These three investigators had noticed that piles of the same number but of different-sized elements sometimes produced strangely variant effects while exhibiting the same degree of electrostatic tension. For example, although both cause physiologically indistinguishable "commotions," a pile of twenty large plates heats metal wires more forcefully than does one of twenty small plates. Haüy reasoned that in physiological "commotions" a balance is established between the greater electrical "mass" discharged in the first moment of the larger pile's operation and the more rapid overall discharge of electricity in the small pile. The pile with small elements obviously requires less time to recharge between successive shocks. But in heating wires the larger pile is more effective due to the intense combustive effect of the greater electrical mass discharged in the first instant.[48]

Haüy's electrostatic comparisons thus corroborate, where the phenomena they were meant to explain might have contradicted, the commission's theory of voltaic electricity. A few holdouts for alternative theories were still in evidence at the *Institut* in 1803, but afterwards they seem to have disappeared almost entirely.[49] Along with Biot's "oxidation" memoir, Haüy's popular textbook, containing some of the best contemporary electrostatic thought on voltaic electricity, no doubt contributed to the rapid acceptance of the commission's views by other French scientists. It was this analysis of voltaic electricity which, though modified in detail, remained unquestioned in essence until 1820.

4. THE ELECTROSTATIC THEORY OF THE PILE
   IN FRENCH PHYSICS, 1804–1820

Just when theoretical homogeneity started to take root, the original intense interest in the principles of voltaic electricity began to wither. By 1804, few papers of theoretical concern were brought to the attention of the *Institut*.[50] A *Société du Galvanisme* was founded in 1803, but

---

[48] Haüy, *op. cit.* (note 43), 29 ff.

[49] The repeated appointment of investigating committees with Biot as the active member may well have been largely responsible for introducing theoretical homogeneity; see note 36 and *Procès-Verbaux, 2,* 579 for typical examples. After 1803, Biot's position seems to be the one generally accepted by French physicists.

[50] In contrast to earlier papers, the few items on voltaic electricity presented to the *Institut* in 1804 seem to have had little theoretical content. These are two typical entries in the *Procès-Verbaux* for 1804: "M. Allizeau annonce qu'il a perfectionné la pile galvanique qu'il avoit présentée l'année dernière. Il présente cette pile nouvelle qui est renvoyée à l'examen de la Commission du Galvanisme" (*3,* 103 [22 *Prairial, An* 12]); "M. Isarn présente son Manuel du galvanisme, Paris, 1804, 1 volume in −8°" (*3,* 105 [13 *Messidor, An* 12]).

its announced purpose had nothing to do with theory; its aim was to investigate voltaic *effects,* particularly medical ones. The Society's new *Journal du Galvanisme* published many articles on chemical experiments performed with the pile, but in all these investigations the pile was used as an instrument and was not itself the object of research.[51] A similar lack of concern for further exploration of the voltaic column can be seen in papers presented to the *Institut* and printed in the well-established French scientific journals. Although the editors of the *Institut's* official *Mémoires* did not print a single article on voltaic electricity during the two decades following Biot's *Rapport,* the *Annales* published almost any relevant study. Yet after 1803 most of the papers in the *Annales* were either letters by foreigners or excerpts from foreign journals.[52] In England, Germany, and Scandinavia, Davy, Ritter, Grotthus, and Berzelius produced important electrochemical researches, which the *Annales* recorded; but in France only two chemists, Gay-Lussac and Thenard, who belonged to the *École Polytechnique,* seemed to follow up any of this research.[53]

The work of Thenard and Gay-Lussac, in fact, belongs to a peculiar middle ground. When they concentrated on the chemical properties of the pile, to which Davy and others directed them, they began to experiment and theorize in ways which forced them to qualify the straightforward electrostatic explanations of Haüy and Biot, whose basic assumptions they nevertheless still wished to maintain. Because of minimal activity elsewhere in France and because of the frequent reporting of their research to the *Institut,* Gay-Lussac's and Thenard's individual and joint publications must be taken as accurate indica-

---

[51] The *Annales* for 1803 contains several notices, *46,* 110–112 and 224, and *47,* 110–111, of a new *Journal du Galvanisme, de Vaccine . . . ,* which is edited by J. Nauche, president of the *Société Galvanique.* The *Journal,* whose contents are frequently summarized in the *Annales,* contains articles primarily on the medical aspects of galvanism. The application of galvanic shock to cases of paralysis is typical. It also prints straightforward accounts of chemical observations; for example, "Notice des expériences faites par la Société Galvanique, sur la decouverte annoncés par M. Pacchioni, de la composition de l'acide muriatique, par M. Riffault, membre de la classe des recherches physiques de cette Société," *Annales, 56* (1806), 152 ff.

[52] A brief catalogue of articles by foreigners in the *Annales* in this period would include works by these authors: Berzelius, *86* (1813), 146 ff. and *87* (1813), 50 ff. and 113 ff.; Bucholz, *66* (1808), 266 ff.; Davy, *44* (1803), 206 ff., *63* (1807), 172 ff., *68* (1808), 203 ff., *70* (1809), 189 ff., *75* (1810), 27 ff. and 129 ff.; Erman, *61* (1807), 113 ff.; Gehlen, *66* (1808), 191 ff.; Grotthus, *58* (1806), 54 ff., *63* (1807), 5 ff.; Heidman, *61* (1807), 70 ff.; Petrini and Cioni, *56* (1806), 269 ff.; Pfaff, *62* (1807), 23 ff.; and Ritter, *64* (1807), 68 ff. In contrast, the only Frenchmen to appear in the *Annales* in these years are: Chompre, *61* (1807), 58 ff.; Chompre and Riffault, *63* (1807), 77 ff.; Guyton, *63* (1807), 113 ff.; Hachette, *65* (1808), 211 ff.; and Gay-Lussac and Thenard, *73* (1810), 197 ff., *78* (1811), 243 ff. and *79* (1811), 36 ff.

[53] Berthollet may have too, since he was charged with reviewing Davy's *Bakerian Lectures* when they arrived at the *Institut* in April 1807; see *Procès-Verbaux, 3,* 516.

tors of whatever active and creative French thought on voltaic electricity there was in the years between 1806, when the second edition of Haüy's *Traité* was issued, and 1816, when Biot's *Traité de Physique* appeared.[54]

In the opening essay of their *Recherches Physico-Chimiques* (1811), Gay-Lussac and Thenard outline and substantiate their qualified version of the electrostatic theory of the pile. Proclaiming general loyalty to the electrostatic approach ("Les diverses circonstances de l'équilibre du fluide électrique dans la pile de Volta ont été parafaitement discutées dans un rapport de la commission nommée par l'Institut"[55]), they go on to raise difficulties for the standard interpretation. The difficulties come from a set of experiments on solutions of acids and salts; these convinced Gay-Lussac and Thenard that the chemical energy of the pile can sometimes vary independently of its directly measurable electrical energy.[56] By collecting volumes of gases and using these volumes as indicators of chemical energy, they were able to show that the electrostatic force, as measured by Biot's apparatus, remains constant in certain circumstances for forty minutes, whereas chemical energy actually declines after twenty. Likewise, different "conducting" substances sometimes produce demonstrably different chemical energies at the same time that they give rise to equivalent electrostatic signs. These results would have posed serious difficulties for the Haüy-Biot theory if Gay-Lussac and Thenard did not themselves fall back on just that theory to account for these phenomena. But the chemists themselves chose to explain differences between chemical and electrostatic energy in terms of discharge velocities and internal conductibilities. Thus, they argued that chemical energy depends on the quantity of electricity discharged in a given time, since the same tension could be imagined to obtain in columns of different discharge rates. And though in this explanation Gay-Lussac and Thenard account for chemical activity in terms of the *rapid movement* rather than the static configuration of electricity, their explanation clearly is a simple variation of Haüy's electrostatic analysis of the effects of large- and small-disc piles.

---

[54] A. M. Ampère was also interested in Davy's work, and his interest was important for the period beginning in 1820. For Ampère's correspondence with Davy, see *Correspondance de Grand Ampère*, ed. L. de Launay (Paris, 1936), *1*, 110, 355–357, and 363. French studies on electrochemistry at this time seem, for all concerned, rather a response to foreign work than a native French development.

[55] Gay-Lussac and Thenard, *Recherches Physico-Chimiques* (Paris, 1811), *1*, 10.

[56] For the distinction between these two "energies," see *ibid.,* 12 ff.

The same advance to and retreat from a critical discussion of the standard electrostatic analysis of the pile can be seen in their individual works. Thenard's discussion of voltaic electricity in his *Traité de Chimie* (1813) combines a careful précis of Biot's *Rapport* with an account of several seemingly anomalous discoveries.[57] He explains how joining the column's poles by a perfect conductor reduces its tension to zero without interfering with certain important chemical effects. But instead of using this observation as a lever with which to topple the electrostatic theory, Thenard concludes with a deliberate comparison of the voltaic apparatus to "une bouteille de Leyde qui aurait la propriété de se recharger d'elle-meme, aussitôt qu'elle serait dechargée." Gay-Lussac's attitudes were similar. Despite his collaboration on the critical sections of the *Recherches Physico-Chimiques,* he falls back on the Haüy-Biot theory in an important article in the *Annales* for 1816; there he explains the insignificance of electrostatic signs in dry piles in terms of the poor conduction of static electricity within them.[58]

The effect of Thenard's and Gay-Lussac's criticisms, therefore, was not to overthrow but to remodel the electrostatic approach to voltaic electricity. If direct comparison of the operating pile to a discharging Leyden jar did not account for all known experimental circumstances, then the actual mechanism of the pile's operation was complicated, but electrostatic just the same. Biot fully realized these implications, and when he wrote the first edition of his four-volume *Traité de Physique, Expérimentale et Mathématique* (1816), he had Gay-Lussac's and Thenard's work clearly in mind.

The eighty closely-printed pages of Biot's discussion of the electric column in his *Traité* can be divided into two distinct but related parts. In one Biot expounds with greater clarity, more extensive numerical detail, and increased mathematical elegance the electrostatic explanations of the pile he had written in 1802 and 1803; but there are no basic conceptual alterations here. In the other part, however, Biot modifies his views in a confrontation with new chemical evidence. Relying heavily on Gay-Lussac and Thenard for information, he shows that his own attitude toward the new phenomena is not fundamentally different from theirs. He, too, claims that variations in chemical activity not correlated with equivalent changes in observable electrical signs require a modification rather than total rejection of the

[57] L. J. Thenard, *Traité de Chimie* (Paris, 1813), *1,* 89 ff.
[58] G(ay)-L(ussac), "Sur les piles seches voltaïques," *Annales,* 2 (1816), 76 ff.

electrostatic theory of the pile. But unlike Gay-Lussac and Thenard, Biot fully recognizes the serious challenge of the chemical phenomena. In response, he deliberately extends the static theory in anticipation of possible experimental refutation, and, as a result, his tone, more than anything else, is very different in 1816 from what it had been a decade before. Thus in the *Traité*, instead of enthusiastically constructing a fresh theory, Biot carefully qualifies and strengthens his old one. While the fundamentals are reexamined and restated, all curious new phenomena are consistently met with *ad hoc* modifications of the basic electrostatic explanation.

Caution is a characteristic of the *Traité*, and Biot is cautious even about the most "fundamental fact" of voltaic electricity: the constant difference of tension between the contacted elements of the pile. This "fact" is now taken as the simplest possible supposition rather than as an indubitable certainty:

> . . . ce n'est qu'une supposition dont les expériences fondamentales . . . ne fournissent aucune preuve. J'ai ouï dire à Coulomb, qu'il avait vérifié cette loi, et qu'elle lui avait paru exacte. Il est clair qu'on ne peut l'établir avec certitude qu'à l'aide de la balance électrique, et en mesurant les quantités d'électricités libres aux diverses hauteurs d'une pile; mais cette observation est influencée par la conductibilité toujours imparfaite des conducteurs humides, et par plusiers autres causes que nous examinerons dans un des chapitres suivans. Quoi qu'il en soit, admettons d'abord la constance de la quantité comme donnant la loi la plus simple, et cherchons à en développer les conséquences par le calcul.[59]

Biot's caution is also apparent when he confronts the difficult phenomena the chemists have revealed. Signs of chemical activity not corresponding to electric tensions force him to abandon another unstated assumption of the old theory, namely, that all static electricity in the pile is directly observable. Some of the electricity might be "hidden," Biot explains, while only a small fraction of the quantity in the pile is actually free and detectable. The voltaic column may therefore be exactly like the electrostatic glass-plate condenser, with most of its electricity normally disguised but ready for an occasional sudden and complete discharge. This electrostatic comparison has multiple advantages: it preserves the basic static analogy; and it ex-

[59] J. B. Biot, *Traité de Physique* (Paris, 1816), 2, 480.

plains "surtout des phénomènes chimiques que nous ne pouvons produire qu'en accumulant des quantités considérables d'électricité, soit par des batteries, soit au moyen de pointes d'une finesse extrême." [60]

Biot's approach towards other chemical actions connected with the pile's operation is similarly electrostatic. Chemical substances either have negligible influence, as he had explained in 1803, or a significant effect that can be accounted for entirely in terms of increased or diminished conductibility. The far more energetic behavior of piles moistened with nitric acid as compared with piles using pure water is due entirely to the greater conductibility of the first solution.[61] Diminishing activity in columns which gradually collect chemical waste is the result of extra *obstacles* that the electricity produced solely by contact must overcome.[62] Regeneration by fresh oxygen is also a conduction effect: contacts are restored by the gas, and electricity again passes freely from element to element.[63] Biot finds, again, that he must modify, but not necessarily abandon, his earlier, relatively straightforward electrostatic theory.

Biot's remarks on "resistance" also deserve mention, because they demonstrate just how completely electrostatic his views remained. The basic concepts he uses to account for the differential transmission of voltaic electricity are Coulomb's electrostatic repulsion and isolating distance, not driving force and mechanical friction, as might be imagined. Immediately after stating that an imperfect conductor offers a sensible resistance to the transmission of electricity, Biot therefore hastens to explain that, if a certain degree of repulsion is obtained, there ought to result "une décharge momentanée dans laquelle toute l'électricité passera librement, comme lorsqu'un corps électrisé se décharge par étincelle à travers l'air. . . ." [64] This direct application of Coulomb's concepts—isolation followed by total static discharge—is a striking indicator of Biot's basically unaltered electrostatic outlook, an outlook which the phenomena of the chemists challenged but obviously did not overthrow.

In 1816, therefore, the electrostatic theory of the voltaic pile would

---

[60] *Ibid.,* 502–503.
[61] *Ibid.,* 515 ff.
[62] *Ibid.,* 519–520.
[63] *Ibid.,* 526–527.
[64] *Ibid.,* 516. Biot also explains on the same page that an imperfect conductor "offrent une résistance sensible à la transmission de l'électricité," but in order to "vaincre" its resistance "un certain degré de force répulsive" is required. This, clearly, is Coulomb's electrostatic language.

seem to be as firmly established as ever. A few concessions and modifications had recently been made to accommodate the chemists, but no revolutionary new phenomena were at hand to disturb the stable equilibrium Biot had recently re-established with his *Traité*. This textbook was well received; as far as voltaic electricity was concerned, everything seemed settled.[65] For the next few years the *Institut* continued to find no study of the pile worthy of discussion or publication in its *Mémoires,* and the *Annales* seems to have found none to publish. But in 1820 everything suddenly changed. A surprising discovery was made by the Danish scientist H. C. Oersted, and French physicists reacted so quickly that soon the whole electrostatic approach to the voltaic apparatus, with its elaborate and interlocking experimental techniques and theoretical explanations, was called in question and then partially abandoned.

5. OERSTED, BIOT, AND AMPÈRE

The phenomenon Oersted discovered was actually quite simple.[66] He noticed that when he joined the poles of a voltaic pile with a connecting wire, a magnetic needle held parallel to the wire would be deflected into a perpendicular position. He attributed this directive turning to a "conflit électrique"[67] whirling about the wire. French scientists, upon hearing of this explanation, were unwilling or unable to understand what "conflit électrique" meant, and they identified the effect as a voltaic one instead. But though there was general consent on this point, disagreements arose over the particular manner of explaining this radically new voltaic effect. A. M. Ampère and J. B. Biot were among the first to offer influential accounts of the new phenomenon. Biot's papers, though actually presented after some of Ampère's, will be considered first, because, unlike Ampère's, they were written from a point of view perfectly consistent with the electrostatic theory of the pile.

On 30 October and 18 December 1820, Biot communicated to the

---

[65] Biot's text was summarized with praise in the "Histoire" section of the *Mémoires* for 1816.

[66] Oersted's discovery was reported to the *Académie* by Arago on 4 September 1820; see *Procès-Verbaux, 7,* 83–84. Oersted's paper, "Experimenta circa effectum . . . ," was printed in *Annales, 14* (1820), 417 ff.

[67] For the meaning of "conflit électrique" and its relationship to Oersted's belief in *Naturphilosophie,* see Robert C. Stauffer, "Speculation and Experiment in the Background to Oersted's Discovery of Electromagnetism," *Isis, 48* (1957), 33–50.

recently re-named *Académie* the results of experiments which he and Savart had completed on Oersted's phenomenon, and on 2 April 1821 he read a paper recounting his principal findings and conclusions.[68] A summary of his views also appeared under the title "Sur l'aimantation imprimées aux metaux" in the widely read *Journal des Savans*. There he begins with a description of the electromotive apparatus; he next itemizes the known effects of voltaic discharge: heating of wires, chemical changes, and intense shocks. By a simple extension of this list, Biot claims, one can include Oersted's phenomenon; i.e., when electricity rapidly traverses metallic bodies, "il leur donne momentanément la vertu magnétique; ils deviennent alors capables d'attirer le fer doux et non aimanté." [69] In other words, something hitherto unobserved happens in a wire when the pile's static electricity discharges through it; magnetic fluids are temporarily disrupted and are rearranged around the wire's circumference, producing an unusual magnetic attraction. But in matters so new, investigators ought to turn to "des recherches purement expérimentales." [70] Biot thus allowed himself to be cautiously convinced that he and his generation of French physicists should continue their gradual extension of the electrostatic theory of the pile.

Biot's next major account of electromagnetism was included in his *Précis Élémentaire* (1824), a shorter textbook based on the *Traité*.[71] Except for extra mathematical details, Biot retains in this version of his theory the backbone of the 1821 article, although by this time Biot's personal attitudes had changed. Rather than referring politely to M. Ampère, as he did in 1821, he accuses Ampère of being a "Cartesian," a devastating rebuke in those days of enlightened Newtonian science.[72] Biot's dramatic change of tone seems to stem from Ampère's explicit rejection of the electrostatic theory of the pile and from the fact that by following Ampère much of French electrical science was changing. Biot was caught out of step. To help understand the position in which Biot found himself in 1824, Ampère's early papers on Oersted's discovery will be considered next.

---

[68] For Biot's papers, see *Procès-Verbaux*, 7, 99 and 118. See also *Mémoires Relatifs, op. cit.* 2, n. 1, p. 80.

[69] Biot, *Journal des Savans* (April 1821), 225.

[70] *Ibid.*, 233.

[71] The 1824 edition of the *Précis* was actually the third French edition. I have used an English version of this edition prepared by John Farrar, *Elements of Electricity, Magnetism and Electromagnetism* (Cambridge, Mass., 1826).

[72] *Ibid.*, 359.

Like Biot, Ampère confronted Oersted's phenomenon with the basic assumptions of the electrostatic theory clearly in mind. But, unlike Biot, Ampère had always considered himself a chemist as well as a physicist, and this made for crucial differences.[73] Chemical sympathies affected his reactions and insights, even though the intellectual heterodoxy Ampère displayed in 1820 far surpassed the restrained ambivalence we have seen already on the part of the chemists. It was also easier for Ampère to entertain unorthodox conceptions because he had no personal investment in the electrostatic theory, nor did he have to justify or defend several years' work on the pile. No conceptual or psychological restraints prevented him from starting with the electrostatic theory and then, almost at once, transforming or abandoning much of what he had started with.

Traces of the old electrostatic theory of the voltaic apparatus can be seen in Ampère's first published paper on Oersted's discovery (December 1820).[74] These residues are peculiar, however, for Ampère appears to rely at least as much on Volta's original explanation as he does on the recent textbook version of the voltaic theory, as though his sensitivity to chemical phenomena impelled him to modify the accepted theory of the voltaic circuit, while his awareness of Oersted's discovery led him to seek the additional authority of Volta to support his somewhat unorthodox designs.[75] For example, as soon as he claims to be breaking new ground by precisely defining two very different effects produced by electromotive action, electric tension and electric current, he alludes to the work of Thenard and Gay-Lussac and then refers his distinction back to Volta.[76] The thrust of his argument—to separate explicitly tension and current phenomena and to classify chemical decomposition and electromagnetic interaction with the latter—is a new one for Ampère's generation of French physicists. Biot, as we have seen, consistently avoided this sort of distinction, but Ampère, relying on Volta's authority, ignores the complicated electro-

---

[73] In the outline for his course at *l'École Centrale*, Bourg (composed 12 March 1802), Ampère made chemistry Part III, after cosmography and mechanics; see Ampère's *Correspondance, op. cit.* (note 54), *1,* 110. He later published a chemical memoir, "Lettre de M. Ampère . . . sur la détermination des proportions dans lesquelles les corps se combinent d'après le nombre et la disposition respective des molécules dont leurs particules intégrantes sont composées," *Annales, 90* (1814), 43 ff. He also had a chemical correspondence; see note 54.

[74] This memoir was published in *Annales, 15* (1821) and was reprinted in *Mémoires Relatifs, op. cit.* (note 22), *2,* 7 ff. This latter is the edition to which reference is made here.

[75] This point will be argued further below.

[76] Ampère, *op. cit.* (note 74), 7–8.

static configuration of the pile; he concentrates instead on the electric fluid, which he insists is subject to a *continual impulsion* just so long as the continuity of conductors remains unbroken.

As a result of again making the continual motion of electricity the center of his concern, Ampère tends from the start to distort, probably unconsciously, the meaning of a few crucial concepts of the accepted electrostatic theory. Electromotive force and resistance most critically reflect his conceptual alterations. In place of the static separation-force and isolating length of Biot's account, Ampère's electromotive force (and resistance) take on something more like Volta's original meaning: "Les courants dont je parle en s'accélérant jusqu'à ce que l'inertie des fluides électriques et la résistance qu'ils éprouvent par l'imperfection même des meilleurs conducteurs fassent équilibre à la force électromotrice, après quoi ils continuent indéfiniment avec une vitesse constante tant que cette force conserve la même intensité; mais ils cessent toujours à l'instant où le circuit vient à être interrompu." [77] Here, current is the continual motion, at constant velocity, of the electric fluid. This conception is clearly different from Biot's. Closely related to this change in the meaning of electromotive force is a new understanding of resistance; for the first time it is imagined to be a frictional sort of impediment to the otherwise smooth motion of impelled electricity.

Despite these alterations, Ampère, in 1820, like Volta and Biot before him, was trying to unify static and current phenomena by using a common electric fluid to explain both. Purely static effects are observed when this fluid is at rest, and current effects, like electromagnetic interaction, occur when it is in motion. This neat sort of classification within the electrostatic framework had almost been made by Gay-Lussac and Thenard for the static and chemical manifestations of voltaic electricity, although it had been submerged again in Biot's theory of "hidden electricities." Ampère in a sense ignored Biot, and turned the chemists' implicit distinction into a fundamental one. This was a theoretical change, but not yet anything like a conceptual revolution.

But once Ampère had gone this far it was relatively easy to go further. In his first paper on electrodynamics he still thought, like Biot, that the center of the circuit was the pile, the place where electricity was first separated and then discharged, inactively for Biot, actively

[77] *Ibid.*, 10.

for Ampère, through the connecting wire; what happened in the wire was some unspecified "reunion" of the electricity. At first, Ampère paid little attention to this reunion although he did explain how current electricity, by returning upon itself and reuniting preparatory to a new separation, could flow in a continuous fashion while normal static discharges last only a short time.[78] In later papers, however, electrical reunions became a major part of Ampère's theory of the circuit. He gradually modified his views in 1821 and 1822, and by 1823 he had submitted his explanation of the electric current to a major overhaul.[79] In a series of articles culminating in his famous electrodynamics memoir of 1823, Ampère completely shifted his attention from the electromotive force in the pile to the reunions resulting from its action inside the wire. The state of the conducting wire became his new starting point for understanding the electric circuit.

Somewhere between 1820 and 1823 Ampère began to think about the reunion of the electrical fluids around the molecules of a magnet and between the particles of a wire. In response to Fresnel's urging (or perhaps again under the influence of the chemists), he started to imagine a neutral, ether-like fluid which resulted from the reunion of positive and negative electricity; it was the passing into and out of this neutral state that now seemed to be the fundamental event in molecular and voltaic currents.[80] While a series of reunions takes place around each particle of a magnet, in a current-carrying wire "les deux fluides électriques parcourent continuellement les fils conducteurs, d'un mouvement extrêmement rapide, en se reunissant et se séparant alternativement dans les intervalles des particules de ces fils . . . mises en mouvement dans les fils conducteurs par l'action de la pile, elles y changent continuellement de lieu, s'y réunissent à chaque instant en fluide neutre, se séparent de nouveau et vont aussitôt se réunir à d'autres molécules du fluide de nature opposée. . . ."[81] A successive separation-reunion-separation of electricity takes place in infinitesimal circles around each molecule of a magnet and passes lengthwise down

---

[78] For Ampère on "reunions" in 1820, see *ibid.*, 14.

[79] L. Pearce Williams has pointed to this shift in Ampère's thought in "Ampère's Electrodynamic Molecular Model," *Contemporary Physics, 4* (1962), 113–123.

[80] Williams, *loc. cit.*, attributes Ampère's conceptual development to Augustin Fresnel's personal influence. It is also possible to imagine, however, that Ampère's interest in chemistry played a role; for the reunion of opposite electricites around a molecule and the production of a neutral fluid seem like conceptions closely connected with electrochemical affinity theories.

[81] Ampère, *Mémoires Relatifs, 3,* 114–115.

the wire. Under this conception, nowhere in a wire or magnet is a burst of static electricity driven by an electromotive force. Instead, the pile works by somehow starting the series of electrical reunions within the wire; and inside the magnet, currents always exist and only have to be aligned by ordinary voltaic electricity or other magnets.[82] The criterion for the existence of current electricity is no longer a discharging voltaic pile with or without attached connecting wires; rather, a current occurs whenever electrical reunions occur in a conductor.

It can now be seen that all of Ampère's serious modifications of the theory of the voltaic circuit were prompted by Oersted's discovery. Prior interest in electrochemistry may have helped, but there was something about the interaction of a wire and a magnet which prevented him from seeing the circuit with Biot's eyes. It is worth trying to probe Ampère's thoughts a little more deeply in order to locate the possible source of his productive shift in viewpoint, especially since the steps he took had an immediate influence on the study of voltaic electricity by other physicists of the French *Académie*.

Oersted's discovery and his own early experiments on the attractive and repulsive forces of current-carrying wires convinced Ampère that electric and magnetic phenomena were really identical.[83] He seems to have believed from the first that, since only like things can interact, either magnets must be electrical or electricity magnetic.[84] The new phenomena persuaded him to accept the first of these possibilities. Once he had made this decision, he knew that he had to try to understand the nature of voltaic electricity, since it played such a large part in his electromagnetic theory. He seems to have asked himself how to conceive of the small voltaic circuits inside a magnet which his theory presupposed. He realized that the magnet's electricity must move continuously, because within a homogeneous metal core there is no break in the continuity of the conductors to interrupt the movement of the electric fluids; thus, the electromotive force must continually drive both positive and negative electricities around a small closed

[82] This conception suggests that another reason for Ampère's change of mind might have been Faraday's discovery of electromagnetic rotations. See the *Mémoires Relatifs* for reprints of papers on this and related problems.
[83] Some of Ampère's work as an experimental physicist is discussed by R. A. R. Tricker, "Ampère as a contemporary physicist," *Contemporary Physics, 3* (1961), 453–468.
[84] The importance of Ampère's metaphysical concern for "homogeneity" considerations has been argued most cogently in Joseph L. Agassi, *Towards an Historiography of Science*, in *History and Theory*, Beiheff 2 (1963), 22 ff.

curve. To prevent the immediate recombination of these separated fluids, this force must impart an enormous velocity. Finally, for a macroscopic voltaic circuit to be perfectly similar to the elementary circuits within a magnet, it must have a perpetual, circular movement of electricity, and an electromotive force which actively drives electricity so long as the continuity of conductors remains uninterrupted. Significantly, these common characteristics of voltaic and magnetic currents are asserted in Ampère's first memoir on electrodynamics:

> ... il ne me paraît guère possible, d'après le simple rapprochement des faits, de douter qu'il n'y ait réellement de tels courants autour de l'axe des aimans, ou plutôt que, l'aimantation ne consiste que dans l'opération par laquelle on donne aux particules de l'acier la propriété de produire, dans le sens des courants dont nous venons parler, la même action électromotrice qui se trouve dans la pile voltaïque. ... Seulement cette action électromotrice se developpant, dans le cas de l'aimant entre les différentes particules d'un même corps, bon conducteur, elle ne peut jamais ... produire aucune tension électrique, mais seulement un courant continu semblable à celui qui aurait lieu dans une pile voltaïque rentrant sur elle-même en formant une courbe fermée. ... [85]

In this early paper, Ampère was obviously reasoning in terms of his first conception of the circuit. Later, however, after he had adopted the reunions theory, he could logically concentrate on the current alone—the alternating transformation of the neutral fluid—while ignoring the nature of the electromotive force and of the conducting substance of the magnet. In fact, only after he had stated the reunions theory for magnetic currents did he apply this conception to the wires of voltaic circuits. Chronology would seem here to confirm this hypothetical reconstruction of the direction of Ampère's thought: from magnetic to voltaic currents.[86]

It should be clear by now that many factors, extraneous from the standpoint of the strict "instrumentalist" French experimental physicist, may have entered into Ampère's thinking. Biot apparently sensed this, and reacted with characteristic caution, though several of his colleagues soon adopted Ampère's views. By 1824, the best experimental

[85] Ampère, *Annales, 15* (1821), 75.
[86] This sort of chronology was first noticed by Williams, *op. cit.* (note 79), although his interpretation of the direction of Ampère's thinking is the reverse of mine.

studies were already being done by men deeply committed to Ampère's theoretical views and associated experimental techniques.

6. AMPÈRE, THE GALVANOMETER, AND THE NEW CURRENT THEORY, 1820–1825

The study of reactions to Ampère's theory of the circuit is complicated by the confusion of converging and diverging channels of inspiration and influence. Ampère's theory was not constructed in one piece and presented to the French scientific community for outright condemnation or praise; rather, it was put together from several parts, few of which were left unchanged from 1820 to 1823. The complicated evolution of Ampère's ideas, whose variations were constantly announced to French scientists, can be roughly characterized in terms of a first "active electromotive force" phase and a later "reunions" one. It will be convenient to maintain this distinction and to consider the multiple responses to Ampère in terms of it.

The *Académie*'s attitudes to Ampère's early papers were mixed, though generally magnanimous. According to Delambre's slightly retrospective summary of the scientific highlights of 1820, the new electrodynamic theory was widely and quickly accepted.[87] He implies that most of the *Académie*'s members immediately adopted Ampère's views and found every recently discovered fact to be in perfect accord with his theory. The immediate enthusiasm of two prominent scientists seems to confirm Delambre's reporting. Late in 1820, Arago tested Ampère's theory by passing static electricity through wires bent into helices.[88] As Ampère had predicted, Arago found that current-current and current-magnet interaction occurs with suitable arrangements of wires carrying *rapidly moving* electricity. Currents can even induce attraction in previously unmagnetized iron filings, and the failure to observe similar forces when the metal chips are replaced with sawdust (which ought to be attracted by unmoving static electricity) appeared

---

[87] Delambre's "Histoire" is in *Mémoires, 4* (for 1819 and 1820; first published, 1824). Ampère's papers are discussed on cxxxvii ff. It is important to note that the major issue was thought to be Ampère's theory of magnetism, not of the electric current; nevertheless, one theory influenced the other.

[88] Arago's experiments before the *Académie* are described in this same "Histoire," *loc. cit.,* cxlix–cli. His papers were read on 25 September and 5 November 1820, and a selection from them can be found in the *Mémoires Relatifs, 2,* 55 ff.

to Arago a definitive confirmation of Ampère's suspicion that forces exerted by electricity in motion are totally different from those produced by the fluids at rest. Augustin Fresnel also tried his hand at some early experiments on electromagnetic interaction, attempting, with indifferent success, to invert current-magnet influences.[89] When he failed to evince clear signs of oxidation or muscular convulsions with a magnetic bar and a helix, Fresnel insisted that his experiments rendered no damage to Ampère's theory, which did not necessarily require these reverse processes. However, the positive reception of Ampère's ideas was not universal, either within or without the *Académie*. Biot did not share in this enthusiastic applause, for, despite his courtesy, he remained the spokesman for the old electrostatic pile theory. And in Ampère's eyes, Biot's opinions, bound up as they were with Coulomb's theory of magnetism, still carried the day. Writing to Roux-Bordier on 21 February 1821, Ampère complained that "l'hypothèse du Coulomb sur la nature de l'action magnétique . . . écartait absolument toute idée d'action entre l'électricité et les prétendus fils magnétiques; la prévention en était au point que, quand M. Arago parla de ces nouveaus phénomènes à l'Institut, on rejeta cela comme on avait rejeté les pierres tombées du ciel, quant M. Pictet vint, dans le temps, lire un mémoire à l'Institut, sur ces pierres. Ils décidaient, tous que c'était impossible."[90]

The pages of the *Annales* manifest similarly mixed reactions to Ampère's early papers outside the official circles of the *Académie*. There is some immediate enthusiasm, some opposition, and, on all sides, there begins to be a general reexamination of voltaic fundamentals. Oersted's original paper was published in French translation in the *Annales, 14* (1820); Ampère's first memoir followed in the next volume; and three papers appear in the volume after that, containing a quickly developed but searching discussion of the electric current. One of the articles in the latter volume is a reprint and translation of Wollaston's twenty-year old memoir, "Expériences sur la production de l'électricité et sur son action chimique."[91] The editors, Gay-Lussac and Arago, point out in a footnote their reasons for re-publishing this paper now:

[89] A. Fresnel, "Note sur des essais ayant pour but de décomposer l'eau avec un aimant," *Mémoires Relatifs, 2*, 76 ff. The facts in the report were announced to the *Académie* on 6 November 1820.

[90] Ampère, *Correspondance, 2*, 566.

[91] Wollaston, "Expériences sur la production de l'électricité et sur son action chimique," *Annales, 16* (1821), 45 ff.

Wollaston's little-known results demonstrate "l'identité d'action chimique des électricités ordinaire et voltaïque," an identity which Arago's experiments had established for magnetic action.[92] By printing Wollaston's memoir and by highlighting its significance, Gay-Lussac and Arago obviously wanted to lend further support to what they took to be Ampère's views. A second article in this volume of the *Annales* is considerably less favorable to the new electrodynamic theory. It is Berzelius' six-page "Lettre sur l'état magnétique des corps qui transmettent un courant d'électricité," which opposes to Ampère's theory an explanation of the new electromagnetic phenomena remarkably similar to Biot's extension of the electrostatic theory.[93] Berzelius argues that in the wire "discharging" the voltaic pile there is an unusual magnetic state, which permits the wire to interact with Oersted's magnetic needle. This easy and natural hypothesis, he claims, avoids the difficulties to which Ampère's theory leads. Ampère's response to Berzelius is the last relevant article in this volume;[94] addressing a letter to Arago, Ampère explains how Berzelius' experiments present no difficulties for his own theory which, with the additional help of Arago's results, successfully "range tous les phénomènes de l'aimant parmi les phénomèns purement électriques." Thus, overall reactions to Ampère's early views seem to fall into the same patterns outside the *Académie* as they did within; there is enthusiasm tempered with caution, with much hard thinking on all sides about the implications of the old and the new theories of voltaic electricity.

The early response to Ampère seems easy to understand. Ampère himself never regarded the "active electromotive force" theory as radically new or fundamentally different from the standard electrostatic account of French textbook physics. He claimed only a new *distinction*, and, except for his views on electromotive force and resistance, his discussion of the circuit was otherwise consistent with Biot's. Ampère quite naturally regarded his own early statements as an articulation and sharpening of an old theory rather than as the enunciation of a new one. By following his lead, French physicists saw nothing illogical in simultaneous commitments to Ampère and Biot. Even the most enthusiastic supporters of the electrodynamic theory introduced odd bits of the static account into their explanation of electromagnetic phe-

---

[92] For the editorial footnote, see p. 45.
[93] Berzelius, *Annales, 16* (1821), 113 ff.
[94] Ampère, *Annales, 16* (1821), 119 ff.

nomena. Arago, for example, when describing his experiments on static discharges through wire helices, refers to "le mouvement imprimé au *fluide magnétique.*"[95] This idea, which Arago attributes to Ampère, could equally well, and in identical language, have been used by Biot, who, in fact, cited Arago's results in support of his own hypothesis.[96] Arago's duplication of current effects with static discharges fits in even better with a completely electrostatic theory of voltaic action than it does with Ampère's modification of it; but Arago did not seem to sense the difference or recognize the issue. Arago, or his co-editor, likewise blurred distinctions between Ampère's and Biot's ideas on another occasion; a footnote to a description of Ampère's experimental apparatus in the *Annales, 18* (1821) reminds the reader that while Ampère correctly realized that he should use wider wires with a greater number of voltaic couples in order to conduct electricity better, the same necessity "avait déjà été observé par M. Biot."[97] The author of this editorial comment apparently discerned no significant difference between Ampère's notion of impulsive electromotive force with frictional resistance and Biot's idea of electrostatic separation force and isolating length.

If Ampère's first hesitant attempts at a new current theory smoothed the way to the acceptance of his electrodynamic theory, while creating ambiguities for others to resolve as best they could, then his clear articulation of the obviously novel reunions hypothesis completely changed the nature of the debate. By 1823, versions of the second current theory had appeared, and old opponents like Biot ceased to be polite in their disagreement. Arago and other early supporters seemed to drop out of the discussion, and the conceptual and experimental relevance of electrical reunions was asserted by a new group.[98] Fourier, Oersted, Seebeck, Becquerel, and De La Rive became participants in this stage of the debate. They all appeared to be acutely conscious of the issues involved in the change-over from Biot's to Ampère's approach to voltaic electricity, and they seemed eager to get on with the neces-

[95] Arago, *Mémoires Relatifs*, 2, 59; the italics are mine.
[96] Biot, *Journal des Savans, op. cit.* (note 69), 233.
[97] See footnote, *Annales, 18* (1821), 94.
[98] It is, in fact, unclear from the printed sources what was underway in French electrical science from mid-1821 to mid-1822, except for the continued development of electromagnetic theory. By mid-1822, the *Académie* was considering Becquerel's experimental papers, and some of these will be discussed below.

sary work of reconstruction. Basing their research on Ampère's latest concepts, they prosecuted their experimental investigations with a new measuring instrument, the galvanometer.

Several researches, based on the use of the new current-measuring device, were published in the *Annales* for 1823. Ampère himself had written earlier of a "galvanometer," consisting of a magnetic needle and a wire, which could register the presence of electricity in motion by means of electromagnetic deflection.[99] This simple instrument was first perfected by Schweigger, who was able to multiply weak current effects with it; it was then fully exploited by Seebeck, who used deflections to detect thermally-produced currents. Oersted discussed both Schweigger's device and Seebeck's discovery in this volume of the *Annales* and also pointed to further studies which could easily be undertaken with galvanometric techniques. In collaboration with Fourier, an early and continuing supporter of Ampère, Oersted also wrote a longer, theoretical memoir;[100] the subject of their paper was the significance of the galvanometer for the exploration of the voltaic circuit.[101] Distinguishing Seebeck's "thermoelectric" from normal "hydroelectric" (or voltaic) circuits, Fourier and Oersted contrasted the greater *quantity* of electric *force* in the former circuits with the greater *intensity* in the latter. Quantity of force was measured by the degree of electromagnetic deviation, while intensity was taken in its usual electrostatic sense. Oersted and Fourier thus attempted to work out a unique set of concepts and experimental techniques specifically applicable to the electric current. Ampère seems to have been pleased with the new instrument and with its rapid experimental and theoretical exploitation, for this same volume of the *Annales* includes a letter by him to Faraday in which he speculatively attributes low-resistance thermoelectric currents to the conducting bodies of magnets.[102]

In addition to these articles on the fundamentals of the circuit, another on a seemingly unrelated subject was printed in this same volume of the *Annales*. It was a discussion by a young physicist, C. A. Becquerel, of electricity produced by pressure. When it had previously

[99] Ampère's remarks on the galvanometer occur in his first electrodynamics memoir, *Mémoires Relatifs*, 2, 12 ff.

[100] Oersted and Fourier, *Annales*, 22 (1823), 375 ff.

[101] It is interesting to note that Ampère mentions Fourier as one of his sympathetic listeners and a faithful attendant at his demonstrations; see Ampère's *Correspondance*, 2, 572.

[102] Ampère, "Une Lettre à M. Faraday," *Annales*, 22 (1823), 389 ff.

been read to the *Académie,* it had elicited high praise from Ampère, and now, in published form, it launched Becquerel's career.[103] By itself Becquerel's paper seems to have had nothing to do with voltaic electricity, Ampère's reunions theory, or Schweigger's electromagnetic multiplier, and in its first conception it may not have had any connection at all.[104] But viewed in the context of his later publications, it looks very much like the first crucial step in Becquerel's reinvestigation of the fundamentals of voltaic electricity, a reinvestigation in which the sequence of logically necessary steps coincided with experiments actually undertaken. With this paper Becquerel began a series of memoirs which, by their sheer quantity and high quality, pushed him to the forefront of French physics as the new authority on voltaic electricity.[105]

Becquerel begins the pressure memoir by considering the previously unrecognized difficulties involved in accurately performing the seemingly most simple electrostatic measurements. He argues that though an observer sometimes finds no signs of electric build-up, this need not mean that static charges are not accumulated. The electric fluids might be *separated,* but the substances in which separation occurs might also be perfect conductors, in which case they would convey the fluids back together for neutralizing recombination before any static measurement could possibly be made. Indeed, to understand properly what electrostatic measurements might mean, one must consider both the velocity with which the static electricities are initially separated and the conducting abilities of the materials in question:

> En général, il paraît que plus les deux corps approcheront d'être bons conducteurs plus il deviendra difficile d'obtenir de l'électricité par pression ... si les corps sont parfait conducteurs, aussitôt qu'une diminution de pression a eu lieu, les deux fluides se combinent instantanément, quelque grande que soit la vitesse de séparation: au contraire, si l'un des deux corps ne jouit pas pleinement de la conducti-

---

[103] Ampère's praise of Becquerel comes in a letter written to G. De La Rive on 12 June 1822; see Ampère's *Correspondance, 2,* 582. Becquerel's memoir, "Expériences sur le développement de l'électricité par la pression," was printed in the *Annales, 22* (1823), 5 ff.

[104] It will be argued below that this first paper was "the first crucial step in Becquerel's reinvestigation of the fundamentals of voltaic electricity," but it is difficult to collect clear evidence for this statement. The first published version of this paper, referred to in note 103, came out long after Ampère's revolutionary work. But Becquerel's earliest public concern for the problem seems to date from March 1820, before Oersted's discovery was even announced; see *Procès-Verbaux, 7,* 33. The resolution of this issue will therefore have to wait for further evidence.

[105] Becquerel's series of papers followed in quick succession in the *Annales.* Two each were published in volumes *22, 23, 24,* and *25.* From then until the end of the decade, Becquerel's papers appeared regularly in the *Annales* and were read frequently to the *Académie.*

bilité, une diminution de pression n'entraîne pas sur-le-champ la recomposition des deux fluides. . . . Cette recomposition mettra plus ou moins de temps à se faire, en raison de la conductibilité des deux corps pressés . . . et il est probable que, dans le cas ou les corps conduissent parfaitement l'électricité, la vitesse de séparation doit être infinie.[106]

After offering this criticism of crude, careless electrostatic measurements, Becquerel turns, in his next paper, to an explanation of the new and more sophisticated galvanometric techniques. In a memoir read to the *Académie* on 16 June 1823 he proposes, as the first task of the new experimental procedure, to calibrate a galvanometric instrument for measuring the electric current:

La première expérience à faire avec cet appareil est de déterminer un rapport entre les écarts de l'aiguille et la force du courant électrique; pour cela, je suspens l'aiguille aimantée à un fil très-fin de platine, d'une balance de torsion, qui m'a servir dans le développement de l'électricité par la pression, à déterminer le rapport entre les pressions et les intensités électriques correspondantes. Ensuite je dispose l'appareil pour que l'aiguille aimantée soit dans la direction du plan du méridien magnétique, quand le fil de suspension est sans torsion. . . . Supposons maintenant que l'aiguille soit chassée de ce plan par l'effet du courant électrique; après plusiers oscillations, elle prendra une position d'équilibre telle que la résultante des actions électriques émanées de chaque point du fil conjonctif fera équilibre. . . . Cette loi [relating force to torsion angle obtained from this data] est précisément la même que celle trouvée par Coulomb entre les forces de torsion et les angles de déviation.[107]

Where the old balance could measure only the apparent intensity of electrostatic forces, the new device, which Becquerel has just calibrated, can measure directly the current intensities themselves. And by referring specifically to "une balance de torsion," Becquerel seems to imply that he expects his measuring instrument to be no less precise than Coulomb's.

Now having a new and accurate experimental technique, Becquerel turns to a study of electricity produced by chemical reactions; the latter's role in the development of voltaic electricity had always been

---

[106] Becquerel, *Annales, 22* (1823), 11–12.
[107] Becquerel, "Du développement de l'électricité par la contact de deux portions d'un même metal, dans un état suffisamment inégal de température," *Annales, 23* (1823), 137–138.

a difficult and troublesome matter for the proponents of the simple, electrostatic theory of the pile.[108] Becquerel's inquiries were part of the general reappraisal of electrochemistry which the editors of the *Annales* helped initiate with their recent publication of Wollaston's twenty-year old paper on oxidation. Becquerel's special contribution was the application of galvanometric techniques to the investigation of chemical reactions. But Becquerel was no pure experimentalist; with Ampère, he believed that where reunions exist, currents exist, and that wherever currents exist the galvanometer can register and measure them directly:

> ... en se servant d'un condensateur pour recueillir une des électricités qui se dégagent, on doit trouver difficilement des traces de ce fluide, puisque le condensateur demandant un certain temps pour se charger, les deux électricités peuvent alors se recombiner; mais si l'on emploie un galvanomètre multiplicateur, tel que celui de M. Schweigger, qui rend sensibles les électricités au moment même où elles se dégagent, et par conséquent, à l'instant où combinaison va s'opérer, on obtiendra des courans plus ou moins forts suivant le degré de conductibilité des substances mises en action et celui de leurs affinités réciproques. . . .[109]

The galvanometer and Ampère's theory were all Becquerel needed; in further studies, he explored all sorts of chemical actions, invariably finding electric currents present.

Becquerel advanced from these preliminary electrochemical studies to a telling criticism of the whole static theory of the pile. Although he never questioned the existence or importance of the electrostatic, tension-producing electromotive force, he nevertheless began to investigate, more radically than any Frenchman had before, the possible roles for chemical reactions in the overall activity of Volta's column. Eventually he concluded that

> ... le cuivre et le zinc, plongés en même temps dans un acide et ne communiquant ensemble que par l'intermédiaire d'une dissolution

---

[108] In his study of thermoelectricity, Becquerel tries to relate chemical to thermal currents by using a common galvanometric technique to measure both. He writes: "Ayant trouvé que les deux bouts d'un fil metallique, dans un état suffisament inégal de température, se constituaient par leur contact mutuel dans deux états électriques contraires, j'ai voulu voir si des effects électriques sembables n'auraient pas lieu lorsque les deux bouts ne seraient pas attaques aussi fortement l'un que l'autre par le même acide" (*op. cit.* [note 107], 152). This seems to have been Becquerel's entrance into electrochemistry.

[109] Becquerel, "Nouveaux resultats électrochimiques," *Annales, 23* (1823), 249.

alcaline, donnent lieu à un courant électrique qui va du zinc au cuivre. L'énergie de ce courant est telle qu'il peut être rendu sensible sur une aiguille aimantée suspendue à un fil de cocon. Ce résultat nous donne sur-le-champ la clef de l'influence de l'action chimique sur la charge de la pile de Volta: en effect, plongeons dans un acide un couple voltaïque, et mettons en communication les deux disques; que va-t-il arriver? Deux courants dans le même sens: le premier sera dû a l'action électromotrice des deux métaux l'un sur l'autre, et le second, à la différence des actions chimiques de l'acide sur les mêmes métaux. Les effets de ces deux courants s'ajouteront donc. Je me borne à signaler ce fait à l'Académie, me réservant d'examiner plus an détail l'influence de l'action chimique sur la pile voltaïque.[110]

Neither impeding nor aiding the distribution of the column's static electricity, chemical reactions now seemed fundamentally important to the basic reunions-starting activity of the pile. For the first time in France, contact force and chemical reactions were on an equal footing.

Becquerel was obviously moving towards a theory of the voltaic apparatus consistent with Ampère's ideas about the electric current. The old problem of electrostatic configuration, however, was a matter which could not easily be dismissed.[111] Becquerel had tried to take into consideration everything he knew about voltaic electricity: the contact force which produced tension; the chemical reactions which produced currents but no tension; and, finally, the influence of the fluid conductors as sources of some unspecified action on the metal plates. He soon stalled in his enterprise, failing to make real progress towards a consistent theory of the pile,[112] and he came to abandon the proj-

---

[110] Becquerel, "De l'état de l'électricité developpée pendant les actions chimiques," *Annales, 24* (1823), 203–204; this memoir was read to the *Académie* on 22 September 1823.

[111] Becquerel and Ampère had to work out the problem of electromotive force among themselves. In his first electrodynamic memoir, Ampère had denied any role to a tension-producing force in the generation of current phenomena. Becquerel, however, never seems to have denied electrostatic contact force, although he also allowed for Ampère's reunions-starting electromotive force. To unravel these difficulties, Becquerel and Ampère collaborated on experimental studies, and together they came around to a position which was closer to Becquerel's prior statements than to Ampère's. Their results were read to the *Académie* on 12 April 1824 and were reported in the *Annales, 27* (1824), 29 ff.

[112] Two of Becquerel's studies of the pile are "Développements relatifs aux effects électriques observées dans les actions chimiques, et la distribution de l'électricité dans la pile de Volta, en tenant compte des actions électromotrices des liquides sur les metaux," *Annales, 26* (1824), 177 ff., and "Des actions électro-motrices de l'eau et des liquides en général sur les métaux," *Annales, 27* (1824), 5 ff. He does not seem to come back to this series of studies, at least in the *Annales*, after 1825.

ect. But Auguste De La Rive, stimulated by Becquerel's failure, started contributing papers to the *Annales*,[113] carrying the reunions theory to its logical conclusion.

De La Rive's first papers on the pile appeared in several issues of the *Annales* for 1825. His principal statement was an ambitiously entitled memoir, "Recherche sur la cause de l'électricité voltaïque." [114] In this he refers specifically to Biot's *Précis,* and, after calling Biot's contact force "occulte" and "mal définie," he advocates a new way of viewing the pile;[115] chemical action rather than electrostatic accumulation should be seen as the basic force of the column:

> . . . quant à l'accumulation de chacun de principes électriques aux extrémités, ou pôles d'une pile composée de plusiers élémens, elle est la résultat de la manière dont l'électricité est produite par l'action chimique. Ainsi le fluide positif, développé par l'action du liquide sur la première lame de zinc, se répand dans ce liquide, rencontre le cuivre de couple suivant, y entre et va neutraliser le fluide négatif de la seconde lame de zinc qui est soudée à ce cuivre. . . . Il y a donc excès a l'une des extrémités de la pile de fluide positif, et a l'autre de fluide négatif; et l'on conçoit que l'énergie de l'action chimique influera sur l'intensité de ces deux fluides accumulés.[116]

Chemical action thus generates the pile's electricity, but De La Rive still has to explain the regular increase in tension resulting from each addition of a copper-zinc couple. He reasons thus: "tous ces effets, qui dependent du nombre des plaques, peuvent s'expliquer par la considération que les deux principes électriques ont toujours, pour se réunir, deux chemins différens qui leur sont offerts: l'un l'appareil voltaïque lui-même, l'autre le conducteur qui en unit les extrémités; la proportion plus ou moins grande d'électricité qui passera par ce conducteur, dépend du rapport qui existe entre sa propre conductibilité et celle de la pile." [117] De La Rive clearly means to say that, as new couples are

---

[113] Auguste De La Rive, like his father, G. De La Rive, was a Swiss scientist who contributed to the French journals, and both were in correspondence with Ampère. For a letter of 1814, see Ampère's *Correspondance, 2,* 483–485.

[114] A. De La Rive, "Recherches sur la cause de l'électricité voltaïque," *Annales, 39* (1828), 297 ff. His other relevant papers are: "Mémoire sur quelques-uns des phénomènes que present l'électricité voltaïque dans son passage à travers les conducteurs liquides," *Annales, 28* (1825), 190 ff.; and "Analyze des circonstance qui déterminent le sens et l'intensité du courant électrique dans un élément voltaïque," *Annales, 37* (1828), 225 ff.

[115] De La Rive, "Recherches sur la cause de l'électricité voltaïque," *loc. cit.*

[116] *Ibid.,* 320–321.

[117] *Ibid.,* 322.

added to the column, "tension" increases. Nevertheless, this increase is not the result of a simple multiplying of the electrostatic separating force but of the increasing width of the column, which prevents the electricity generated in chemical reactions from easily recombining.

Through Becquerel's and De La Rive's work, Ampère's reunions explanation of the voltaic current was thus ultimately made a part of a fundamental critique of the electrostatic theory of the pile. And within the space of a few years, Biot's account of voltaic electricity had been confronted by a cogent, complete, and contrary theory of the pile and circuit. The reunions theory was the work of Ampère, Becquerel, and De La Rive, all scientists new to French experimental physics in the 1820's. Those of the old Coulombian stamp, like Biot, had first taken only a negative interest in the new developments. The reunions theory, however, had the special advantage of being closely associated with the most recent and presumably the most precise electrical measuring instrument, and France's instrumentalist physicists could not ignore the new theory and still leave their methodological commitments intact. They had somehow to accomodate the new theory, because it seemed that electrostatic experimental techniques, though not quite discredited, were somewhat out of date. Fortunately, France's Coulombian physicists proved flexible, and their active assimilation of the reunions idea, which seems to have been well underway by the late 1820's, must be considered as the last episode in the history of the new circuit theory.[118]

7. POUILLET—ASSIMILATION BY THE COULOMBIANS

The assimilation process will be treated here only in the barest and most tentative way. The work of C. S. M. Pouillet, an early collaborator of Biot,[119] will be briefly discussed. Pouillet's attitudes will be taken as representative of the reactions of flexible Coulombians.

---

[118] The merger of the electrostatic and reunions theories of current in the mid- to late 1820's is a complicated event well worth a separate study in itself. Here I will just mention one paper and use it as an indicator of what seems to be the general reconsideration of voltaic fundamentals going on at this time: Colladon, "Déviation de l'aiguille aimantée par le courant d'une machine électrique ordinaire," *Annales, 33* (1826), 62 ff.

[119] Pouillet collaborated with Biot in October 1815 on a study of "toutes les lois de la diffraction de la lumière . . ."; see *Procès-Verbaux, 5,* 557. In August 1822 he read a memoir to the *Académie* on "les phénomènes électro-magnétiques," in which Biot and Poisson, both advocates of Coulomb's theory of magnetism, reported; see *Procès-Verbaux, 7,* 364. By 1825, Pouillet was engaged in a series of studies on atmospheric electricity, using electrostatic instruments; see, for example, *Procès-Verbaux, 7,* 220 and 238.

Pouillet discussed the fundamentals of voltaic electricity in his text-book, *Élémens de Physique* (1828), maintaining here a remarkably balanced point of view.[120] Pouillet still accepted Biot's main ideas about voltaic electricity: the electromotive force segregates the electrical fluids and then holds them apart; the pile is essentially electrostatic in its properties; and the current itself is essentially a discharge. This does not mean that Pouillet was reactionary. He was writing in 1828, not in 1816, and as a result terms like "discharge," "current," and "electromotive force" meant something different to him than what they had meant to Biot. The notion of discharge current, for example, is fundamentally changed; ordinary electrostatic discharges are said to pass along conductors rather as a series of *decompositions and recombinations* than as a simple translation of electricity.[121] Voltaic currents, though said to be discharge currents, are therefore meant to be understood as the perpetual reunion of the two electrical fluids. An altered meaning for electromotive force follows as a natural corollary. Pouillet asks the reader to imagine

... une plaque de cuivre, ou un élément négatif, communiquant au sol par le fil conducteur non métallique. . . . Sur sa surface supérieur posons une plaque de zinc de même dimension, à l'instant du contact la force électromotrice exerce son action, le fluide résineux qu'elle développe passe sur le cuivre, et s'écoule dans le sol; le fluide vitré au contraire passe sur le zinc, et s'y accumule jusqu'à ce qu'il y ait acquis la tension maximum que la force électromotrice soit capable de retenir; il ne faut pour cela qu'un moment inappréciable cette tension, ou plutôt l'épaisseur électrique qui la produit, étant prise unité, nous dirons que le cuivre est à l'état naturel, tandis que le zinc est couvert d'une épaisseur *l* d'électricité vitrée. Si par quelque moyen nous allions enlever au zinc une partie du fluide qui le couvre, il n'aurait plus l'épaisseur *l* qu'il doit avoir; la force électromotrice le reproduirait à l'instant par un nouveau développement, qui réparerait exactement la perte, et par un égal développement de résineux qui

---

[120] The full title of Pouillet's textbook was *Élémens de Physique Expérimentale et de Météorologie* (Paris, 1828).

[121] Pouillet's notion of successively recombining static electricity might have been more directly borrowed from Gay-Lussac, with whom he lectured in a course in the "Faculté des Sciences." Gay-Lussac and Thenard had referred to static discharge in this way, though somewhat cryptically, in their *Recherches Physico-Chimiques, op. cit.* (note 55), 47. All this is further evidence for believing that bits and pieces of a special view of the voltaic apparatus were advocated by the chemists, even during Biot's period of dominance at the *Académie,* and may have been brought forward by Oersted's discovery, which served mainly to crystallize these views.

s'écoulerait dans le sol. A chaque portion de fluide que l'on viendrait de la sorte enlever au zinc, il y aurait une réparation subite pour reproduire sans cesse l'épaisseur *l*, qui est l'état de l'équilibre galvanique; et si l'on établissait par example la communication du zinc avec le sol par un fil *non metallique*, son fluide vitré s'écoulerait sans cesse, et serait sans cesse réparé; en même temps le fluide résineux developpé sur le cuivre s'écoulerait de même; tellement, que si l'on approachait, l'un de l'autre deux fils non metallique qui touchent le zinc et le cuivre, les fluides se recomposeraient à leur point de contact; et l'on aurait *une circulation* électrique continue; les fluides seraient séparés au contact des metaux, et recomposés au contact des fils conducteurs non électromoteurs qui communiquent avec eux.[122]

Pouillet is here constructing a theory of a pile which simultaneously distributes electrostatic tensions in a regular order and continually supplies new electricity to the conducting wires for perpetual reunion.[123] This, in effect, is Biot's pile set in motion, with a dynamic equilibrium taking the place of the static. And in the wire, which Pouillet, unlike Biot, fully and sympathetically considers, there is energetic activity:

Les pôles de la pile isolée étant des sources *indéfinies* d'électricités contraires, il est évident que si l'on met chacun d'eux en communication avec un fil de métal, le fil partagera le fluide du pôle qu'il touche, et l'on aura ainsi deux conducteurs, l'un positif, l'autre négatif, qui étant mis en présence devront donner une recomposition *continuelle* ... le deux fils ... étant approchés à une petite distance, on voit jaillir une étincelle, une autre la suit de près, puis une autre, et ainsi de suite; c'est un courant de feu continu, c'est une baterie inépuisable qui, se décharge toujours sans être jamais déchargée. Lorsqu'on met les fils conducteurs en contact immédiat, et que l'on ferme ainsi le circuit de la pile, les étincelles disparaissent, mais tous les effets électriques ne sont pas détruits. Les fluides se développent encore dans tous les couples, entre tous les élémens, et ne cessent de venir se recomposer dans tous les points des fils conducteurs qui rejoignent les deux pôles de la pile. Ainsi au-dehors tout paraît immobile, et au-dedans tout est en activité et en mouvement. Une des preuves les plus frappantes de cette rapide circulation de l'électricité est le phénomène que présent un fil métallique un peu fin, que l'on

---

[122] Pouillet, *Élémens, 1*, Par. 2, 631–632.
[123] Pouillet clearly realizes his dual intentions; he distinguishes between one sort of electrostatic, tension-producing electromotive force and another sort of continuous, dynamic "production."

interpose entre les conducteurs, pour fermer le circuit: si ce fil est un peu long, il devient chaud subitement; un peu plus court, il devient rouge, et plus court encore, il devient rouge-blanc. . . .[124]

It seems clear from these passages that Pouillet appropriated the best from the electrodynamic theory and applied it to the electrostatic. His circuit theory considered both a pile with an arithmetic series of tensions that, as a Coulombian, he knew how to calculate and measure, and a wire carrying a true Ampèrian current. At the same time, his exposition seems cogent enough to convince even overly enthusiastic reunionists to allow a place for electrostatic tensions within the pile. When, several years later, Pouillet was still pursuing this hybrid approach, his studies culminated in an analysis of the total resistance in the circuit which, very much like Ohm's, joined Coulombian, tension-dependent isolating length with more contemporary notions of variable conductibility.[125] Pouillet's textbook thus merged past traditions and contained the seeds of future progress.

The *Élémens* is therefore something more than a convenient terminus for this paper. It shows clearly and concretely the importance of Ampère's electromagnetic theory and of the closely related work of Becquerel and De La Rive for the study of the electric current by a large community of French physicists. It also shows, specifically, how early supporters of Biot's theory had to acknowledge the sharpened distinction between the static and dynamic effects of electricity, and how they had to accept the galvanometric evidence for reunions which Becquerel had neatly collected and persuasively expounded. Finally, the *Élémens* demonstrates how new ideas about the electric current were assimilated partly through a slight modification and finer articulation of already accepted ideas and partly through a major change in experimental techniques.

The physicists of the *Académie,* in other words, transformed their ideas without all of them realizing explicitly that a conceptual revo-

---

[124] Pouillet, *op. cit.* (note 122), 635.

[125] Several roughly contemporaneous studies, notably those of Becquerel (for example, *Annales, 32* [1826], 420 ff.), concentrated on galvanometrically-measured conductibility to the exclusion of the electrostatic force needed to overcome resistance. Pouillet himself tried consistently to combine both galvanometric and electrostatic measures for current-flow. His work beginning in the late 1820's culminated in an important memoir in 1837, which, in many respects, was parallel to studies by Georg Ohm. (Pouillet, "Mémoire sur la pile de Volta et sur la loi générale de l'intensité prennent les courants," *Comptes Rendus, 4* [1837], 267 ff.) Ohm's classic "Die galvanische Kette" (1827) is summarized in Whittaker, *op. cit.* (note 2), 90–93.

lution was in progress. To some, the introduction of new ideas probably seemed more like a late modification of old notions, and the contours of a deeper conceptual shift were hidden behind the deceptive facade of a continued commitment to the methodology of instrumental physics. But at the end of the 1820's, when Becquerel's studies earned him a place within the official circle of the *Académie*,[126] other French physicists could look back to Oersted's discovery as the source of both conceptual and experimental innovation. To them, it was the crucial mutation on which the evolution of circuit theory so fundamentally depended.

---

[126] Becquerel narrowly defeated Pouillet for his position in the *Académie*, winning 29–28 on 20 April 1829 after several drawn ballots; see *Procès-Verbaux, 9*, 229. In what looks like a sign of formal recognition, the official *Mémoires* of the *Académie* published Becquerel's "Mémoire sur l'électrochimie et l'emploi de l'électricité pour opérer des combinaisons" in *9* (1830), 551 ff. This was the first article on voltaic electricity printed in the *Mémoires* since Biot's "Rapport." In his introductory remarks to this memoir, Becquerel cites Oersted's discovery and the galvanometric techniques it made possible as the turning points in the study of electrochemistry.

Maxwell, Osborne Reynolds, and the Radiometer

BY S. G. BRUSH* AND C. W. F. EVERITT**

Many window-shoppers are familiar with the radiometer, a sidewalk attention-getting device frequently displayed by jewellers. It is a simple instrument, consisting of a small glass bulb, partially evacuated, inside of which a set of vanes is mounted on a spike. Each vane is silvered on one side and blackened on the other. When the radiometer is placed in sunlight or near a source of heat, the vanes spin around.

Some rather involved physics is needed to explain completely how a radiometer works; we shall not attempt it here.[1] This essay will recount how physicists sought an understanding of the phenomenon nearly a hundred years ago, and in so doing laid the foundations of the modern science of rarified gas dynamics. This account relies heavily on referees' reports preserved in the Royal Society Archives; they contain valuable evidence of the relations between the protagonists. Referees' reports in general are a useful primary source which has been too much neglected by historians of science.[2]

In 1873 the English chemist William Crookes was attempting to determine the atomic weight of thallium, an element he had discov-

* Department of History, and Institute for Fluid Dynamics and Applied Mathematics, University of Maryland, College Park, Maryland 20742.

** Physics Department, Stanford University, Stanford, California 94305.

[1] See for example L. Loeb, *Kinetic Theory of Gases,* 2nd. ed. (New York, 1934), E. Kennard, *Kinetic Theory of Gases with an Introduction to Statistical Mechanics* (New York, 1938), A. E. Woodruff, *Phys. Teacher, 6,* No. 7 (1968), 358–363.

[2] We wish to acknowledge the generous cooperation of the staffs of the Royal Society of London and the Cambridge University Library in providing us with copies of the relevant documents. We also thank Brigadier Wedderburn-Maxwell, the Royal Society, and Cambridge University for permission to quote. We intend to publish the complete texts of these and other letters and manuscripts of Maxwell relating to kinetic theory at a later time.

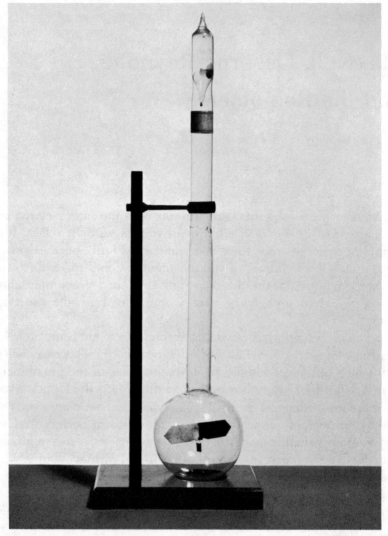

FIG. 1. Crooks Radiometer. Crown Copyright Reserved Photograph, Science Museum, London (courtesy of Dr. F. Greenaway).

ered spectroscopically a few years before. Realizing that his delicate pan-balance might be disturbed by irregular air currents, he made his weighings in a vacuum. To his great surprise, he found that heat seems to diminish the weight of bodies in a vacuum. Further investigation showed that while the body moves up when the heat source is

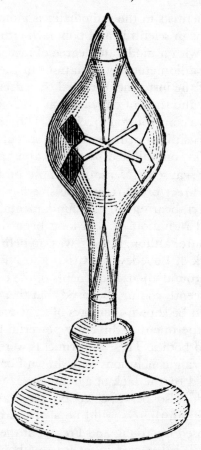

FIG. 2. Radiometer or Light-Mill, constructed by Dr. Geissler, of Bonn, as shown in Reynold's paper in the *Philosophical Transactions, 166* (1876), 725.

below it, it moves down when the source is above. Thus the effect is not simply a change of weight, but rather a directed force of repulsion associated with heat.[3] It was to illustrate this force more conveniently that Crookes constructed his radiometer (Fig. 1).

Although "light-mills" similar to the radiometer had been made and studied in Germany (Fig. 2), it was not until Crookes began to publicize his discovery that the radiometer attracted the attention of the scientific world at large. The intense, though short-lived, interest

[3] For a comprehensive account of the work of Crookes, see A. E. Woodruff, *Isis, 58* (1966), 188–198.

of contemporary scientists in the radiometer is indicated by the hundreds of publications in scientific and popular journals between 1875 and 1877. Anyone with a moderate degree of mechanical skill could construct his own radiometer, try variations in the size, shape, and color of the parts of the instrument, record the effects of temperature, pressure, humidity, and time of day, and have the satisfaction of publishing his results as a contribution to science. Those with a more theoretical bent could rush into print with their speculations about how the radiometer works, uninhibited by any established theory.

The radiometer caused a sensation, most probably because it seemed to prove a direct pressure or repulsive force of light or heat. It thus had a direct bearing on the fundamental problem of the nature of light and radiation. It had long been thought that light might exert a pressure, although there was no definite experimental proof until the work of Lebedew in Russia and Nichols and Hull in the United States around 1900. Proponents of the corpuscular theory of light in the eighteenth century believed that their theory would be confirmed if it could be shown that rays of light possess momentum. Several tried the experiment of directing powerful beams of light on delicately suspended bodies. These experiments were generally inconclusive—rather, varying conclusions were drawn from them. The issue was further confused by the lack of agreement as to whether or not light pressure would also be expected on the wave theory. If the waves are longitudinal, there should be a pressure; but if they are transverse, as proposed by Young and Fresnel at the beginning of the nineteenth century, it is less certain, at least without Maxwell's electromagnetic theory.

A somewhat skeptical observer, the physician W. B. Carpenter, reported the initial reaction of British scientists in 1873: when the radiometer "was first exhibited at the soirée of the Royal Society, there was probably not one who was not ready to believe with its inventor that the driving round of its vanes was effected by the direct mechanical energy of that mode of radiant force which we call Light." [4] Carpenter criticized this "readiness to admit a novelty" as evidence of an unscientific attitude, and compared it with the gullibility of scientists who accepted accounts of supernatural phenomena. [5]

[4] W. B. Carpenter, *The Nineteenth Century*, *1* (1877), 242–256.
[5] W. B. Carpenter, *Mesmerism, Spiritualism, Etc. Historically and Scientifically Considered* (New York, 1877), 7.

This was an unkind dig at Crookes himself, who had been involved in investigations of spiritualism in the years immediately preceding.

According to Carpenter, "two of the most distinguished among British mathematical physicists" agreed with Crookes that the motion of the vanes was a direct effect of radiant energy. One of these may have been James Clerk Maxwell; in any case, Maxwell's first impressions of the radiometer are significant since he played an important role in the later development of its theory. In a letter to his friend William Thomson (later Lord Kelvin), Maxwell gave the following account of Crookes's demonstration at the Royal Society soirée:

> They whip spirits all to pieces. A candle at 3 inches acts on a pith disk as promptly as a magnet does on a compass needle. No time for air currents and the force is far greater than the weight of all the air left in the vessel. Attraction by a bit of ice very lively. All this at the best attainable vacuum . . .[6]

Maxwell had made two major contributions to theoretical physics by this time, both relevant to the radiometer problem. First, in his 1867 paper on kinetic theory, he had developed a powerful method for dealing with viscosity, heat conduction, and diffusion in gases, based on transport equations for the flow of momentum, energy, and matter. Second, on the basis of his electromagnetic theory of light, Maxwell produced a quantitative estimate of the pressure of light. In his *Treatise on Electricity and Magnetism* (1873), Maxwell wrote that concentrated radiation from an electric lamp, "falling on a thin metallic disk, delicately suspended in a vacuum, might perhaps produce an observable mechanical effect," although he estimated that the force due to ordinary sunlight would be only about a tenth of the horizontal magnetic force in Britain (§793). Thus he would certainly be interested in the results of Crookes's experiments.

When Crookes submitted a written account of his work to the Royal Society, it was sent to Maxwell to review for publication in the *Philosophical Transactions*. Maxwell made his report on 24 February 1874. Fortunately the referees' reports on this and other papers on the radiometer (except for the ones published by Larmor in 1907 in the *Memoir and Scientific Correspondence* of G. G. Stokes) have been preserved at the Royal Society; they give a valuable insight into the development of the theory.

[6] Letter No. 108 in the Maxwell Collection at Cambridge University.

Maxwell wrote that Crookes "has made a great discovery with respect to the mode in which a body placed in air of different densities is acted on when bodies of different temperatures are placed near it. . . ." He agrees that the observed "repulsion from a heat-emitting body" is "due to radiation." Although, as he notes, he had suggested "a probable repulsive action of radiation" in his treatise on electricity, "the effects observed by Mr. Crookes seem to indicate forces of much larger value." Maxwell seems willing to accept this quantitative refutation of his own theory, and recommends publication of the paper.[7]

In the summer of 1874, Coggia's comet made a spectacular appearance in the skies of Europe. Its tail, visible for several weeks, showed the familiar behavior of pointing away from the sun. In the past, observations of comets had stimulated astronomers to suggest that some kind of repulsive action of the sun, perhaps light pressure, acts on the volatile matter of the comet and pushes it out on the opposite side. The new comet of 1874 must have reopened the discussion of light pressure in many European scientific circles, and helped to prepare the way for the radiometer fad of 1875–1877.

Crookes himself suggested in his first paper that experiments on the radiometer might help to reveal something about the nature of the sun's repulsive action on comets. Maxwell had discussed comets' tails in earlier correspondence with astronomers, but had not adopted the light-pressure explanation. That comets' tails were a subject of conversation in Maxwell's house during the visit of the comet in 1874 is evident from an anecdote of Garnett: the frequent repetition of the word "tail" caused Maxwell's terrier to run round in circles chasing his own.[8]

One of the first scientists to challenge the light-pressure theory of the radiometer was Osborne Reynolds, Professor of Engineering at Owens College, Manchester. Reynolds thought comets' tails were either an electrical effect or a "negative shadow."[9] In 1873 he had published a paper on the steam engine where he discussed the condensation of a mixture of steam and air on cold surfaces; this study may have inspired his suggestion, a year later, that the radiometer effect depends on the condensation and evaporation of gas molecules

---

[7] Royal Society Archives, 1874, No. 295.
[8] Reported by A. Ferguson, *Nature, 128* (1931), 604–608.
[9] O. Reynolds, *Papers on Mechanical and Physical Subjects* (Cambridge, 1900–1903), *1*, 8. Most of the other papers by Reynolds mentioned in this article may be found here.

at the surface of the vanes. Although his radiometer theory in its original form did not prove successful, it did prefigure in a general way later explanations based on gas-surface interactions. More important at the time, however, was the work of a young physicist, Arthur Schuster, who came to Manchester in 1875. Schuster, with Reynolds' encouragement, carried out what was subsequently regarded as the "crucial experiment" on the radiometer: the observation of the rotation of the radiometer case in a direction opposite to that of the vanes.[10] This was taken to mean that no momentum is brought into the system by the radiation, only energy; the actual turning of the vanes must then depend on some internal mechanism, presumably associated with the residual gas in the bulb.

By this time it had also been pointed out by several observers that light ought to produce a greater force on the reflecting, silvered side of the vanes than on the blackened side where it is absorbed; in fact they soon noted that the vanes move in just the opposite way.

By 1876, the British scientists who were actively working on the radiometer problem—Crookes, Schuster, Reynolds, Stoney, Dewar, and Tait—agreed that the rotation was not a direct effect of radiation striking the vanes, but that it depended in some way on the residual gas and the temperature difference between the two sides of the vanes. They thought that gas molecules striking the hot, black side would somehow exert more pressure than those striking the cold, silvered side. However, they disagreed about the precise reason for this pressure difference. They recognized that the ordinary methods of the kinetic theory of gases based on the Clausius mean-free-path concept were inadequate to deal with rarified gases. The mean free path of a molecule might be larger than the distance a molecule travels between successive collisions with solid surfaces; in this case equilibration of temperature, which usually takes place rapidly as a result of inter-molecular collisions, would be ineffective, and it might be possible for an unequally heated vane to establish a large temperature gradient in the surrounding gas. However, no one was able to calculate accurately the net force on the vanes due to this temperature gradient without arbitrary assumptions.

[10] A. Schuster, *Proceedings of the Royal Society, 24* (1876), 391–392; *Philosophical Transactions, 166* (1877), 715–724. See also his *Biographical Fragments* (London, 1932), 230–232: Schuster says he first proposed the experiment to Reynolds and others during the winter of 1873–1874 but was reluctant to do it himself for fear of "cutting into what I consider to be other people's work." Finally Reynolds set up the apparatus and they did it together.

This was clearly a problem for an expert in kinetic theory, and Maxwell was frequently asked for his opinion. Realizing that the radiometer problem was very difficult, he did not want to commit himself in public until he had considered it carefully. His reports to the Royal Society on the papers of Crookes, Reynolds, and Schuster show him skeptical of the existing theories and, indeed, more concerned with the details of experimental apparatus than with theoretical explanations. However, on one occasion his reputation as an authority on physics forced him to appear before an imperious audience; in May of 1876 Maxwell wrote to his uncle, Robert Cay:

> ... I was sent for to London, to be ready to explain to the Queen why Otto von Guericke devoted himself to the discovery of nothing, and to show her the two hemispheres in which he kept it, and the pictures of the 16 horses who could not separate the hemispheres, and how after 200 years W. Crookes has come much nearer to nothing and has sealed it up in a glass globe for public inspection. Her Majesty however let us off very easily and did not make much ado about nothing, as she had much heavy work cut out for her all the rest of the day....[11]

For the next year and a half there is little evidence of Maxwell's thinking about the radiometer problem. However, his official Cambridge University report reveals that the Cavendish Laboratory had acquired four radiometers by May 1877.[12] It was also during this period that Arthur Schuster came to Cambridge from Manchester. While the flurry of excitement about radiometers was generally dying down, since light pressure seemed not to be directly involved after all, Maxwell was just beginning to define the real problem for himself and to set to work on it seriously.

It was not until just before his death in 1879 that Maxwell's paper "On stresses in rarified gases arising from inequalities in temperature" was completed: it was worth waiting for.[13] The main part of the paper was devoted to calculating the force resulting from temperature inequalities in the interior of a gas. Maxwell found that the stress in

---

[11] Maxwell to Robert Cay, 15 May 1876, Peterhouse Library, Cambridge University.

[12] J. C. Maxwell, *Cambridge University Reporter*, 15 May 1877, p. 434. There are no radiometers in the list of apparatus for the previous year. (*Ibid.*, 20 May 1876, p. 496.)

[13] J. C. Maxwell, *Phil. Trans.*, *170* (1879), 231–256; reprinted in Maxwell's *Scientific Papers* (Cambridge, 1890), 2, 681–712. The paper is marked "Received March 19—Read April 11, 1878," but additional notes were added, dated May and June 1879. Abstracts appeared in *Proc. Roy. Soc.*, 27 (1878), 304–308 and *Nature*, *18* (1878), 54–55.

gases is proportional to the second spatial derivative of the temperature. Thus, contrary to the results of the earlier theories of FitzGerald and Stoney, a constant temperature gradient (constant first spatial derivative) would not by itself produce an inequality of pressure. However, a small object in the gas, acting as a heat-source, could produce a changing temperature gradient and a resulting stress which might be sufficient to account for the motion of the radiometer vanes. If the vane itself is the source of the heat flow, the stress will be found only near the edges of the vane.

As Maxwell was careful to point out, his kinetic theory gives only the *normal* stress on the surface of the vane. The theory cannot be used to compute tangential stresses without further assumptions; in fact, the usual assumption in kinetic theory is that the physical state of the gas next to a surface is the same as that of the surface itself, in which case there can never be a tangential stress. This in turn involves two assumptions, one being that there is no "slipping" (discontinuous velocity change) of the gas relative to the surface, the other that there is no temperature-discontinuity. Both are of dubious validity in rarified gases in view of the discovery of slip phenomena by Kundt and Warburg.[14]

Maxwell now recognized a fatal defect in his original theory: when any number of solid spheres are maintained at various temperatures in a gas, a state of equilibrium will eventually be reached in which there is a steady flow of heat, and in this case there can be no forces on any of the spheres as long as only normal stresses are considered. Nevertheless, Crookes's experiments showed that forces do act "between solid bodies immersed in rarified gases, and this, apparently, as long as inequalities of temperature are maintained" (§§11–12). Maxwell proposed to attribute the effect to "the phenomenon discovered in the case of liquids by Helmholtz and Piotrowski[15] and for gases by Kundt and Warburg, that the fluid in contact with the surface of a solid must slide over it with finite velocity in order to produce a finite tangential stress." Kundt and Warburg had found that the velocity of sliding of a gas over the surface due to a given tangential stress varies inversely as the pressure, so that this

[14] A. Kundt and E. Warburg, *Annalen der Physik,* ser. 2, *155* (1875), 337–366 and *156* (1875), 177–211; *Philosophical Magazine,* ser. 4, *50* (1875), 53–62.
[15] H. Helmholtz and G. von Piotrowski, *Sitzungsberichte der kaiserlichen Akademie der Wissenschaften in Wien, 4* (1860), 607–658.

effect should be stronger in raified gases. The existence of such currents sweeping along the surfaces of solid bodies immersed in a rarified gas would completely destroy the simplicity of the kinetic-theory solution of the problem.

In an Appendix added to his paper in May 1879, Maxwell said that while he had previously been reluctant to give any quantitative treatment of the effect of slipping, one of the referees of the paper had encouraged him to do so and had even suggested various possible hypotheses about the gas-surface interaction. Maxwell now proposed a set of equations which "express both the fact that the gas may slide over the surface with a finite velocity . . . and the fact that this velocity and the corresponding tangential stress are affected by inequalities of temperature at the surface of the solid, which give rise to a force tending to make the gas slide along the surface from colder to hotter places." [16]

Here is the basis of the modern theory of the radiometer effect, and of many other phenomena that take place in rarified gases. Maxwell does not claim the credit for discovering the tendency of a gas to creep along the surface from colder to hotter places. Instead, he states that the phenomenon was discovered by Osborne Reynolds, who called it "thermal transpiration," and who showed that it is a necessary consequence of the kinetic theory of gases. Maxwell reveals that "it was not till after I had read Professor Reynolds' paper that I began to reconsider the surface conditions of gas, so that what I have done is simply to extend to the surface phenomena the method which I think most

---

[16] Maxwell assumed that "of every unit of area [of the surface] a portion f absorbs all the incident molecules, and afterwards allows them to evaporate with velocities corresponding to those in still gas at the temperature of the solid, while a portion $1-f$ perfectly reflects all the molecules incident upon it." He derived, as a first approximation, the following equation, relating the velocity of the gas near a flat surface in the yz plane to the temperature θ, to its gradients, and to the viscosity μ and density ρ of the gas:

$$v - G\left(\frac{dv}{dx} - \frac{3}{2}\frac{\mu}{\rho\theta}\frac{d^2\theta}{dx\,dy}\right) - \frac{3}{4}\frac{\mu}{\rho\theta}\frac{d\theta}{dy} = 0,$$

where G is the "Gleitungs-coefficient" (slip coefficient) previously introduced by Helmholtz and Piotrowski. G is related to the mean free path l of a molecule by the equation

$$G = \frac{2}{3}\left(\frac{2}{f} - 1\right)l.$$

Kundt and Warburg found that for air at different pressures p flowing through glass capillary tubes from $17°C$ to $27°C$, $G = 8/p$ centimeters. (Appendix, added May 1879, to the paper cited in note 13.)

A rough qualitative explanation of the origin of the radiometer force based on the surface flow set up by a temperature gradient was given by Maxwell in a referee's report (Fig. 4, below).

suitable for treating the interior of the gas. I think that this method is, in some respects, better than that adopted by Professor Reynolds, while I admit that his method is sufficient to establish the existence of the phenomena, though not to afford an estimate of their amount."

Thus the most important part of Maxwell's paper, the treatment of gas-surface interactions, owes its origin to the suggestion of an unnamed referee, and its success to Maxwell's acquaintance with a paper by Osborne Reynolds—a paper which had not yet been published! The crucial importance of the referees' reports for disentangling the history of the radiometer begins to emerge.

Maxwell's paper was sent to the Royal Society in the middle of March 1878. It was referred to William Thomson, who submitted his report on 15 June 1878. Thomson had received a summary of the theory in a letter from Maxwell on 7 March 1878, [17] and he says in his report that he has already discussed the paper with its author. It seems probable that Maxwell knew that Thomson was the referee; he could hardly have failed to recognize the literary style of Thomson, with whom he frequently corresponded (there might have been some pretense of anonymity since the Secretary of the Royal Society, Stokes, had sent a typewritten copy of portions of the report to Maxwell rather than the original). The most important part of Thomson's report (Fig. 3) reads:

> If the surface were like the sketch, so that all collisions would be against the projecting gills and only an infinitely small proportion of them against the bottoms of the hollows (b in the sketch) would it act as would the same space occupied by the gas? Suppose the surface of the solid to be covered with square pyramids or with simple harmonic hills ($z = h \sin mx \sin ny$) as in the second sketch. How would it be if the vertical angle (θ) of each pyramid is infinitely small? (The same answer here of course as to the preceding question.) If $\theta = 180°$ and the surface of the solid is supposed to be frictionless (that is to say if we have simply a smooth frictionless solid) how will it be in this case? (I suppose in this, as in the other extreme, the resultant action on the solid will be equilibrant.) [18]

It was apparently this problem proposed by Thomson that induced Maxwell to throw caution to the winds and formulate a theory of gas-

---

[17] Published by J. Larmor, *Proceedings of the Cambridge Philosophical Society, 32* (1936), 743–744.
[18] Royal Society Archives, 1878, Maxwell (70) Thomson's Report (marked "123" in the upper right-hand corner of the first page).

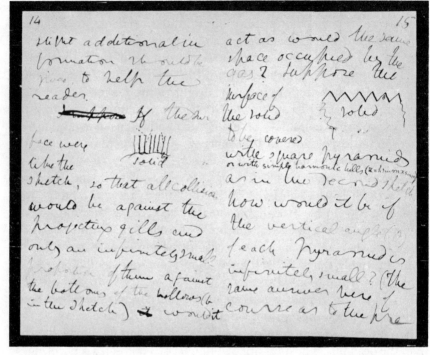

FIG. 3. A page from William Thomson's report on Maxwell's paper (1878).

surface interactions. Some insight into the early stages of this theory can be gleaned from Maxwell's report,[19] dated 23 October 1878, on the sixth of Crookes's series of papers "On repulsion resulting from radiation." Crookes submitted the paper to the Royal Society on 27 June 1878, so Maxwell presumably received it only a few weeks after seeing Thomson's report on his own paper. Maxwell is interested mainly in a new instrument constructed by Crookes, "a small fly with clear mica vanes which can be placed in different parts of the radiometer bulb to detect differences of pressure on its vanes." Explorations with this device would permit a direct test of Maxwell's prediction that if the isothermal surface adjacent to the surface of the solid is not parallel to that surface, there will be a tangential stress of the gas on the solid. (The force would be directed from places where consecutive isothermal surfaces are close together to places where they are further apart.) "At the surface of the solid," Maxwell writes in his report, "the

[19] Royal Society Archives, 1878, No. 88.

gas cannot support this stress but must yield to it. This causes a current sweeping over the surface." The theory is confirmed qualitatively by Crookes's observations.

Three months later, in January 1879, Osborne Reynolds submitted his paper "On certain dimensional properties of matter in the gaseous state" to the Royal Society. It contains a general theory of the flow of rarified gases, with applications to the radiometer and to the new phenomenon of "thermal transpiration," one of the first major discoveries stimulated by the radiometer. Maxwell was asked to report on Reynolds' paper; he made good use of the information contained in it in extending his own theory.

Thermal transpiration, as Reynolds defines it, is the flow of gas through porous plates caused by a temperature difference on the two sides of the plate. To illustrate the magnitude of the effect, Reynolds stated that a temperature difference of 160°F between two sides of a porous plate, with hydrogen gas maintained at atmospheric pressure on one side, would sustain a permanent pressure difference of one inch of mercury, the higher pressure being on the hotter side. The rate of transpiration depends on the porosity of the plate and the density of the gas in such a way that similar results are obtained "so long as the density of the gas is inversely proportional to the lateral dimensions of the passages through the plates." In other words, similar effects could be observed in rarified gases with channels of larger diameter. Now it was just this "scaling" property of the phenomenon which led Reynolds to discover it in the first place, according to his own account. He had recognized that the radiometer effect depends on the ratio of the size of the vanes to the mean free path of the molecules in the gas: there would be no effect for very large vanes, or for very small mean free paths. Since the mean free path decreases as the pressure (or density) increases,

> it appeared that by using vanes of comparatively small size the force should be perceived at comparatively greater pressures of gas. . . . On considering how this might be experimentally tested, it appeared that to obtain any result at measurable pressures the vanes would have to be very small indeed; too small almost to admit of experiment. And it was while thinking of some means to obviate this difficulty that I came to perceive that if the vanes were fixed, then instead of the movement of the vanes we should have the gas moving past the vanes—a sort of inverse phenomenon—and then instead of having small vanes, small spaces might be allowed for the gas to pass. Whence it was at

117

once obvious that in porous plugs I should have the means of verifying these conclusions. . . .[20]

In his first report on Reynolds' paper, Maxwell approved heartily of the experiments described, but was less satisfied with the theoretical deductions.[21] He felt that Reynolds had not clearly stated what happens at the surface, and suggested the hypothesis "that the surface has on it small prominences of various shapes from which the molecules rebound, with a velocity which is greater or less than the impinging molecules, accordingly as the temperature of the solid is greater or less than that of the gas." Maxwell even worked out the beginnings of a quantitative theory based on this idea, remarking that "conditions of this kind might perhaps assist Prof. Reynolds in forming a theory about the rebounding molecules." Reynolds did make some use of this suggestion—he says in his published paper that Section VII "was revised and somewhat enlarged in August, 1879, in accordance with a suggestion made by one of the referees," but he does not adopt the particular formulation given by Maxwell. Nevertheless there is a clear train of influence running through the referees' reports from Thomson to Maxwell to Reynolds, leading eventually to the modern theory of gas-surface interactions in rarified gases.[22]

Maxwell's report concludes (Fig. 4) with his own explanation of the mechanism of thermal transpiration (and incidentally of the radiometer effect):

We may use this method directly to explain the fact of thermal transpiration, etc.

Let P be an element of the surface of a solid and let us suppose that the temperature is the same for gas and solid along any vertical line but that it increases from left to right.

Let us also suppose that the pressure is at first the same everywhere. Consider the molecules moving in the direction AP. They come

---

[20] O. Reynolds, *Phil. Trans.*, *170* (1879), 727–845 (received 17 January, read 6 February 1879, with an Appendix added December 1879).

[21] Royal Society Archives, 1879, No. 188 (28 March 1879).

[22] Not much has been added to the theory since the work of these men. Although the concept of "accommodation coefficient" was introduced in the twentieth century by Knudsen and others as a further development of Maxwell's ideas, it remains for the most part an empirical parameter representing the probability that a molecule will be absorbed temporarily by the surface to acquire an average energy corresponding to the surface temperature before bouncing off again. Attempts to calculate it from molecular theory have not been particularly successful. As recently as 1961, L. Waldmann remarked in a symposium that "for the calculation of the slip coefficients, at present only Maxwell's theory of 1879 is available." (*Rarefied Gas Dynamics*, ed. L. Talbot, [New York, 1961], 323.)

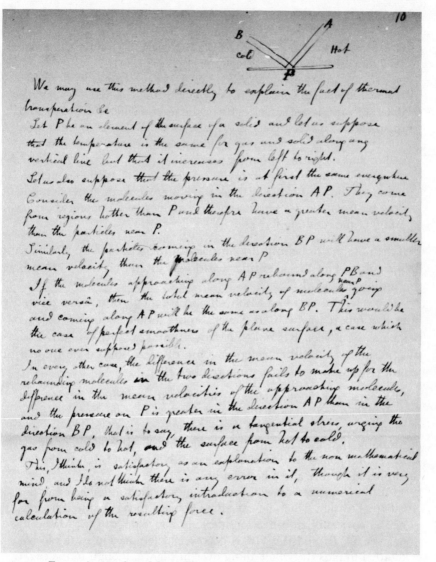

We may use this method directly to explain the fact of thermal transpiration &c

Let P be an element of the surface of a solid and let us suppose that the temperature is the same for gas and solid along any vertical line but that it increases from left to right.

Let us also suppose that the pressure is at first the same everywhere Consider the molecules moving in the direction AP. They come from regions hotter than P and therefore have a greater mean velocity than the particles near P.

Similarly, the particles coming in the direction BP will have a smaller mean velocity than the molecules near P

If the molecules approaching along AP rebound along PB and vice versâ, then the total mean velocity of molecules going up and coming along AP will be the same as along BP. This would be the case of perfect smoothness of the plane surface, a case which no one ever supposed possible.

In every other case, the difference in the mean velocity of the rebounding molecules in the two directions fails to make up for the difference in the mean velocities of the approaching molecules, and the pressure on P is greater in the direction AP than in the direction BP, that is to say, there is a tangential stress, urging the gas from cold to hot, and the surface from hot to cold.

This, I think, is satisfactory as an explanation to the non mathematical mind, and I do not think there is any error in it, though it is very far from being a satisfactory introduction to a numerical calculation of the resulting force.

FIG. 4. A page from Maxwell's report on Reynold's paper (1879).

from regions hotter than P and therefore have a greater mean velocity than the particles near P.

Similarly the particles coming in the direction BP will have a smaller mean velocity than the molecules near P.

If the molecules approaching along AP returned along PB and

vice versa, then the total mean velocity of molecules near P going and coming along AP will be the same as along BP. This would be the case of perfect smoothness of the plane surface, a case which no one ever supposed possible.

In every other case, the difference in the mean velocity of the rebounding molecules in the two directions fails to make up for the difference in the mean velocities of the approaching molecules, and the pressure on P is greater in the direction AP than in the direction BP, that is to say, there is a tangential stress, urging the gas from cold to hot and the surface from hot to cold.

This, I think, is satisfactory as an explanation to the nonmathematical mind, and I do not think there is any error in it, though it is very far from being a satisfactory introduction to a numerical calculation of the resulting force.[23]

Unfortunately, this beautifully simple argument did not appear in print at the time when the scientific world was interested in knowing how the radiometer works.

The relations between Maxwell and Reynolds deteriorated rapidly after this report was written. Several more letters about Reynolds' paper passed between Reynolds, Maxwell, Stokes, and William Thomson (acting as the other referee). One of these should be quoted at length (Maxwell to Stokes, 2 September 1879):

Of course I cannot profess to follow with minute attention the course of an acrobat [Reynolds] who drives 24 in hand, but as on more than one occasion he throws up the reins and starts a new team, it is probable that the results will be sufficiently flexible to adapt themselves to the facts, whatever the facts may be. But O. R. says he has improved all this, and I hope he has. . . .

But to criticize Stoney when he comes to the Radiometer and the Stoney Stratum which he calls the Crookes' Layer required hermeneutic powers of an order considerably above dp/dt [i.e., Maxwell] or even O. R. and this fact is a powerful temptation to the feeble minded to pass the whole thing over in silence, though to the eagle eye of Thomson a happy expression, even though in the midst of erroneous words, may illuminate the whole conglomeration of blunders with a meaning which the author himself could never be made to recognize.

With respect to Graham's experiments, O. R. is right and Thomson wrong, for there is considerable difficulty, perceived by Graham

[23] Maxwell, *op. cit.* (note 21).

himself, in getting any simple law out of them. The result was that Graham was driven to the excessively fine pores of graphite plate, whereby he approximated to the law of effusion and got rid of transpiration almost entirely. On the other hand, O. R. is impervious to Thomson's lucid statement of the difference between effusion and transpiration, namely that what restrains the flow of gas is collisions but that in effusion the collisions between two molecules of gas are very few compared with those between a molecule and a solid surface whereas in transpiration the collisions between two molecules mostly preponderate, so that in effusion the velocity of the molecules governs the velocity of effusion, whereas in transpiration viscosity is the ruling consideration. I am afraid I have not answered your letter at all, except about O. R. being the discoverer of dimensional properties in gases. I have always felt inclined to give him leave to practise at his "mean range" till he has qualified himself to go in among the all comers for the R. S. meetings.[24]

This is the letter which Horace Lamb referred to when he wrote to Larmor in 1905 regarding the latter's edition of the Stokes letters: "Maxwell's letter is certainly amazing and characteristic, and hits off very happily and good-naturedly some of O. R.'s peculiarities. The gaiety of the world will lose by its omission, but I cannot urge its publication at present. I remember that O. R. long bore a grudge against Maxwell on account of this paper, and he might still be sensitive to the brilliant raillery." [25]

We must now say something about the reasons for Reynolds' "grudge against Maxwell on account of this paper." There is more to it than the normal resentment of authors against the referees who criticize and delay the publication of their papers. In this case Maxwell had made use of Reynolds' discovery of thermal transpiration (with proper acknowledgment) in his own paper, which was published before Reynolds'. Here Maxwell had cast doubts in public on the value of Reynolds' work before Reynolds was in a position to defend himself. The situation was further aggravated by the fact that any criticism of Maxwell's role in the affair was soon impossible because of the sympathy called forth by his painful illness (cancer of the abdomen) and death. This illness may also account for Maxwell's erratic

[24] Royal Society Archives, 1879, No. 57.
[25] See J. Larmor, *Memoir and Scientific Correspondence of the late Sir George Gabriel Stokes* (Cambridge, 1907); letter from Lamb to Larmor, 1 October 1905, in Stokes papers, Add. 7618, Box 3, Cambridge University.

behavior during this period. Tait, writing to Stokes in December 1879, suggested that "Maxwell's hereditary malady [his mother had died of the same cause] had begun to affect him some months before his death," as shown by his conflicting recommendations of Chrystal and Garnett for the Edinburgh chair of mathematics:

> I began to fear [Tait continued] that his mind was affected; but, happily, this phase was very transient. I had full details of his malady from my colleague Sandars, who attended him and who was an old fellow-student of his in Edinburgh, but I considered it much too painful and delicate a subject to be even hinted at in public. No man could have behaved more tenderly or nobly than Maxwell did under long trials of the most overwhelming nature. How he could manage to do splendid mathematical work all the time is inconceivable.[26]

Here was a trap waiting for Reynolds—the circle of Maxwell's powerful friends (Tait, Thomson, Stokes, and many others) would rush to his defense if anyone should dare criticize the dead hero.

On 23 October 1879, just two weeks before Maxwell's death, Reynolds sent a letter to Stokes to be communicated to the Royal Society. He protested the slur on his work in Maxwell's paper, and ventured "to request those interested in the subject to withhold their opinion until they have an opportunity of reading my paper. In the meantime I can only express my opinion that Professor Maxwell is mistaken in supposing that the results which are obtained from his method are more definite than those to be obtained by mine."[27] Apparently Reynolds wanted the letter to be read to the Royal Society immediately, but Stokes was reluctant to do so. In a telegram to Reynolds on 5 November 1879, just after Maxwell had died, Stokes demanded that Reynolds should either rewrite his letter or allow Stokes to add his own comments after presenting it.[28] Reynolds chose the second alternative, and his letter was read the following April with a note by Stokes appended. In this note, Stokes mentioned that Maxwell's work had been partly based on a suggestion of William Thomson.

In the meantime, Thomson was working to undermine the originality of Reynolds' discovery of thermal transpiration. He wrote to Stokes about a paper by W. Feddersen on the subject, based on a the-

---

[26] Stokes papers, Box 13, Cambridge University.
[27] O. Reynolds, *Proc. Roy. Soc., 30* (1880), 300–302; reprinted in *Papers, 1,* 391.
[28] Telegram from Stokes to Reynolds, 5 November 1879, in Stokes-Kelvin papers, Add. MS. 7618, Box 1, Cambridge University.

oretical prediction of Carl Neumann in 1872.[29] Stokes passed on the reference to Reynolds, who added a note to his own paper in December 1879, in which he stated that M. J. Violle[30] had attributed Feddersen's results to the presence of water vapor. But in attempting to repel the threatened German invasion of his priority, Reynolds had hastily grasped at a French straw; he seems to have misunderstood Violle's remarks. To discuss the work of Feddersen, Neumann, Dufour, and others would be out of place here; we mention this incident only because it throws another sidelight on the relations between Reynolds and Maxwell's friends.

FitzGerald, in a note in the *Philosophical Magazine* for February 1881, recalled Maxwell's criticism of Reynolds' work, and further complained that "Prof. Reynolds' paper is very elaborate, and necessarily somewhat difficult, not only from the nature of the subject, but also, in parts, owing to the inelegant method that Prof. Reynolds has pursued." Reynolds replied:

> With regard to Professor Maxwell's remarks on my paper, and his own work on the same problem, of course, the sad circumstance of his death occurring, so that this was about the last work he did, renders it very difficult to approach the subject; but with reference to what I have already said, and in explanation of the apparently imperfect idea at which he arrived as to the scope and purpose of my method, it may be stated that, before writing his own paper, Professor Maxwell had only seen my paper in manuscript in the condition in which it was first sent in to the Royal Society, when the preliminary part was very much compressed, and, as I fear, somewhat vaguely stated, besides being founded on different assumptions from the present.[31]

By now scientists were no longer concerned primarily with the radiometer itself. The radiometer had served to direct attention to the phenomena of rarified gases; this subject could now develop independently without being tied to a particular instrument. Crookes

[29] Thomson to Stokes, 11 April 1880, in Stokes-Kelvin letters, Add. MS. 7618, Box 1, Cambridge University; see W. Feddersen, *Ann. Physik, 148* (1873), 302–311; C. Neumann, *Berichte über die Verhandlungen der Königlich Sächsischen Gesellschaft der Wissenschaften zu Leipzig*, Mathematisch-Physische Classe, 24 (1872), 49–64.

[30] J. Violle, *Journal de Physique, 4* (1875), 97–104.

[31] O. Reynolds, *Phil. Mag.*, ser. 5, *11* (1881), 335–342; *Papers, 1,* 384.

moved on to study electrical discharges in rarified gases, opening a path to experimental atomic physics; Reynolds turned to other topics in fluid dynamics, such as the problem of the onset of turbulence.

As often happens in science, a phenomenon which can be reproduced experimentally fairly easily, such as the radiometer effect, turns out to be complicated from a theoretical viewpoint; and then a very refined experimental technique is needed to isolate individual aspects which can be analyzed theoretically. Almost no research on the radiometer itself was done for about forty years after Maxwell's death. G. D. West, in 1920, was probably correct in attributing this to the fact that "the majority of physicists felt that the matter had been placed beyond their grasp, rather than because some well-understood explanation had been given which rendered further research superfluous." [32] William Sutherland, attempting to revive interest in the subject in 1896 on the grounds that Reynolds' theory was unsatisfactory, gave a vivid retrospective account:

The comparative neglect into which the radiometer has fallen is probably the natural compensation for the exalted interest of its two or three years' reign over the scientific imagination twenty years ago. In reading amongst the papers about it published at that time, one gets an impression of the laboratory of Crookes as of an arsenal where night and day the equipment of a great expedition into the unknown was being pushed on under the sleepless eye of a patriot leader; but in the answering bustle outside, Stokes, Schuster, Stoney, FitzGerald, Pringsheim, Reynolds and others soon showed that the new conquest was simply an outlying part of the Kinetic Theory of Gases. Or, to vary the figure, Crookes appears as a friendly counsel subjecting Nature to a passionate and eloquent cross-examination with his fellow physicists as judge and jury bringing in a verdict for the Kinetic Theory. And then the interest died rapidly away, perhaps mostly on account of Reynolds' great paper . . . which was probably held to settle consequences of the kinetic theory of gases, especially as the same train of reasoning had led him to his discovery of Thermal Transpiration with the beautiful establishment of its simple quantitative laws, simple in the illumination of his theory, but complex enough without it.[33]

[32] G. D. West, *Proceedings of the Physical Society of London, 32* (1920), 222.
[33] W. Sutherland, *Phil. Mag.,* ser. 5, *42* (1896), 373–391 and 476–492; see also Reynolds, *Phil. Mag.,* ser. 5, *43* (1896), 142–148.

That, in outline, was the "public history" of the radiometer; we have attempted in this paper to give some indication of the private history of the interactions between the major participants.

Acknowledgment: the work of one of us (S. G. B.) was supported in part by a research grant from the U. S. National Science Foundation.

Gibbs on Clausius

BY MARTIN J. KLEIN*

"The plan which you particularly mention, that of historical and critical articles, seems one where there is quite an opening for useful activity."

J. W. Gibbs

This paper is dedicated to my teacher, Professor Laszlo Tisza

1. Early in June 1889 Josiah Willard Gibbs received a letter asking him to write a review of the scientific career of Rudolf Clausius, who had died the previous August. The request came from Josiah Parsons Cooke, Professor of Chemistry at Harvard, writing on behalf of the American Academy of Arts and Sciences at Boston.[1] Clausius had been elected to membership in the Academy in 1873 and it was therefore customary and proper that he be commemorated in the pages of its *Proceedings* by a fellow member. This posthumous privilege of membership would prove to be the greatest, if not the only, benefit Clausius derived from his otherwise purely honorary association with the Academy at Boston: no learned society in the world, and Clausius belonged to some twenty-five of them, could have offered a person better qualified than Gibbs to write about Clausius.

Gibbs had never met Clausius, so far as we know. Clausius never visited the United States, and there is no record that Gibbs crossed his path at any time during the three years of postdoctoral study that he spent in Europe right after the American Civil War. Nor does the catalogue of Gibbs's scientific correspondence show any letters from

* Department of History of Science and Medicine, Yale University, New Haven, Conn. 06520.
[1] Lynde Phelps Wheeler, *Josiah Willard Gibbs*, revised ed. (New Haven, 1952), Appendix III. Catalogue of Gibbs's "Scientific Correspondence," p. 226.

Clausius, even though Gibbs regularly sent him reprints of his papers.[2] (Clausius apparently took little interest in the later developments of the thermodynamics he helped to create. Max Planck has described his own failure to get any response from Clausius to his early work on thermodynamics. Clausius did not answer Planck's letters and he was not at home when Planck called on him.[3]) Despite this lack of any personal relationship with Clausius, Gibbs had the best of reasons for accepting the invitation to write about him. He had been living with Clausius' ideas for the more than fifteen years since he had started his own work in thermodynamics.

It is no surprise, then, that Gibbs responded promptly to the invitation to write on Clausius' work.[4] "I will look the matter up and see what I can do," he wrote to Cooke. He found the prospect "in many ways a pleasant one," though he quickly disclaimed any "facility at that kind of writing, or indeed at any kind." He did not "expect to do justice to the subject," but he thought he "might do something," when he could "get a little relief from some pressing duties." All this is about what one might expect, but the central paragraph of Gibbs's letter sounds another note.

"There are some drawbacks: of course it has not escaped your notice that it is a *very* delicate matter to write a notice of the work of Clausius. There are reputations to be respected, from Democritus downward, which may be hurt, if not of the distinguished men directly concerned, at least of their hot-headed partisans. Altogether I feel as if I had to take my life in my hands."

Even though we recognize the obvious hyperbole in the Yale professor's last statement to the Harvard professor concerning the writing of an obituary for the late professor at Bonn, we are still likely to wonder what was so "*very* delicate" about this task. The bitter disputes over the assignment of credit for the key developments in thermodynamics are now largely forgotten. Scientists brought up on a thermodynamics which seems the very model of a staid and settled discipline will probably find it hard to believe that this same thermodynamics could ever have been the occasion of serious debate, fundamental disagreements, and harsh and wounding words. But Gibbs had

2 *Ibid.,* Appendix IV. Gibbs's Mailing Lists for Reprints, p. 239.
3 Max Planck, *Scientific Autobiography and Other Papers,* transl. Frank Gaynor (New York, 1949), p. 19.
4 J. W. Gibbs to J. P. Cooke, 10 June 1889. Quoted in full in J. G. Crowther, *Famous American Men of Science* (London, 1937), p. 292.

lived through the controversies, and while he had long since quietly settled the issues in his own mind, he knew that the "hot-headed partisans" of the anti-Clausius camp were still alive and active. The writing of an obituary notice of Clausius would require Gibbs to take a public stand on the issues, but this would have to be done without any hint of a polemic tone, in view of the occasion.

The obituary Gibbs proceeded to write[5] certainly met all the demands of a difficult situation. It has been aptly described by J. G. Crowther as "one of the most remarkable obituary notices in scientific literature." It sets Clausius in his rightful place, in Gibbs's "classical statement of the origin of thermodynamics and the extent of the contributions by the various founders."[6]

Gibbs's article on Clausius is remarkable for more than just that "classical statement of the origins of thermodynamics." He did not hesitate to discuss Clausius' most controversial thermodynamic concept, the disgregation, and to give it his unqualified approval. Clausius himself had considered disgregation, which has long since been forgotten, to be a concept of greater physical significance than entropy. What Gibbs had to say on this subject is particularly interesting for the insight it gives into his views on the relationship between thermodynamics and "molecular science." Gibbs's brief comments on disgregation, which I discuss in detail later in this essay, make his Clausius obituary an essential document for the understanding of his scientific thought.

2. The first and most important issue that Gibbs had to settle, in writing on Clausius, was the significance of Clausius' contributions to thermodynamics. It is a long time now since that significance has been questioned, but when Gibbs wrote the question was far from closed. Gibbs left no room for doubt on where he stood. "This memoir marks an epoch in the history of physics," he wrote, referring to Clausius' first memoir on the mechanical theory of heat, published in 1850.[7] "If we say," Gibbs continued, "in the words used by Maxwell some years

[5] J. W. Gibbs, "Rudolf Julius Emanuel Clausius," *Proc. Amer. Acad.*, *16* (1889), 458. Reprinted in *The Scientific Papers of J. Willard Gibbs* (New York, 1906; reprinted 1961), 2, 261–267. All my quotations from Gibbs not otherwise identified are from this source.

[6] J. G. Crowther, *op. cit.* (note 4), 293.

[7] R. Clausius, "Über die bewegende Kraft der Wärme, und die Gesetze, welche sich daraus für die Wärmelehre selbst ableiten lassen," *Pogg. Ann.*, *79* (1850), 368, 500. This along with Clausius' other principal publications in thermodynamics, was reprinted in English translation in R. Clausius, *The Mechanical Theory of Heat*, ed. T. A. Hirst (London, 1867), pp. 14–69.

ago, that thermodynamics is 'a science with secure foundations, clear definitions, and distinct boundaries,' and ask when those foundations were laid, those definitions fixed, and those boundaries traced, there can be but one answer. Certainly not before the publication of that memoir."

Gibbs's statement was certainly justified. It had been only the previous year, 1849, that William Thomson had seen a real conflict, as yet irreconcilable, between the new experimental results of James Prescott Joule and the successful existing theory based on Sadi Carnot's memoir of 1824. "The extremely important discoveries recently made by Mr. Joule of Manchester, that heat is evolved in every part of a closed electric conductor, moving in the neighborhood of a magnet, and that heat is *generated* by the friction of fluids in motion, seem to overturn the opinion commonly held that heat cannot be *generated*, but only produced from a source. . . ."[8] Thomson thought, nevertheless, that "the fundamental axiom adopted by Carnot may be considered as still the most probable basis for an investigation of the motive power of heat." He had good reason to think so; his brother James had just used Carnot's theory to predict that the freezing point of water would be lowered by the application of pressure.[9] William Thomson confirmed his brother's "very remarkable speculation" of "an entirely novel physical phenomenon," by experiment, not only qualitatively but in "remarkably close agreement" with the quantitative prediction of the lowered freezing point under pressure.[10] But this successful theory was based on the conservation of the caloric fluid, and it was put in doubt by Joule's results. Thomson found it a "very perplexing question"; giving up Carnot's axiom would lead to "innumerable other difficulties—insuperable without farther experimental investigation." He was sure of only one thing: "It is in reality to experiment that we must look—either for a verification of Carnot's axiom, and an explanation of the difficulty we have been considering; or for an entirely new basis of the Theory of Heat."[11]

[8] W. Thomson, "An Account of Carnot's Theory of the Motive Power of Heat," *Trans. Roy. Soc. Edinburgh, 16* (1849), 541. Reprinted in Sir William Thomson, *Mathematical and Physical Papers, 1* (Cambridge, 1882), 113–155. Quoted passages on pp. 116–117.
[9] J. Thomson, "Theoretical Considerations on the Effect of Pressure in Lowering the Freezing Point of Water," *ibid.,* 156–164.
[10] W. Thomson, "The Effect of Pressure in Lowering the Freezing Point of Water Experimentally Demonstrated," *ibid.,* 165–169. Quoted passages on pp. 165 and 169.
[11] W. Thomson, *op. cit.* (note 8), footnote on p. 119.

Thomson had been wrong, as Gibbs implied when he wrote: "The materials indeed existed for such a science, [in 1850], as Clausius showed by constructing it from such materials, substantially, as had for years been the common property of physicists." What was lacking was not experiment but analysis, analysis that Clausius supplied in his memoir, "On the Moving Force of Heat and the Laws of Heat Which May Be Deduced Therefrom." [12] Clausius showed that one did not have to make a choice between Carnot and Joule. The essence of Carnot's axiom, i.e., the requirement that some heat pass from a warm body to a colder one whenever work is done in a cyclic process, could be maintained even if one dropped Carnot's additional statement that no heat is lost in the process. In this revised form Carnot's axiom was compatible with Joule's statement that whenever work is produced by heat, a quantity of heat proportional to the work done is consumed, and conversely. Clausius saw that the problem of choosing *the* right law of thermodynamics was falsely put: there were two independent laws to be formulated from which the subject could then be developed.

This was the crux of the matter. As Gibbs put it, "the science of thermodynamics came into existence" with the publication of Clausius' memoir. And Gibbs added, "it might have been said at any time since the publication of that memoir, that the foundations of the science were secure, its definitions clear, and its boundaries distinct."

Thomson himself had immediately recognized the importance of Clausius' work and paid generous tribute to it in his own fundamental paper written a year later.[13] He, too, had realized that Carnot's axiom could be modified so as to be consistent with Joule's work and a mechanical theory of heat. But while he claimed that his own work was independent of Clausius', he gave Clausius clear priority. "The merit of first establishing the proposition [Carnot's theorem on reversible cycles] upon correct principles is entirely due to Clausius, who published his demonstration of it in the month of May last year. . . ." Thomson referred to the two laws of thermodynamics as the propositions of Joule and of Carnot and Clausius, respectively. Although his formulation of the second law differed from that given by Clausius, Thomson stated flatly that the two forms were logically equivalent.

The controversy over the importance of Clausius' work did not

---

[12] R. Clausius, *op. cit.* (note 7).

[13] W. Thomson, "On the Dynamical Theory of Heat," *Trans. Roy. Soc. Edinburgh,* 20 (1851), 261. Reprinted in W. Thomson, *op. cit.* (note 8), 174–316. Quoted passage on p. 181.

break out until 1872. The previous year James Clerk Maxwell published his textbook, *Theory of Heat*.[14] Clausius was barely mentioned in Maxwell's discussion of thermodynamics, and he promptly wrote to the *Philosophical Magazine* to complain and to set the record straight.[15] Maxwell obviously realized that he had been unfair to Clausius, and made the changes Clausius requested in the next edition of his book.[16] But Clausius also objected in the same article to the treatment his work had received at the hands of Peter Guthrie Tait in the latter's *Sketch of Thermodynamics*,[17] which had appeared several years earlier. Tait most emphatically did not admit the justness of Clausius' criticisms and claimed that it was William Thomson "who first *correctly* adapted Carnot's magnificently original methods to the true Theory of Heat." [18] Tait dismissed Clausius' contribution in these words: "Now it is one thing to rush into print with a proof which has afterwards to be explained and patched up, and quite another thing to wait till one hits on a complete and irrefragable demonstration." [19] Tait even went so far as to take Thomson to task: "I think Thomson has done mischief as regards scientific history by giving Professor Clausius undue credit. . . . Moreover Thomson is certainly mistaken when he asserts that even this [Clausius' statement of the second law] is equivalent to his own Axiom." [20]

This fierce polemic continued through two volumes of the *Philosophical Magazine*. It finally broke off, unresolved, with Clausius writing that "the tone in which Mr. Tait has written renders it impossible for me to continue the discussion." [21] Tait considered this "a trifle too much" and hoped that "in future [Clausius] will leave offensive and unjust charges unmade." [22] Both men restated their cases at

[14] J. C. Maxwell, *Theory of Heat* (London, 1871).

[15] R. Clausius, "A Contribution to the History of the Mechanical Theory of Heat," *Phil. Mag., 43* (1872), 106.

[16] See, for example, the first American edition of Maxwell's *Theory of Heat*, published in New York in 1872.

[17] P. G. Tait, *Sketch of Thermodynamics* (Edinburgh, 1868).

[18] P. G. Tait, "On the History of the Second Law of Thermodynamics, in reply to Professor Clausius," *Phil. Mag., 43* (1872), 517.

[19] *Ibid.*

[20] *Ibid.*

[21] R. Clausius, "A Necessary Correction of one of Mr. Tait's Remarks," *Phil. Mag., 44* (1872), 117. Also see his earlier "On the Objections Raised by Mr. Tait against my Treatment of the Mechanical Theory of Heat," *Phil. Mag., 43* (1872), 443.

[22] P. G. Tait, "Reply to Professor Clausius," *Phil. Mag., 44* (1872), 240. Also see his earlier "Reply to Professor Clausius," *Phil. Mag., 43* (1872), 338.

length and with unabated emotions some years later in subsequent editions of their books.[23] Tait withdrew the one sentence of warm praise for Clausius which had appeared in his first edition and Clausius explained how and why he considered Tait's views to be motivated by scientific chauvinism.

All these bitter words had prompted Gibbs's feeling that he was taking up a *"very* delicate matter" in writing about Clausius. Yet Gibbs did not hesitate to praise Clausius warmly for the qualities he had shown in his 1850 memoir. "The constructive power thus exhibited, this ability to bring order out of confusion, this breadth of view which could apprehend one truth without losing sight of another, this nice discrimination to separate truth from error,—these are qualities which place the possessor in the first rank of scientific men."

Gibbs had not waited until Clausius was dead before indicating, quietly but decisively, where he stood in the controversy. His first scientific memoir,[24] which appeared in the spring of 1873, immediately after the Clausius-Tait exchanges, made his position clear. Gibbs began this study of graphical methods in thermodynamics by listing the fundamental quantities of thermodynamics. One of those on his list was the entropy, defined by Clausius in 1865. Gibbs wrote as though it were obvious that entropy was a quantity whose significance was on a par with that of energy. He introduced casually, without emphasis, the idea of the fundamental thermodynamic equation of a system as the equation for its energy expressed in terms of its volume and entropy, and devoted particular attention to the volume-entropy diagram for graphical representation.

Just how far from obvious all this was in 1873 can be inferred from a footnote[25] in which Gibbs pointed out simply that he was using the term entropy "in accordance with the original suggestion of Clausius, and not in the sense in which it has been employed by Professor Tait

---

[23] (a) P. G. Tait, *Sketch of Thermodynamics*, 2nd ed. (Edinburgh, 1877). See especially the Preface and compare section 49 in the two editions. Also see P. G. Tait, "On the Dissipation of Energy," *Phil. Mag.*, 7 (1879), 344 and W. Thomson's "Note on the Preceding Letter," *Phil. Mag.*, 7(1879), 346.
(b) R. Clausius, *Die Mechanische Wärmetheorie*, 2nd ed. (Braunschweig, 1876), *1*, 354–359, 387–388; (Braunschweig, 1879), *2*, 306–334.
[24] J. W. Gibbs, "Graphical Methods in the Thermodynamics of Fluids," *Trans. Connecticut Acad.*, 2 (1873), 309. *Scientific Papers, 1*, 1–32.
[25] *Ibid.*, footnote on p. 2.

and others after his suggestion." Tait had misused the term "entropy" in the first edition of his book:[26] he had made a serious error in stating Clausius' definition, and then compounded it by arbitrarily deciding to use the same word "entropy" to mean the negative of the quantity he incorrectly attributed to Clausius. This error was propagated by Maxwell,[27] who gave Tait's incorrect version of Clausius' definition and then adopted Tait's sign change. It is worth noting that Maxwell saw the point as soon as he read Gibbs's paper. "It is only lately," he wrote to Tait, "under the conduct of Professor Willard Gibbs that I have been led to recant an error which I had imbibed from your [*Thermodynamics*], namely that the entropy of Clausius is *unavailable energy* while that of [Tait] is *available energy*. The entropy of Clausius is neither one nor the other." [28] Later editions of Maxwell's book were modified accordingly.[29]

Gibbs went a step further two years later when he submitted his long memoir, "On the Equilibrium of Heterogeneous Substances," [30] to the Connecticut Academy. He decided to put a motto at the head of his great work. To use a motto at all is something of a gesture in a scientific paper, and Gibbs's choice of this particular motto can certainly be viewed as a showing of his colors since he chose to quote Clausius' version[31] of the two laws of thermodynamics: "Die Energie der Welt ist constant. Die Entropie der Welt strebt einem Maximum zu." That Gibbs quoted Clausius in the original language may not be irrelevant either, for Tait had made quite a point of his own "too British point of view." Gibbs also saw to it that his statements went directly to those who could appreciate their implications. He sent reprints of his works to Tait, Thomson, Maxwell, Clausius, and others

[26] P. G. Tait, *op. cit.* (note 17), 100.
[27] J. C. Maxwell, *op. cit.* (note 16), 186.
[28] J. C. Maxwell to P. G. Tait, 1 December 1873. Quoted in Cargill Gilston Knott, *Life and Scientific Work of Peter Guthrie Tait* (Cambridge, 1911), p. 115. See also Maxwell's letter to Tait of 13 October 1876, quoted on p. 222, where Maxwell wrote: "When you wrote the Sketch your knowledge of Clausius was somewhat defective. Mine is still. . . ." This book also has much information about the Tait-Clausius controversy, as well as the other controversies in which Tait was involved, including his attack on Robert Mayer's priority in the formulation of energy conservation.
[29] See, for example, the 4th ed. (London, 1875), p. 162.
[30] J. W. Gibbs, "On the Equilibrium of Heterogeneous Substances," *Trans. Connecticut Acad., 3* (1875–1878), 108, 343. *Scientific Papers, 1,* 55–349.
[31] R. Clausius, "Über verschiedene für die Anwendung bequeme Formen der Hauptgleichungen der mechanischen Wärmetheorie," *Pogg. Ann., 125* (1865), 353. Gibbs's motto consisted of Clausius' concluding words, on p. 400. English version in *The Mechanical Theory of Heat,* pp. 327–365.

on a list that looked like a Who's Who of contemporary physical scientists.[32]

When in 1889 Gibbs expressed his high opinion of the importance of Clausius' work and his great respect for the genius of its author, he was only making explicit a position which his published work had demonstrated implicitly to his readers from the beginning of his career.

3. A modern reader of Gibbs's essay on Clausius will probably be puzzled by the passage that follows his appraisal of Clausius' role in the creation of thermodynamics. In this passage Gibbs described Clausius' concept of disgregation and the use he made of it. "Disgregation" sounds archaic now, while Gibbs's own writings on thermodynamics seem hardly to have aged. One is therefore surprised to see him giving a relatively lengthy discussion of disgregation, and even more surprised to find him praising *this* aspect of Clausius' work for its "remarkable insight." Gibbs had his reasons, however, and they are worth looking for. But we have to begin by seeing what Clausius meant by disgregation and why he introduced it into his theory.

In his 1850 memoir Clausius formulated his basic assumptions as cautiously as possible. He assumed the equivalence of heat and work and, by fixing his attention on cyclic processes, avoided the need for saying anything explicit about changes within the system itself. "We shall forbear," he wrote, "entering at present on the nature of the motion which may be supposed to exist within a body, and shall assume generally that a motion of the particles does exist, and that heat is the measure of their *vis viva*. Or yet more generally, we shall merely lay down one maxim [equivalence of work and heat] which is founded on the above assumption." [33] Clausius' mention of his "forbearance" was not just an empty phrase. In 1857, stimulated by the appearance of a paper by August Karl Krönig, Clausius published his own views on "The Nature of the Motion which we call Heat." [34] He wrote that, even before 1850, he "had already formed a distinct conception of the nature of this motion, and had even employed the same in several investigations and calculations." He had

---

[32] L. P. Wheeler, *op. cit.* (note 2).

[33] R. Clausius, *The Mechanical Theory of Heat*, p. 18.

[34] R. Clausius, "Über die Art der Bewegung, welche wir Wärme nennen," *Pogg. Ann., 100* (1857), 353. Reprinted in S. G. Brush, *Kinetic Theory* (Oxford, 1965) *1*, 111–134. Quoted passage on p. 112.

"intentionally avoided mentioning this conception" in his papers before 1857 because he "wished to separate the conclusions which are deducible from certain general principles from those which presuppose a particular kind of motion." This comment is very characteristic of Clausius' way of handling the relationship between the general principles of the new thermodynamics and the more special assumptions about the underlying molecular motions. He did not hesitate to picture the molecular motions for himself and to use these pictures freely as a guide to his reasoning, but he tried to keep his public presentations free of assumptions that could be attacked as being inadequately founded or even gratuitous.

Clausius extended this protective attitude to even his most general ideas on the nature of internal processes in bodies. While he made passing reference to such matters as internal work, the heat in a body, and the effect of heat on molecular arrangements in his earlier papers,[35] he was very careful to keep these remarks outside the logical structure he was building up. By 1862, however, Clausius felt that his insight into the general thermodynamic aspects of what went on inside a system, when heat was supplied to it or work was done on it, had developed enough to justify a detailed memoir.[36] He looked upon this work as the culmination of his thermodynamics, but, aware that others might well demur at accepting his new assumptions, he emphasized that it did not affect the status of his previous results. ("I will, however, at once distinctly observe that, whatever hesitation may be felt in admitting the truth of the following principles, the conclusions arrived at in my former paper, in reference to circular processes, lose thereby none of their authority." [37]) The concept of disgregation was one of the key ideas in Clausius' new discussion.

A general consideration of the various kinds of "processes by which heat can perform mechanical work" had led Clausius to conclude that "the effect of heat always tends to loosen the connection between the molecules, and so to increase their mean distances from one another." He described the *disgregation* of a body as the measure of "the degree in which the molecules of the body are dispersed," so

---

[35] In his 1850 memoir, for example, on pp. 20, 23, 29 of *The Mechanical Theory of Heat*. Also in the 1854 memoir, *ibid.*, 112, 113.
[36] R. Clausius, "Über die Anwendung des Satzes von der Äquivalenz der Verwandlungen auf die innere Arbeit," *Pogg. Ann.*, *116* (1862), 73. *The Mechanical Theory of Heat*, pp. 215–250.
[37] *Ibid.*, 216.

that his general conclusion could be re-expressed simply as "heat tends to increase the disgregation." [38] The problem now was to find the appropriate mathematical representation of this quantity.

In his earliest paper Clausius had expressed the first law in the form,

$$dQ = dU + dW, \tag{1}$$

where dQ represents the heat added to the system in an infinitesimal process, dW represents the external work done by the system in this process, and dU is the change in a property of the system later called the internal energy. While Clausius did not use the symbol d to indicate an inexact differential, he did stress the difference between Q and W—quantities defined for a process and dependent on the path traversed—and U, a function of the state of the system only.

Clausius' new step directly violates the spirit and the letter of present-day thermodynamics. He resolved the state function, U, the internal energy, into the sum of two other state functions: the heat in the body, H, and the internal work function, I,

$$dU = dH + dI. \tag{2}$$

Combining these two equations, Clausius rewrote the first law in the form,

$$dQ = dH + dL, \tag{3}$$

where

$$dL = dI + dW \tag{4}$$

is the total work done in the process, the sum of internal and external work. (Clausius used another sign convention and explicitly included the mechanical equivalent of heat, but nothing is distorted by writing his equations in the slightly more intelligible form I have used here.)

All of this could be look upon as just a formal rearrangement. Clausius had an additional assumption to make, however: "The mechanical work which can be exerted by heat in any alteration of the arrangement of a body is proportional to the absolute temperature at which this alteration occurs." [39] This statement as it stands is not completely clear, but the mathematical form that Clausius gave it shows what he really had in mind. For this is the assumption which

[38] *Ibid.*, 220.
[39] *Ibid.*, 223.

Clausius used to give disgregation a quantitative meaning. "Since the increase of disgregation is the action by means of which heat performs work, it follows that the quantity of work must bear a definite ratio to the quantity by which the disgregation is increased; we will therefore fix the still arbitrary determination of the magnitude of disgregation so that, at any given temperature, the increase of disgregation shall be proportional to the work which the heat could perform at that temperature." [40] Combining this with his additional assumption given above, Clausius could write the equation

$$dL = T\,dZ, \tag{5}$$

where Z is the disgregation, a state function.

If this result is used in equation (3), Clausius' new form of the first law, one readily obtains the equation,

$$\frac{dQ}{T} = \frac{dH}{T} + dZ. \tag{6}$$

Clausius knew from his earlier analyses that dQ/T is an exact differential, or, equivalently, that the integral of dQ/T around any closed reversible path must vanish. Since Z is a state function, (because Clausius assumed it to be one), dZ is an exact differential, too, and therefore dH/T must also be exact. But if H is a state function, (the "heat in the body"), and dH/T is an exact differential, it can easily be shown that H must be a function of the temperature only.

Clausius' discussion led to the definite conclusion: "The quantity of heat actually present in a body depends only on its temperature, and not on the arrangement of its component particles." [41] And while the "quantity of heat actually present in a body" is not directly accessible to experiment, its temperature derivative is a specific heat which Clausius called the "true specific heat." According to his result, this true specific heat depends only on temperature and not, for example, on whether the substance is in its solid, liquid, or gas phase, or on its state of chemical combination. Since the actual specific heats of liquid water, steam, and ice are evidently different, Clausius had to argue that this was because the measured specific heat contained an additional term having to do with changing the state of aggregation. Thus,

[40] *Ibid.*, 227.
[41] *Ibid.*, 236.

if one differentiates Clausius' form of the first law with respect to temperature at constant volume one obtains the equation,

$$\left(\frac{dQ}{dT}\right)_v = \frac{dH}{dT} + T\left(\frac{\partial Z}{\partial T}\right)_v, \tag{7a}$$

or, equivalently,

$$c_v = c_{\text{true}} + T\left(\frac{\partial Z}{\partial T}\right)_v. \tag{7b}$$

Here c_v is the specific heat at constant volume and c_{true} is Clausius' true specific heat.

Since this true specific heat is independent of volume, phase, and chemical structure, it must be identified with the value of c_v when the substance is a gas at low pressure, that is, with the specific heat of a monatomic ideal gas at constant volume:

$$c_{\text{true}} = \frac{3}{2}R \tag{8}$$

for one mole, where R is the gas constant. The function H is then also universally determined,

$$H = \frac{3}{2}RT, \tag{9}$$

where the possible additive constant is chosen to be zero. (I must point out that these last two equations do not appear explicitly in Clausius' memoir.)

Clausius recognized that this result of his on the true specific heat was "considerably at variance with the ideas hitherto generally entertained of the heat contained in bodies,"[42] and that his new assumptions would hardly command universal assent. That was why he was so careful to prevent this work from interfering with the acceptance of his better-founded results. Thus, when Clausius reworked his papers into a systematic exposition of thermodynamics for the second

[42] *Ibid.*, 216.

edition of his book in 1876, he carefully omitted any mention of disgregation,[43] although he promised to treat it in a third volume, which he did not live to complete.

For his own part, he was thoroughly convinced of both the correctness and the importance of his work on disgregation and the heat in a body. In an article he wrote in 1865 Clausius introduced the term "entropy," denoted by S, for the state function whose differential is the quantity dQ/T, which he had shown to be exact many years earlier. He found the physical meaning of the entropy, however, by resolving it with the help of (6), derived in his 1862 paper:

$$dS = \frac{dH}{T} + dZ. \tag{10}$$

Since H is a function of temperature only this equation is directly integrable. The entropy is the sum of the "transformation-value of the heat present in the body," Clausius' characterization of the function whose differential is dH/T, and the disgregation, which measures the arrangement of the particles. (In the case of an ideal gas these two terms are familiar in their explicit form as $c_v \ln T$ and $R \ln V$, respectively.)

Clausius saw the disgregation as a concept more fundamental than the entropy, since entropy was to be interpreted physically with the help of disgregation. He found entropy most useful as a summarizing concept. This viewpoint comes out particularly clearly in a lecture Clausius gave in 1867 on the Second Law to the German Scientific Association.[44] He explained and analyzed several aspects of the law in considerable detail, with disgregation playing an important part in the discussion. It is only in the concluding paragraphs of his talk that Clausius brought in entropy at all, to pull the ideas together. "I have endeavoured to express the whole of this process by means of one simple theorem, whereby the condition towards which the universe is gradually approaching is distinctly characterized." Entropy is described as "a magnitude which represents the sum of all the transformations which must have taken place in order to bring any body or system of bodies into its present condition." And the "one simple

---

[43] R. Clausius, *op. cit.* (note 23b). The third volume, which appeared posthumously, was put together by Max Planck and Carl Pulfrich. It contained no treatment of disgregation.
[44] R. Clausius, "On the Second Fundamental Theorem of the Mechanical Theory of Heat," *Phil. Mag., 35* (1868), 405. Quoted passages on p. 419.

theorem" is, of course: "The entropy of the universe tends towards a maximum."

4. At least one of Clausius' contemporaries was willing to express himself on the subject of disgregation in no uncertain terms. One of the harshest remarks Tait made about Clausius in the course of their controversy concerned just this subject: "Professor Clausius has rendered many services to science, especially in the Kinetic Theory of Gases; but he has done, and seems to take credit in doing, uncompensated mischief by his introduction of what he calls *innere Arbeit* and *Disgregation*. In our present ignorance of the nature of matter, such ideas can do only harm; and no one will dispute his full claim to originality as regards *them*." [45] (The italics are most definitely Tait's.)

Tait stated his objections to disgregation and its allied concepts in a clearer and more temperate form in his book, particularly in the 1877 edition.[46] He thought it was "pure assumption" to resolve the entropy into the sum of disgregation and a term coming from the heat in the body. He blamed "this sort of speculation" on "the assumption that bodies must contain a certain quantity of actual, or thermometric, heat." Tait considered such speculations ill-advised. "We are quite ignorant of the condition of energy in bodies generally. We know how much goes in, and how much comes out, and we know whether at entrance or exit it is in the form of heat or work. But that is all."

Maxwell quoted these words with evident approval in a review of Tait's book for *Nature*.[47] He even twitted Clausius a little on his own, writing that Clausius "allows a certain quantity [of the heat entering

[45] P. G. Tait, "Reply to Professor Clausius," *Phil. Mag., 43* (1872), 338. Actually Clausius' "claim to originality" for the disgregation was disputed by William John Macquorn Rankine. Rankine had introduced what he called the "metamorphic function" in 1851 (basing his work on a vortex atom). It differed from Clausius' disgregation by a term depending only on the temperature. See W. J. M. Rankine, "On Thermodynamic and Metamorphic Functions, Disgregation, and Real Specific Heat," *Phil. Mag., 30* (1865), 407 and R. Clausius, "On the Determination of the Disgregation of a Body, and on the True Capacity for Heat," *Phil. Mag., 31* (1866), 28.

A recent paper by Edward E. Daub, "Atomism and Thermodynamics," *Isis, 58* (1967), 293, discusses Rankine's work in some detail and compares Rankine and Clausius on disgregation. I discuss this paper further in notes 48 and 51 below.

[46] P. G. Tait, *op. cit.* (note 23a), 137.

[47] J. C. Maxwell, "Tait's *Thermodynamics*," *Nature, 17* (1878), 257. Reprinted in *The Scientific Papers of James Clark Maxwell*, ed. W. D. Niven (Cambridge, 1890), 2, 660–671. Quoted passages on pp. 664–665.

a body] to remain in the body as heat, and this remnant of what should have been utterly destroyed lives on in a sort of smouldering existence, breaking out now and then with just enough vigour to mar the scientific coherence of what might have been a well compacted system of thermodynamics." Maxwell summed up with a classic statement on the nature of the subject: "If we define thermodynamics, as I think we may now do, as the investigation of the dynamical and thermal properties of bodies, deduced entirely from what are called the First and Second laws of Thermodynamics, without any hypotheses as to the molecular constitution of bodies, all speculations as to how much of the energy in a body is in the form of heat are quite out of place."

But Maxwell saw further into Clausius' ideas and was more sympathetic to them than his rather irascible friend in Edinburgh. He pointed out that Clausius had actually said what he meant by the heat in a body, even if only in a footnote added to his 1862 paper when it was reprinted in book form. "In the middle of a sentence," wrote Maxwell, "we read: ' . . . the heat actually present in a unit weight of the substance in question—in other words, the *vis viva* of its molecular motions.' Thus the doctrine that heat consists of the *vis viva* of molecular motions, and that it does not include the potential energy of molecular configuration—the most important doctrine, if true, in molecular science—is introduced in a footnote under cover of the unpretending German abbreviation 'd.h.'." Maxwell recognized that disgregation, internal work, and heat in a body, while improper and superfluous in thermodynamics, might well have a significance in molecular science. He was not convinced, however, that Clausius had proven his case.

5. If we now return to Gibbs's essay on Clausius, we find a remarkably different tone. Once he had established Clausius' position as the creator of thermodynamics, Gibbs observed that Clausius had probably done less than Thomson in developing the consequences and applications of the subject. "His attention, indeed, seems to have been less directed toward the development of the subject in extension, than toward the nature of the molecular phenomena of which the laws of thermodynamics are the sensible expression. He seems to have very early felt the conviction, that behind the second law of thermodynamics . . . there was another law of similar form but relating to the

quantities of heat (i.e. molecular *vis viva*) absorbed in the performance of work, external or internal."

In describing the ideas proposed by Clausius in 1862, Gibbs referred to the revised version of the first law, (3) above, by saying: "The element of heat may evidently be divided into two parts, of which one represents the increase of molecular *vis viva* in the body, and the other the work done against forces, either external or internal." Gibbs, whose mastery of thermodynamics was unparalleled, was quite prepared to take Clausius' step as "evident," even though he obviously realized how "unthermodynamic" it had been considered by Maxwell, and, indeed, it was.

He described "the proposition of which Clausius felt so strong a conviction" as the statement that the total work differential, dL, could be written as $T\,dZ$ where Z was a state function, the disgregation. Here, too, Gibbs had an illuminating remark to make. He pointed out that Clausius regarded the disgregation "as determined by the position of the elementary parts of the body without reference to their velocities." Now as a matter of fact Clausius had never discussed this point at any length. He had repeatedly said that the disgregation was fully determined by the arrangement of the constituent molecules of the body, but he had not emphasized that it was *not* determined by molecular velocities. Considered as a thermodynamic state function, Z would, in general, depend on the temperature as well as on the volume.[48] Gibbs's remark referred to the level at which molecular properties determined the nature of the thermodynamic functions, and not to this thermodynamic level itself. His statement that molecular velocities played no part in determining the disgregation suggested that he was in a position to go beyond Clausius in this discussion.

This is confirmed by the rest of what Gibbs had to say. After

---

[48] Even though Clausius always treated the disgregation of a general system as a function of both temperature and volume Tait wrote that, "the very definition of Z makes it a function of T alone" (*op. cit.* [note 46]). Tait did not give any reasons for this statement, a particularly remarkable one since the disgregation of an ideal gas is a function of volume only. Daub, however, has recently argued, (*op. cit.* [note 45]), that disgregation should always depend only on volume and be independent of temperature. His claim is based on the assertion that disgregation, according to Clausius, was "fully determined by the arrangement of the constituent particles," whereas "temperature represented the kinetic energy of the particles and not their configuration." I think this is an erroneous argument, based on a confusion over the admittedly loosely used word "arrangement." See the discussion of equations (11)–(17) in the text of this paper.

briefly restating Clausius' results on disgregation and the heat in a body, Gibbs commented that Clausius had only been able to advance them "rather as a hypothesis than as anything for which he could give a formal proof." But Gibbs could proceed to write: "The substantial correctness of these views cannot now be called in question." He did not give any detailed justification for this statement. He simply referred to the demonstrations by Maxwell and Boltzmann that average molecular kinetic energies are proportional to the absolute temperature, and especially to Boltzmann's determination of "the precise nature of the functions which Clausius called entropy and disgregation." Gibbs was sure of the "substantial correctness" of Clausius' views, where Maxwell had been at best skeptical, largely on the strength of Boltzmann's work. It is a little startling then to note that the relevant memoir by Boltzmann[49] appeared in 1871, years before Maxwell's remarks were written. How could Maxwell have rejected Boltzmann's analysis which so thoroughly convinced Gibbs?

One cannot give a final answer to such a question, but I would conjecture that Maxwell never read Boltzmann's 1871 paper. Although he had considerable respect for Boltzmann's work, and carried on a dialogue with him on the kinetic theory of gases, Maxwell's intellectual style was very different from his Viennese colleague's. In 1873 he wrote to Tait: "By the study of Boltzmann I have been unable to understand him. He could not understand me on account of my shortness, and his length was and is an equal stumbling block to me. Hence I am very much inclined to join the glorious company of supplanters and to put the whole business in about six lines."[50] But Maxwell had a more specific reason than just his distaste for Boltzmann's many lengthy memoirs. In 1866 Boltzmann[51] had tried to prove that the second law of thermodynamics was a purely mechanical theorem, and announced a "purely analytical, completely general proof of the second law of thermodynamics" from mechanical

---

[49] L. Boltzmann, "Analytischer Beweis des zweiten Hauptsatzes der mechanischen Wärmetheorie aus den Sätzen über das Gleichgewicht der lebendigen Kraft," *Wiener Berichte, 63* (1871), 712. Reprinted in L. Boltzmann, *Wissenschaftliche Abhandlungen,* ed. F. Hasenöhrl (Leipzig, 1909), *1,* 288–308.

[50] J. C. Maxwell to P. G. Tait, August 1873. Quoted in Knott, *op. cit.* (note 28), 114.

[51] L. Boltzmann, "Über die mechanische Bedeutung des zweiten Hauptsatzes der Wärmetheorie," *Wiener Berichte, 53* (1866), 195. *Wiss. Abh., 1,* 9–33. It should be pointed out that Boltzmann gave mechanical interpretations of both entropy and disgregation in this paper, too. Daub (*op. cit.* [note 45]) discusses this briefly, but I think he overstates his case when he claims that Boltzmann "had been guided in his analysis" by Clausius' splitting of the entropy. I do not find any direct evidence in Boltzmann's paper to support this claim.

principles. Several years later, quite unaware of this work, Clausius had offered a similar theorem,[52] and a small priority dispute resulted.[53] Maxwell, however, was convinced that all these efforts were misguided and that the second law was essentially statistical in nature. He had used statistical reasoning from the beginning in his own work on the theory of gases. Maxwell commented to Tait that "it is rare sport to see those learned Germans contending for the priority of the discovery that the 2nd law of thermodynamics is the *Hamiltonsche Princip.*"[54] Now Boltzmann's 1871 memoir was also called "Analytical Proof of the Second Law of Thermodynamics from the Theorems on the Equilibrium of Kinetic Energy." Maxwell might well have assumed that it was simply more of the same, another attempt "to degrade [Hamilton's principle] into the 2nd law of thermodynamics, as if any pure dynamical statement would submit to such an indignity."[55] Ironically enough, Boltzmann had already adopted Maxwell's statistical approach and his 1871 memoir gave the second law of thermodynamics a basis in statistical mechanics—for the first time.

Gibbs *had* studied Boltzmann's work, and it had left him with renewed respect for Clausius' intuition: "But the anticipation, to a certain extent, at so early a period in the history of the subject, of the ultimate form which the theory was to take, shows a remarkable insight, which is by no means to be lightly esteemed on account of the acknowledged want of a rigorous demonstration." Clausius' splitting of the entropy into disgregation and a term related to the heat in a body was "the ultimate form" of the theory, according to Gibbs. It is time to look at the corresponding theorem in "molecular science," and to see what Gibbs was really talking about.

I shall use a modern notation here since there is nothing in particular to be gained by going back either to the notation of Boltzmann's memoir or to the notation of Gibbs's book on statistical mechanics. (Gibbs did include these results in his book, carefully referring to

[52] R. Clausius, "Über die Zurückführung des zweiten Hauptsatzes der mechanischen Wärmetheorie auf allgemeine mechanische Principien," *Pogg. Ann., 142* (1871), 433.

[53] L. Boltzmann, "Zur Priorität der Auffindung der Beziehung zwischen dem zweiten Hauptsatze der mechanischen Wärmetheorie und dem Prinzip der kleinsten Wirkung," *Pogg. Ann., 143* (1871), 211.
R. Clausius, "Bemerkungen zu der Prioritätsreclamation des Hrn. Boltzmann," *Pogg. Ann., 144* (1871), 265.

[54] J. C. Maxwell to P. G. Tait, 1 December 1873. Quoted in Knott, *op. cit.* (note 28), 115.

[55] J. C. Maxwell to P. G. Tait, 13 October 1876. Quoted in Knott, *op. cit.* (note 28), 222.

Clausius and Boltzmann in one of the few literature citations in the whole work.[56]) I shall also take for granted the fundamental equation which relates averages over the molecular quantities to the thermodynamic functions. It was established first in Boltzmann's 1871 memoir and is discussed in every textbook on statistical mechanics.

Suppose we have a system of n particles whose energy E can be written in the form,

$$E = \sum_{i=1}^{n} \frac{1}{2m_i} (p_{xi}^2 + p_{yi}^2 + p_{zi}^2) + \Phi(x_1, y_1, z_1, \ldots, x_n, y_n, z_n). \quad (11)$$

The x_i, y_i, z_i are the Cartesian coordinates of particle i whose mass is m_i and whose momentum has components p_{xi}, p_{yi}, p_{zi}. The function Φ is the potential energy of interaction among the particles. Let us define the quantity $\mathfrak{Z}$, the partition function, by the equation,

$$\mathfrak{Z} = \int \cdots \int dp_{x1} \cdots dp_{zn} \, dx_1 \cdots dz_n \exp(-E/kT). \quad (12)$$

The integrals run from $-\infty$ to ∞ for the momentum components and over the volume V, within which the system is confined, for the coordinates. As usual T is the temperature and k is a universal constant. One result of the basic theorem of statistical mechanics is that the entropy S of this system (considered as a function of T and V) can be found from the equation,

$$S = k \ln \mathfrak{Z} + kT \frac{\partial \ln \mathfrak{Z}}{\partial T}. \quad (13)$$

Now, because the kinetic energy terms in (11), (the molecular *vis viva*), are simply quadratic and add on to the potential energy, and because the integrals over momenta run out to infinity, one can always do the momentum integrals in $\mathfrak{Z}$. The partition function of any system whose energy has the form (11) can be written as

$$\mathfrak{Z} = (\text{constant}) \; T^{3n/2} \, \mathfrak{Q}, \quad (14)$$

where $\mathfrak{Q}$, the configuration integral, has the form

$$\mathfrak{Q} = \int d\tau \exp(-\Phi/kT), \quad (15)$$

where $d\tau$ is simply an abbreviation for $dx_1 \ldots dz_n$.

[56] J. W. Gibbs, *Elementary Principles in Statistical Mechanics* (New Haven, 1902), 69.

The entropy S can, then, always be written in the form,

$$S = (\text{const.}) + \frac{3nk}{2} \ln T + k \ln \mathfrak{D} + kT \frac{\partial \ln \mathfrak{D}}{\partial T}. \qquad (16)$$

This is precisely Clausius' decomposition of the entropy into the transformational value of the heat content, $(3nk/2) \ln T$, and the disgregation, $k \ln \mathfrak{D} + kT \, \partial \ln \mathfrak{D} / \partial T$.

In an exactly parallel way one can show that the internal energy U of the system is given by the equation, analogous to (2),

$$U = \frac{3nkT}{2} + \frac{\int d\tau \, \Phi \exp(-\Phi/kT)}{\int d\tau \exp(-\Phi/kT)}. \qquad (17)$$

The first term is the "heat in the body," H, i.e., the molecular kinetic energy, and the second is the internal work function, I, that is, the average of the intermolecular potential energy, Φ.

Clausius' resolution of both internal energy and entropy into separate terms, associated respectively with the molecular kinetic energy and the molecular configuration, has become a simple theorem in (classical) statistical mechanics. It is also a very general theorem, since it depends only on classical statistical mechanics and on the assumed form, (11), of the energy. The theorem can be paraphrased by saying that the molecular kinetic energy contributes in exactly the same way to the thermodynamic properties of all systems, and that systems differ in their thermodynamic properties only because of the differences in their intermolecular potential energies.

It is true that Clausius' terms, "the heat in the body" and the "true specific heat," do not seem to add anything to the situation. But the result is so general, and so closely related to the basic equation relating molecular quantities to thermodynamic properties, that one can easily see why Gibbs referred to Clausius' anticipation of "the ultimate form which the theory was to take."

6. When Gibbs undertook to write on Clausius, he had been quick to disclaim any "facility at that kind of writing," but his diffidence was clearly unfounded. One of the skills he exhibited in his essay was that of neatly avoiding a negative comment where there was nothing to be gained by making one. In discussing the origins of the laws of thermodynamics, for example, Gibbs had to say something

about the work of W. J. Macquorn Rankine, who had proposed versions of both laws in the same period as Clausius and Thomson. Rankine's work was based on an elaborate model of molecular vortices. His formulations of the laws, though suggestive and influential, sometimes strained his readers' "powers of deglutition" as Maxwell phrased it in the course of a long and sometimes hilarious passage on Rankine's work.[57] Gibbs was briefer and kinder, as he dismissed Rankine in a sentence: "Meantime Rankine was attacking the problem in his own way, with one of those marvellous creations of the imagination of which it is so difficult to estimate the precise value."

Gibbs was equally deft in dealing with some aspects of Clausius' work. He gave Clausius proper credit for his early papers on the kinetic theory of gases, and for introducing the concepts of the mean free path and the virial. But it was characteristic of Clausius that he never really adopted the statistical point of view introduced by Maxwell and used to such advantage by both Maxwell and Boltzmann.[58] The velocity distribution function never took its place in Clausius' work. Gibbs wrote that "Clausius was concerned with the mean values of various quantities," and that "Maxwell occupied himself with the relative frequency of the various values which these quantities have," as did Boltzmann. "In reading Clausius," Gibbs went on, "we seem to be reading mechanics; in reading Maxwell, and in much of Boltzmann's most valuable work, we seem rather to be reading in the theory of probabilities. There is no doubt that the larger manner in which Maxwell and Boltzmann proposed the problems of molecular science enabled them in some cases to get a more satisfactory and complete answer, even for those questions which do not at first sight seem to require so broad a treatment."

After very brief comments on Clausius' researches in electrodynamics Gibbs referred to the sheer extent of his work, the number of papers bearing his name. This made a natural link to Gibbs's final words, as applicable to their author as to his subject.

"But such work as that of Clausius is not measured by counting titles or pages. His true monument lies not on the shelves of libraries, but in the thoughts of men, and in the history of more than one science."

---

[57] J. C. Maxwell, *op. cit.* (note 47).
[58] See the Ph.D. dissertation of Elizabeth Wolfe Garber, *Maxwell, Clausius and Gibbs: Aspects of the Development of Kinetic Theory and Thermodynamics* (Case Institute of Technology, 1967).

Author's Note

I had hoped to have this paper ready in time to present it to Professor Tisza on his sixtieth birthday, 7 July 1967, but that proved to be impossible. My own ideas on thermodynamics and statistical mechanics, as well as on many other subjects, owe a great deal to my studies with Tisza and to our occasional discussions over a period of twenty years. I thought the subject of this paper might be particularly appropriate since Professor Tisza is such a devoted Gibbsian, as he so clearly indicated in his recent book, *Generalized Thermodynamics,* where he, too, quotes several times from Gibbs's obituary of Clausius. As a matter of fact, I remember an occasion about ten or twelve years ago, when I had been reading Wheeler's biography of Gibbs and the first volume of Jones's *Freud,* and remarked to Tisza how pale and thin Gibbs seemed next to Freud. He very properly attributed some of this to the difference in the biographers and referred me to Gibbs's obituary of Clausius, (and to Gibbs's letters on Quaternions), for a dimension of Gibbs that Wheeler did not capture. I hope he enjoys this delayed result of his reference.

My ideas on disgregation owe much to Dr. Charles Weiner, who wrote a seminar paper on this subject for me at Case Institute of Technology in the fall of 1962. His paper, "Clausius and the 'Internal' Explanation of Entropy" was presented to the History of Science Society when it met at Indiana University in the spring of 1963, but it was unfortunately never published.

I should like to thank Case Institute of Technology, Yale University, and the John Simon Guggenheim Memorial Foundation for support.

Origins of Lorentz' Theory
of Electrons and the Concept
of the Electromagnetic Field[1]

BY TETU HIROSIGE*

1. INTRODUCTION

Lorentz' theory of electrons, developed at the end of the nineteenth century, showed the limitations of classical physics and prepared the way for its transformation at the beginning of the twentieth. Its historical significance in modern physics may be considered in several aspects. First, it laid a firm foundation for the investigation of the structure of atoms by giving a theoretical explanation of the Zeeman effect. Second, Lorentz' discussion of the properties of electrons, viz., his consideration of the electromagnetic mass of the electron and of the self-reaction of the electron to its field, gave rise to a development of the theory of elementary particles.[2] The modern subject of solid-state physics also originates in part in Lorentz' theory of electrical conduction in metals based on the theory of electrons. Finally, the most immediate and important significance of the theory of electrons for the development of theoretical physics is that it completed the prerelativistic theory of the electromagnetic field, and thus paved the way for the theory of relativity.

* Department of Physics, College of Science and Engineering, Nihon University, Kanda-Surugadai, Tokyo.
[1] This is a revised and enlarged version of my earlier paper written in Japanese: "Lorentz Densiron no Keisei to Denziba Gainen no Kakuritu," *Kagakusi Kenkyu,* ser. 2, *1* (1962), 9-19 and 59-74. An abridged English version, "Lorentz's Theory of Electrons and the Development of the Concept of Electromagnetic Field," has been published in *Japanese Studies in the History of Science,* No. 1 (1962), pp. 101-110.
[2] See, for example, Abraham Pais, *Developments in the Theory of the Electron* (Princeton, 1948).

The fundamental merit of Lorentz' achievement in the development of electromagnetic theory is the separation of the electromagnetic field from matter, rendering the field an independent physical reality. In 1920 Einstein, being appointed to a special professorship in Leiden University, stated, in his inaugural address, that in Maxwell's and Hertz's theories the ether did not differ essentially from ponderable matter; it was Lorentz who "attained this most important progress in the theory of electricity since Maxwell by taking away from the ether its mechanical qualities and from matter its electromagnetic ones." [3] Einstein also declared that "the Maxwell-Lorentz theory of the electromagnetic field has served as a model for the space-time theory and the kinematics of the special theory of relativity." [4]

There are several accounts of the development of Lorentz' theory of electrons. To take some examples: E. Whittaker devotes the last chapter of his *History of the Theories of Aether and Electricity, I. The Classical Theories* to the theory of electrons, stating that Lorentz' stationary ether "is simply space endowed with certain dynamical properties; its introduction was the most characteristic and most valuable feature of Lorentz' theory." [5] This chapter of Whittaker's book is, however, an interpretation in terms of modern ideas and notation of the results obtained by Lorentz; it does not contain a thorough analysis of the process by which Lorentz reached those results. Much more acute remarks are found in P. Ehrenfest's short article "Professor H. A. Lorentz as Researcher." [6] Ehrenfest notices that Lorentz, under Helmholtz' influence, at first subscribed to the viewpoint of action at a distance in his approach to Maxwell's theory. He further states that Lorentz' dissertation "already contains the preparations for the great idea which will later so characteristically distinguish Lorentz' electron theory from its competitors and will make it win out. We mean the clear division of roles, in any given electromagnetic or optical event taking place in a piece of glass or metal, between the 'aether' on the one hand and 'ponderable matter'

---

[3] Albert Einstein, *Äther und Relativitätstheorie* (Berlin, 1920), p. 7. "Er erzielte diesen wichtigsten Fortschritt der Elektrizitätstheorie seit Maxwell, indem er dem Äther seine mechanischen, der Materie ihre elektromagnetischen Qualitäten wegnahm."

[4] *Ibid.*, 7–8. "Der Raum-Zeit-theorie und Kinematik der speziellen Relativitätstheorie hat die Maxwell-Lorentzsche Theorie des elektromagnetischen Feldes als Modell gedient."

[5] Edmund T. Whittaker, *A History of the Theories of Aether and Electricity, I. The Classical Theories* (London, 1951), p. 393.

[6] Paul Ehrenfest, *Collected Papers* (Amsterdam, 1955), pp. 471–478. This apparently was an address delivered at Rotterdam in 1923.

on the other hand."[7] He also suggests that Lorentz' atomistic view of electricity would have helped him to reach this division of roles. However, being a short eulogy, Ehrenfest's article does not trace the course of the development of Lorentz' ideas. The question of the connection of Lorentz' early viewpoint of action at a distance with his theory of electrons is not considered either.

T. Takabayasi has made a penetrating conceptual analysis of the bearing of the theory of electrons on the quantum theory of atomic structure.[8] Concerning the relation between the development of electromagnetic theory and the theory of electrons, Takabayasi states that "the monism of ether . . . obstructed the way to the distinction between the electric charge and the ether and to the conception of charge as a substance. Rather, electricity should have been conceived as particles endowed with mass, that is, as substantialized." He explains that "at first Lorentz, as an adherent of action at a distance, assumed the ions of Weber's theory and on this hypothesis explained dispersion. But he soon denied the doctrine of action at a distance, preferring Maxwell's theory. He preserved, however, the notion of ions which had been the reasonable core of Weber's theory."[9] However, it was not the primary purpose of Takabayasi to analyze the formation of the theory of electrons, and he does not deal further with it.

Generally speaking, in the historical study of a scientific achievement it is of cardinal importance to distinguish carefully the significance which the achievement bore in the contemporary conditions of scientific cognition from the part which it occupies in the present-day scheme of scientific doctrine. It is not rare in the history of science that an idea or a result which will be judged to be erroneous or of little importance in light of present-day understanding has indeed been an important moment in the development of scientific cognition. It is therefore necessary, in the analysis of the history of science, to pay due attention both to the historical factors underlying the scientific achievement and to the historical conditions arising from it and preparing the way for subsequent developments.

---

[7] *Ibid.*, 472.

[8] Takehiko Takabayasi, "History of Quantum Mechanics. I. The Discovery of the Electron (in Japanese)," *Shizen* (June 1949), pp. 55–62, (July 1949), pp. 48–57, (Sept. 1949), pp. 42–50, (Oct. 1949), pp. 62–71; "History of Quantum Mechanics. II. The Decipherment of Spectra," *Shizen* (Jan. 1950), pp. 71–79, (March 1950), pp. 49–55, (July 1950), pp. 74–80, (Dec. 1950), pp. 70–79, (Feb. 1951), pp. 72–80, (May 1951), pp. 95–105.

[9] *Ibid.*, (Sept. 1949), pp. 43 and 44.

With such a view in mind, this study proposes to trace the whole process of the formation of the fundamental ideas of Lorentz' theory of electrons, and to elucidate his route to the recognition that the electromagnetic field is an independent physical reality. The theory of electrons is accessible to everyone through Lorentz' masterly *Theory of Electrons*, which has recently been reprinted in paper-back edition.[10] But before Lorentz' theory could be cast into the mature form of the *Theory of Electrons*, a number of problems had to be solved. By inquiring into these problems, on the one hand, Lorentz obtained a clear picture of the microscopic structure of matter, and, on the other hand, he shaped the modern concept of the electromagnetic field. In this process the introduction of an atomistic viewpoint was not a contingent but a necessary and immanent condition for the establishment of the concept of the field as an independent physical reality. The following contains an analysis 1) of the interconnection of Lorentz' twin inheritances, Maxwell's theory and Continental electrodynamics, 2) of the historical significance of the hypothesis of a stationary ether, and 3) of the relation of Lorentz' theory of electrons to Hertz's electrodynamics.

2. FIELD AND CHARGE IN MAXWELL'S THEORY[11]

According to Faraday's contiguous-action theory of electromagnetic phenomena the electromagnetic action is exerted by the intervention of a medium of polarized particles. Maxwell's theory also describes electromagnetic phenomena as being caused by changes in the mechanical state of a medium. In the first chapter of Maxwell's *Treatise on Electricity and Magnetism*, which may be considered the final exposition of his conception of electromagnetic theory, he explains that "the peculiar features of the theory are:—That the energy of electrification resides in the dielectric medium, whether that medium be solid, liquid, or gaseous, dense or rare, or even what is called a vacuum, provided it be still capable of transmitting electrical action." [12] Ex-

---

[10] Hendrik Antoon Lorentz, *The Theory of Electrons and Its Applications to the Phenomena of Light and Radiant Heat* (Leipzig, 1909; New York; Dover reprint, 1952).

[11] Part of this section is included in my earlier paper, "Koten-denzikigaku to Sôtaisei-riron" ("Classical theory of electromagnetism and the theory of relativity"), *Kagakusi Kenkyu*, No. 52 (1959), pp. 1–8.

[12] James Clerk Maxwell, *A Treatise on Electricity and Magnetism*, 3rd ed. (Cambridge, 1891), *1*, Part I, Chap. 1, §62, p. 68.

plaining his intention, Maxwell states the following in a section entitled "Plan of This Treatise":

> If we now proceed to investigate the mechanical state of the medium on the hypothesis that the mechanical action observed between electrified bodies is exerted through and by means of the medium, as in the familiar instances of the action of one body on another by means of the tension of a rope or the pressure of a rod, we find that the medium must be in a state of mechanical stress.
>
> . . . .
>
> Electric tension, in this sense, is a tension of exactly the same kind, and measured in the same way, as the tension of a rope, and the dielectric medium, which can support a certain tension and no more, may be said to have a certain strength in exactly the same sense as the rope is said to have a certain strength. Thus, for example, Thomson has found that air at the ordinary pressure and temperature can support an electric tension of 9600 grains weight per square foot before a spark passes.[13]

As to the void space—the so-called vacuum—Maxwell supposes that it is filled with ether. He contributed an article entitled "Ether" to the ninth edition of the *Encyclopaedia Britannica,* apparently written in 1878 or 1879, the year he died, in which he describes the ether as "a material substance of a more subtle kind than visible bodies, supposed to exist in those parts of space which are apparently empty."[14] In the second-to-last paragraph of this article he says that "whatever difficulties we may have in forming a consistent idea of the aether, there can be no doubt that the interplanetary and interstellar spaces are not empty, but are occupied by a material substance or body, which is certainly the largest, and probably the most uniform body of which we have any knowledge."[15] One should note that the ether is defined as a *material substance.* The ether in Maxwell's theory, therefore, was supposed to be a dielectric medium which in many respects resembles ponderable dielectrics such as paraffin and glass.

From these considerations we may conclude that in Maxwell's conception the electromagnetic field is a kind of mechanical state of dielectric media which include the ether as a special case. Contrary

[13] *Ibid.,* §35, pp. 63–64.
[14] Maxwell, *The Scientific Papers of J. C. Maxwell* (Cambridge, 1890), 2, 763.
[15] *Ibid.,* 775.

to this, today the electromagnetic field is regarded as an independent dynamical system distinct from, and on the same footing as, ponderable matter. Separating the electromagnetic field from ponderable matter in this way, one investigates the interaction between the field and matter. Maxwell did not conceive of the electromagnetic field in this manner. For him it was not an independent dynamical reality, but one of the mechanical states of a material substance.

Such a conception was common among those who accepted Maxwell's theory in the late nineteenth century as is evidenced by the well-known fact that the followers of Maxwell in Britain devoted themselves to the construction of mechanical models of the ether. Although popular books on the history of physics often state that Hertz's discovery of electromagnetic waves in 1888 verified the theory of the electromagnetic field, what was commonly believed proved at the time was the existence of the ether as a mechanical substance.

The earliest physicist to show a deep insight into the possibility of generating electromagnetic radiation, one of the most remarkable consequences of Maxwell's theory, was evidently G. F. FitzGerald. Oliver Lodge once wrote that "although we knew all about electric oscillation, no one but G. F. FitzGerald in 1883 had thought of their [sic] being competent to emit waves into the ether." [16] It therefore is interesting to see what FitzGerald said, immediately after Hertz's experiment, in his opening address at the 1888 meeting of the British Association:

> The year 1888 will be ever memorable as the year in which this great question has been experimentally decided by Hertz in Germany and, I hope, by others in England. It has been decided in favour of the hypothesis that these actions take place by means of an intervening medium.[17]
>
> . . . .
>
> Rowland's experiment proving an electro-magnetic action between electric charges depending on their absolute and not relative velocities has already proved the existence of a medium relative to which the motion must take place, but the connection is rather metaphysical and is too indirect to attract general attention.[18]

. . . .

[16] Oliver Lodge, *Advancing Science* (London, 1931), p. 88.
[17] George Francis FitzGerald, "Address to the Mathematical and Physical Section of the British Association," *Report of British Association,* 1888, pp. 557–562, esp. p. 558; *The Scientific Writings of the Late George Francis FitzGerald* (Dublin-London, 1902), pp. 229–240.
[18] *Ibid.,* 559.

As I have endeavoured to impress upon you, no *experimentum crucis* between the hypotheses is possible except an experiment proving propagation in time either directly, or indirectly by an experiment exhibiting phenomena like those of the interference of light. . . . there is no doubt that [Hertz] has observed the interference of electro-magnetic waves quite analogous to those of light, and that he has proved that electro-magnetic actions are propagated in air with the velocity of light.[19]

. . . .

Hertz's experiment proves the ethereal theory of electro-magnetism. It is a splendid result. Henceforth I hope no learner will fail to be impressed with the theory—hypothesis no longer—that electro-magnetic actions are due to a medium pervading all known space. . . .[20]

After these discussions, reviewing some of the existing mechanical theories of the ether, he concluded that "we seem to be approaching a theory as to the structure of the ether. . . . Anyway we are learning daily what sort of properties the ether must have." [21]

It is clear that FitzGerald considered Hertz's experiment as proof of the existence of the ether as a mechanical substance. Lodge also considered it as such. He contributed a report of the 1888 meeting of the British Association to *The Electrician* where, referring to FitzGerald's address, he wrote that by Hertz's experiment "the existence of an ether is raised out of the rank of hypothesis, which it has long been, into the domain of demonstrated facts." [22]

To sum up, Maxwell and his followers conceived of the electro-magnetic field as a special state of matter, an "accident" of dielectric substances and not a "substance" in itself.

The concept of the electromagnetic field as a mechanical state of dielectric media has its root in the circumstances in which Maxwell's theory was formed. In the first place, Maxwell took his fundamental ideas from Faraday's conception of contiguous action. Second, the following circumstance should be noted: at first Maxwell constructed his fundamental equations on the basis of a model which consists of magnetic vortices and electric particles.[23] In this model, the electric particle, which is conceived of as the element of electric current and

19 *Ibid.,* 560.
20 *Ibid.,* 561.
21 *Ibid.*
22 Lodge, *op. cit.* (note 16), 101.
23 Maxwell, "On Physical Lines of Force," *Phil. Mag.,* ser. 4, *21* (1961), 161, 231, 338; *23* (1862), 12, 85; *Scientific Papers, 1,* 451–513.

charge, is regarded both as a constituent of the ether in a vacuum and a constituent of ponderable matter. Owing to this model Maxwell arrived at the idea of "electric displacement" as a true displacement of electric particles, an idea which was indispensable for the construction of his theory. At the same time, however, such a model could not help giving rise to the notion that the electromagnetic field is a mechanical state of the ether or ponderable matter; this mechanical state was assimilated to an elastic strain. Maxwell explained in his *Treatise* that a displacement D is produced in a medium by an electric force E, and that its magnitude is, as in the case of an elastic body, proportional to the force E: $D = (K/4\pi)E$.[24]

Although Maxwell himself later discarded the model of vortices and idle-wheel particles, he did not replace it by any alternative conception of the electromagnetic field and electric charge. Rather he declared that he would not raise the question of the nature of the electric charge, though he admitted that the charge is a physical quantity: "While admitting electricity, as we have now done, to the rank of a physical quantity, we must not too hastily assume that it is, or is not, a substance, or that it is, or is not, a form of energy, or that it belongs to any known category of physical quantities."[25] By avoiding the question of the nature of the electric charge, Maxwell's theory was rendered vague. The relations between the electric charge as the origin of the Coulomb force and as the entity that is displaced within a dielectric substance were confounded. In the "Einleitende Uebersicht" to his collected papers on electromagnetic waves, *Untersuchungen über die Ausbreitung der Elektrischen Kraft*, Hertz asserts that the confusion has arisen from residues of Maxwell's earlier mode of thought:

> quite a number of expressions remained which were derived from his earlier ideas. And so, unfortunately, the word "electricity," in Maxwell's work, obviously has a double meaning. In the first place, he uses it (as we also do) to denote a quantity which can be either positive or negative, and which forms the starting-point of distance-forces (or what appear to be such). In the second place, it denotes that hypothetical fluid from which no distance-forces (not even apparent ones) can proceed, and the amount of which in any given space must, under all circumstances, be a positive quantity.[26]

[24] Maxwell, *op. cit.* (note 12), *1*, Part I, Chap. 1, §60, pp. 64–67; Chap. 2, §68, pp. 75–76.
[25] Maxwell, *op. cit.* (note 12), *1*, Part I, Chap. 1. §35, p. 38.
[26] Heinrich Hertz, *Gesammelte Werke von Heinrich Hertz. 2. Untersuchungen über die Ausbreitung der Elektrischen Kraft* (Leipzig, 1894), p. 29. The quotation is from the English translation, *Electric Waves, Being Researches on the Propagation of Electric Action with Finite Velocity Through Space* (London, 1893; New York; Dover reprint, 1962), p. 27—hereafter cited as *Electric Waves*.

To avoid this confusion Hertz proposes always to interpret the meaning of the word "electricity" in a suitable way; and he himself confines the use of the word to the first sense stated above. But at the same time he admits that he has not succeeded in doing this to his entire satisfaction.

To summarize: in Maxwell's conception the ether and ponderable matter remain the same kind of entity inasmuch as electricity in a rather vague sense is contained within the interior of both the ether and ponderable matter and inasmuch as a displacement of the electricity is considered to be a dielectric polarization of both media. The next step was to inquire into the nature of electricity, to open a way to distinguish the ether from ponderable matter. If the ether and ponderable matter were distinguished from each other, a move would be made toward establishing the independence of the electromagnetic field. In addition, to inquire into the nature of electricity would prompt an elucidation of the concept of the field from another direction. For, in the fourth quarter of the nineteenth century, any inquiry into the nature of electric charge would almost certainly involve suppositions about the microscopic structure of matter. Insofar as matter is treated macroscopically, the electromagnetic field in the interior of a material body could be nothing but a state of that body; consequently the field in general would continue to be considered as a mechanical state of continuous media; in particular, electromagnetic phenomena in a vacuum could only be looked upon as being borne by a dielectric medium, the ether. But once one considers matter as an aggregate of microscopic particles and supposes that the ether pervades the intermolecular spaces, one could readily suppose that the electromagnetic field within a material body is borne by the ether in the intermolecular spaces. This would be the first step toward the separation of the electromagnetic field from matter. The great merit of Lorentz' theory of electrons is that it carries out this very task.

3. INTERPRETATION OF MAXWELL'S THEORY IN TERMS OF ACTION AT A DISTANCE

Lorentz' theory of electrons originates in his treatment of optical phenomena by means of the electromagnetic theory of light. The problem which he had to solve first of all was how to construct a consistent theory of the reflection and the refraction of light, the problem which had brought the elastic-ether theory of light to a standstill.

The difficulty of the elastic theory was how to reconcile the boundary conditions with the experimentally confirmed sine and tangent laws of Fresnel and with the fact that longitudinal waves had not been found in light.

Between 1870 and 1872, early in his undergraduate days at Leyden University, Lorentz carefully studied the papers of Maxwell on electromagnetism, and later he even tried to generate electric waves by discharging a Leyden jar.[27] At the same time, he studied Fresnel's wave theory of light and was deeply impressed by its penetrating lucidity. In 1927, at the occasion of Fresnel's centenary, he told the French Physical Society that "I can say that Fresnel has been one of the teachers to whom I owe most."[28] It was, however, Helmholtz' article "Ueber die Bewegungsgleichungen der Elektricität für ruhende leitende Körper"[29] which stimulated Lorentz to take up the problem of the reflection and refraction of light from the viewpoint of the electromagnetic theory of light. This task was completed in 1875 and submitted to Leyden University as his doctoral thesis, "Over de theorie der terugkaatsing en breking van het licht"[30] ("On the Theory of the Reflection and Refraction of Light"). In the introduction of this thesis Lorentz cites Helmholtz: "it was Helmholtz' remark that induced me to study the extent to which the phenomena of reflection and refraction of light lead to the theory of Maxwell in preference to the undulatory theory, which has hitherto been adopted."[31]

It surprises us to see Lorentz take action at a distance as the point of departure of his investigation, despite his conviction of the superiority of Maxwell's electromagnetic theory of light. He declares: "In deriving the equation of motion of electricity, I largely follow Helmholtz. Like this physicist I will start from action at a distance; we shall thus have

[27] G. L. de Haas-Lorentz, ed. *H. A. Lorentz—Impressions of His Life and Work* (Amsterdam, 1957), pp. 27, 28, 31, 32.

[28] Lorentz, "Centenaire d'Augustin Fresnel (1788–1827)," *Revue d'optique, 6* (1927), 514; *Collected Papers* (The Hague, 1935–1939), *9*, 340–342, esp. 341. "Pour ma part, je puis dire que Fresnel a été un des maîtres auxquels je dois le plus. . . ."

[29] Hermann von Helmholtz, "Ueber die Bewegungsgleichungen der Elektricität für ruhende leitende Körper," *Borchardt's Journ. f. die reine u. ang. Math., 72* (1870), 57–129; *Wissenschaftliche Abhandlungen, 1,* 545–628.

[30] Lorentz, "Over de theorie der Terugkaatsing en breking van het licht," *Academisch Proefschrift* (Leiden, 1875); *Collected Papers, 1,* 1–192. This volume also contains the French translation of this thesis.

[31] Lorentz, *ibid.,* 2. "Het was deze opmerking van Helmholtz, die mij aanleiding gaf, te onderzoeken, in hoeverre de verschijnselen der terugkaatsing en breking van het licht aanleiding geven, om de theorie van Maxwell boven de tot nu toe aangenomen undulatietheorie te verkiezen."

the advantage of founding the theory on the most direct interpretation of the facts." [32] This seemingly curious fact, however, can be explained by considering the following historical circumstances.

At that time, Maxwell's theory was, in general, considered to be very difficult to understand. When Maxwell's *Treatise* was first published, it was considered, partly because of its advanced mathematics, to be "a kind of intellectual primeval forest, almost impenetrable in its uncleared fecundity." [33] Lorentz himself later said: "It was not always easy to understand Maxwell's thoughts." [34] It was Helmholtz' great service to re-formulate Maxwell's theory in terms of action at a distance and to make it comprehensible to contemporary physicists. Even Hertz, the discoverer of electrically generated electromagnetic waves, observed that

> many a man has thrown himself with zeal into the study of Maxwell's work, and, even when he has not stumbled upon unwonted mathematical difficulties, has nevertheless been compelled to abandon the hope of forming for himself an altogether consistent conception of Maxwell's ideas. I have fared no better myself. Notwithstanding the greatest admiration for Maxwell's mathematical conceptions, I have not always felt quite certain of having grasped the physical significance of his statements. Hence it was not possible for me to be guided in my experiments directly by Maxwell's book. I have rather been guided by Helmholtz's work. . . .[35]

Another testimony pointing to the role of Helmholtz in promulgating Maxwell's theory is found in the autobiography of M. I. Pupin, who achieved fame with his inventions in telecommunication engineering. From 1883 to 1885 Pupin studied physics at Cambridge University, where Maxwell had recently occupied a professorship. But it was only after he moved to Berlin in the autumn of 1885 and began to study under Helmholtz that he came to understand clearly the physical significance of Maxwell's theory.[36]

Helmholtz originally set out to examine the relation of W. Weber's

[32] Lorentz, *ibid.*, 30. "Bij de afleiding van de bewegingsvergelijkingen der electriciteit zal ik grootendeels Helmholtz volgen. Even als deze natuurkundige zal ik daarbij uitgaan van de onmiddelijke werking op een afstand; aldus toch hebben wij het voordeel, dat aan de theorie meest rechtstreeksche opvatting der feiten ten grondslag ligt."

[33] Ehrenfest, *op. cit.* (note 6), 472.

[34] De Haas-Lorentz, *op. cit.* (note 27), 32.

[35] Hertz, *op. cit.* (note 26), 22. The quotation is from *Electric Waves,* p. 20.

[36] Michael Pupin, *From Imigrant to Inventor* (New York, 1923), pp. 234–241.

electrodynamics, based on action at a distance,[37] to the principle of conservation of energy. Helmholtz found that Weber's force between two elements of electric current implies an unstable equilibrium, whereas F. E. Neumann's force did not lead to this difficulty. For the case of two closed circuits, Weber's and Neumann's laws gave one and the same induced electromotive force. The difference between the two theories could be reduced to terms which vanish by integration over a closed circuit. But since an experiment could only be performed with closed circuits, there always remained an undetermined term in the expression for the force. In view of these circumstances, Helmholtz advanced a general theory which included, as special cases, both Weber's and Neumann's theories as well as the law of induced electromotive force of Maxwell's theory. The expression which Helmholtz found for the electrodynamic potential between two elements of current iDs and $jD\sigma$ is

$$-\frac{1}{2}A^2\frac{ij}{r}[(1+k)\cos(Ds, D\sigma) + (1-k)\cos(r, Ds)\cos(r, D\sigma)]DsD\sigma,$$

where r is the distance between two elements of current and A is a constant depending on the units employed, and k is an undetermined constant. If one used the electrostatic unit of current, the value of $1/A$ was, according to Weber's and Kohlrausch's measurement, equal to $31074 \cdot 10^6$mm/sec, which agreed with the velocity of light. In Helmholtz' formula $k = 1$ and $k = -1$ yield Neumann's and Weber's laws respectively, and $k = 0$ corresponds to Maxwell's theory.

On the basis of this generalized formula Helmholtz derived an equation expressing a wave-propagation of states of polarization through a medium capable of magnetic and dielectric polarizations. According to Maxwell electric disturbances propagate themselves in a dielectric in the form of transverse waves whose velocity in the atmosphere is equal to that of light; Helmholtz thought that Maxwell's result "might have a remarkable significance for the future development of physics." Moreover, since "the problem of the velocity of propagation of electric actions has recently been discussed," it seemed to him important "to investigate what could be concluded from the generalized law of induction in the case that there is magnetizable and

---

[37] Cf. A. E. Woodruff, "Action at a Distance in Nineteenth Century Electrodynamics," *Isis, 53* (1962), 439–459.

dielectrically polarizable media."[38] Helmholtz summed up the result of his investigation in these words: "In dielectric insulators, even when they are not magnetizable, electric motions can propagate themselves as transversally and longitudinally oscillating waves";[39] he added a footnote suggesting the possibility that this "analogy between the motions of electricity in dielectrics and the motion in the luminiferous ether" might solve the difficulty of the elastic-solid theory of reflection and refraction of light.[40] It was this footnote that Lorentz quoted in his thesis.

Helmholtz' derivation of the equation of propagation of electricity will be outlined: he begins with the supposition that, in analogy with the case of an induced magnetic moment, the electric polarization P (whose components Helmholtz denotes by $\mathfrak{x}, \mathfrak{y}, \mathfrak{z}$) is proportional to the electric force that acts at the point under consideration. The constant of proportionality, which Helmholtz calls the *Polarisationsconstant*, is denoted by ϵ. The total current at each point is a sum of the displacement current and the conduction current:

$$\iota = \frac{\partial P}{\partial t} + \frac{1}{\epsilon \kappa} P, \tag{1}$$

where κ is the specific resistance of the medium.

The electric force that determines the polarization is composed of four parts:
1. The static force:

$$-\operatorname{grad} \phi.$$

On a surface where the polarization and ϕ have discontinuities, the condition

$$n \cdot (P - P_1) = \frac{1}{4\pi} \left[\frac{\partial \phi}{\partial n} - \frac{\partial \phi_1}{\partial n} \right] \tag{2}$$

should be satisfied, where n is a unit vector in the normal direction.
2. The electromotive force induced by a distribution of electric current:

$$-A^2 \frac{\partial A}{\partial t}.$$

---

[38] Helmholtz, *op. cit.* (note 29). *Wissenschaftliche Abhandlungen, 1*, 557.
[39] Helmholtz., *ibid.*, 557. "In dielektrischen Isolatoren, selbst wenn sie nicht magnetisirbar sind, können sich elektrische Bewegungen in transversal und longitudinal oscillirenden Wellen fortpflanzen."
[40] Helmholtz, *ibid.*, 558.

A (which Helmholtz denotes by U, V, W) satisfies the following equations:

$$\Delta A = (\mathrm{1} - k)\,\mathrm{grad}\,\frac{\partial\phi}{\partial t} - 4\pi\iota, \tag{3}$$

and

$$\mathrm{div}\,A = -k\,\frac{\partial\phi}{\partial t}. \tag{4}$$

3. The electromotive force induced by changes in distribution of the magnetic moment M (which Helmholtz denotes by λ, μ, ν):

$$A\,\frac{\partial}{\partial t}\,\mathrm{rot}\,\mathbf{G}.$$

G (which Helmholtz denotes by L, M, N) satisfies

$$\Delta G = -4\pi M, \tag{5}$$

and

$$\mathrm{div}\,G = -\chi. \tag{6}$$

4. The external electromotive force E_{ext} (Helmholtz' $\mathfrak{x}$, $\mathfrak{y}$, $\mathfrak{z}$).
   Adding these forces one obtains

$$P/\epsilon + \mathrm{grad}\,\phi + A^2\,\frac{\partial A}{\partial t} - A\,\frac{\partial}{\partial t}\,\mathrm{rot}\,G - E_{\mathrm{ext}} = \mathrm{o}. \tag{7}$$

The magnetic polarization too is proportional to the magnetic force. Helmholtz writes the proportionality constant as θ. The magnetic force is composed of two parts:
1. The static force:

$$-\mathrm{grad}\,\chi.$$

On a surface where free magnetism is found, the condition

$$n\cdot(M - M_1) = \frac{\mathrm{1}}{4\pi}\left[\frac{\partial\chi}{\partial n} - \frac{\partial\chi_1}{\partial n}\right] \tag{8}$$

should be satisfied.
2. The magnetic force that is produced by a distribution of electric current:

$$-A\,\mathrm{rot}\,A.$$

Adding these forces gives

$$M/\theta + A \operatorname{rot} A + \operatorname{grad} \chi = 0. \tag{9}$$

The quantities A and G are defined by integrals of functions of the current ι and the moment M. In order to eliminate these integrals from the equations, Helmholtz made use of the following theorem:

If in a space S, a vector A is everywhere continuous, and on the surface of S, $n \cdot A = 0$, then the two conditions

$$A = 0$$

and

$$\operatorname{rot} A = 0 \quad \text{and} \quad \operatorname{div} A = 0$$

are equivalent to each other.

We take as S the whole space, and as the quantity A in this theorem the expressions on the left-hand side of equations (7) or (9). If we take the rotation of equation (7) and transpose terms, we obtain

$$\operatorname{rot}\left(\frac{P}{\epsilon}\right) = \frac{1 + 4\pi\theta}{\theta} A \frac{\partial M}{\partial t} + \operatorname{rot} E_{\text{ext}}. \tag{10}$$

Taking the divergence of equation (7),

$$\operatorname{div}\left(\frac{P}{\epsilon}\right) = -\Delta\phi + A^2 k \frac{\partial^2\phi}{\partial t^2} + \operatorname{div} E_{\text{ext}}. \tag{11}$$

In a similar way, we obtain from equation (9)

$$\operatorname{rot}\left(\frac{M}{\theta}\right) = A \left[\operatorname{grad} \frac{\partial\phi}{\partial t} - 4\pi \frac{\partial P}{\partial t} - \frac{4\pi}{\kappa\epsilon} P \right], \tag{12}$$

$$\operatorname{div}\left(\frac{M}{\theta}\right) = -\Delta\chi. \tag{13}$$

Then, according to the theorem, in order that equations (10) and (11), or (12) and (13), are equivalent respectively to equations (7) or (9),

$$P = M = \phi = \chi = 0 \tag{14}$$

must be satisfied at infinity, and the expressions on the left-hand sides of equations (7) and (9) should be everywhere continuous. Since A and G and their first derivatives are by definition everywhere continuous, the conditions that must be satisfied on a surface where ϵ, θ, and κ have discontinuities are

$$\frac{\boldsymbol{P}}{\epsilon} + \operatorname{grad} \phi - \boldsymbol{E}_{\text{ext}} = \left(\frac{\boldsymbol{P}}{\epsilon}\right)_1 + (\operatorname{grad} \phi)_1 - (\boldsymbol{E}_{\text{ext}})_1 \qquad (15)$$

and

$$\frac{\boldsymbol{M}}{\theta} + \operatorname{grad} \phi = \left(\frac{\boldsymbol{M}}{\theta}\right)_1 + (\operatorname{grad} \phi)_1. \qquad (16)$$

In the case that there is no external electromotive force and the medium is a homogeneous, perfect insulator, ϵ and θ are constant, and $1/\kappa$ can be put equal to 0. Then, taking the rotation of (10) and making use of (12), one obtains

$$\Delta \boldsymbol{P} = 4\pi(1 + 4\pi\theta)A^2 \frac{\partial^2 \boldsymbol{P}}{\partial t^2} + \left[1 - \frac{(1 + 4\pi\theta)(1 + 4\pi\epsilon)}{k}\right] \operatorname{grad} \operatorname{div} \boldsymbol{P}.$$

$$(17)$$

In a similar way

$$\Delta \boldsymbol{M} = 4\pi\epsilon(1 + 4\pi\theta)A^2 \frac{\partial^2 \boldsymbol{M}}{\partial t^2},$$

$$\operatorname{div} \boldsymbol{M} = 0 \qquad (18)$$

are obtained.

Equation (17) has exactly the same form as the equation that expresses the vibration of solid elastic bodies in which the velocity of propagation is given by

$$\frac{1}{A\sqrt{4\pi\epsilon(1 + 4\pi\theta)}} \qquad \text{for transverse waves,}$$

or

$$\frac{1}{A}\sqrt{\frac{1 + 4\pi\epsilon}{4\pi\epsilon \cdot k}} \qquad \text{for longitudinal waves.}$$

However, equations (18) for the magnetic polarization correspond to the equations for an incompressible elastic body in which only transverse waves can be propagated. The velocity of propagation is the same as above. Helmholtz thus concluded that "the results obtained so far render plausible a propagation with finite velocity of the electric action-at-a-distance force without any essential change in the foundation of the received theory of electrodynamics." [41]

[41] Helmholtz, op. cit. (note 29). Wissenschaftliche Abhandlungen, 1, 628. "Die bisher vorliegenden Erfahrungen auch ohne wesentliche Aenderungen in den Grundzügen der accepirten Theorie der Elektrodynamik eine Ausbreitung der elektrischen Fernwirkungen mit endlichen Geschwindigkeiten als möglich erscheinen lassen."

L. Rosenfeld, in his highly interesting analysis of the development of the recognition of the identity of light vibrations and electromagnetic waves, has criticized Helmholtz, saying that "in laying on them [Maxwell's conceptions] a rather heavy hand he [Helmholtz] completely spoiled their subtle harmony." He added that "not only was such an approach to Maxwell's theory entirely alien to its spirit, but it tended to obscure its characteristic features." [42] As far as the physical meaning of Helmholtz' theory is concerned, this criticism may be admitted without reservation. But in historical context, Helmholtz' theory may be judged differently. In fact, most physicists at that time gained their understanding of Maxwell's theory through Helmholtz' version of it; in particular, Lorentz' exploration of the electromagnetic theory of light was motivated by it. And it should not be overlooked that, against the background of the Maxwellian conception of the electromagnetic field, Helmholtz' theory could not be considered a *heresy*. In a sense, it was adapted to Maxwell's conception; for, its concrete model of polarization offered an intelligible representation of Maxwell's view of polarization as a special condition of a material medium.

4. ELECTROMAGNETIC THEORY OF LIGHT

Since these were the features of Helmholtz' theory, it is no surprise that in Lorentz' theory of 1875 the electromagnetic field is not yet conceived of as a dynamical entity independent of all material substances. Lorentz in those days considered light as a propagation in a material medium of variations of the state of polarization. In the second chapter of his thesis, having reproduced the derivation of Helmholtz' wave equation, he concludes: "Now in an elastic body, transverse and longitudinal vibrations can propagate themselves; we therefore deduce that, in an insulator, a propagation of transverse and longitudinal *electric vibrations* can similarly take place, where, by *electric vibration*, we mean that the dielectric polarization is a periodic function of time." [43] The same view was also expressed in his next paper, written three

---

[42] L. Rosenfeld, "The Velocity of Light and the Evolution of Electrodynamics," *Nuovo Cimento*, Supplement to ser. 10, *4* (1956), 1630–1669. The quotation is on page 1664.
[43] Lorentz, *op. cit.* (note 30); *Collected Papers, 1,* 78. "Nu kunnen zich in een elastich lichaam transversale en longitudinale trillingen voortplanten en wij besluiten daaruit, dat in den isolator op dezelfde wijze een voortplanting van transversale en longitudinale *electrische trillingen* mogelijk is, waarbij wij onder *electrische trilling* het verschijnsel te verstaan hebben, dat de dielectrische polarisatie een periodieke functie van den tijd is."

years later, advancing a theory of the refractive index (to be discussed more fully in the next section). At the beginning of this paper, Lorentz reiterates his position regarding Maxwell's electromagnetic theory of light: "I still hold that its main principle, i.e., the hypothesis that vibrations of light are movements of the same character as electric currents, can hardly be questioned." [44]

Though the equations of motion of electricity of Helmholtz and Lorentz have the same form as the equations of elastic-solid theory, they require different boundary conditions, making it possible to overcome the difficulties of the elastic-solid theory of light. The solution comes about as follows. Choosing the x-axis perpendicular to the boundary surface, Lorentz writes the components of polarization P as ξ, η, ζ and the components of M/θ as L, M, N. Then equations (15) and (16) lead to[45]

$$\frac{\xi_1}{\epsilon_1} - \frac{\xi_2}{\epsilon_2} = -\left[\left(\frac{\partial\phi}{\partial x}\right)_1 - \left(\frac{\partial\phi}{\partial x}\right)_2\right], \frac{\eta_1}{\epsilon_1} = \frac{\eta_2}{\epsilon_2}, \frac{\zeta_1}{\epsilon_1} = \frac{\zeta_2}{\epsilon_2}, \quad (19)$$

and

$$L_1 - L_2 = -\left[\left(\frac{\partial\chi}{\partial x}\right)_1 - \left(\frac{\partial\chi}{\partial x}\right)_2\right], M_1 = M_2, N_1 = N_2. \quad (20)$$

To these should be added the conditions of continuity for ϕ and χ, from equations (2) and (8):

$$\xi_1 - \xi_2 = \frac{1}{4\pi}\left[\left(\frac{\partial\phi}{\partial\chi}\right)_1 - \left(\frac{\partial\phi}{\partial x}\right)_2\right],$$

$$\theta_1 L_1 - \theta_2 L_2 = \frac{1}{4\pi}\left[\left(\frac{\partial\chi}{\partial x}\right)_1 - \left(\frac{\partial\chi}{\partial x}\right)_2\right]. \quad (21)$$

The latter equation allows χ to be eliminated from the first of equations (20). The result is

$$(1 + 4\pi\theta_1)L_1 = (1 + 4\pi\theta_2)L_2. \quad (22)$$

If we put $P/\epsilon = E$, $E + 4\pi P = D$, $M/\theta = H$, and $H + 4\pi M = B$, then Lorentz' conditions represent the continuity of the normal components of D and B, and of the tangential components of E and H.

[44] Lorentz, "Over het verband tusschen de voortplantings snelheid en samenstelling der midden-stofen," *Verh. Kon. Akad. Wet. Amsterdam, 18* (1878), 1; *Collected Papers, 2,* 1–119. This volume includes the English translation. The quotation is from p. 1.
[45] Lorentz, *op. cit.* (note 30); *Collected Papers, 1,* 90.

These conditions are well-known in present-day textbooks, so that there is no need to reproduce Lorentz' derivation of the laws of reflection and refraction and of Fresnel's sine and tangent laws. But a remark should again be made here: Lorentz does not consider the oscillation of the field intensity E, but the vibration of the polarization P of a dielectric body.

That Lorentz in those days did not start from Maxwell's equations themselves but from Helmholtz' theory of dielectric polarization can also be seen from his arguments about the velocity of propagation of electric vibrations.[46] Following Helmholtz, Lorentz remarks that the value which can be determined experimentally is not the constant A but instead $\sqrt{(1 + 4\pi\epsilon_0)(1 + 4\pi\theta_0)}A$, where ϵ_0 and θ_0 are the values of ϵ and θ for the air or the free ether. Hence if one writes A' for $\sqrt{(1 + 4\pi\epsilon_0)(1 + 4\pi\theta_0)}A$, the velocity of propagation of transverse vibrations is given by

$$v = \frac{1}{A'} \sqrt{\frac{(1 + 4\pi\epsilon_0)(1 + 4\pi\theta_0)}{4\pi\epsilon(1 + 4\pi\theta)}} .$$

For the air or the free ether it becomes

$$v = \frac{1}{A'} \sqrt{\frac{1 + 4\pi\epsilon_0}{4\pi\epsilon_0}} .$$

However, Maxwell, in his *Treatise*, demonstrated the velocity of propagation of electromagnetic waves to be

$$v = \frac{1}{A'} .$$

And experimental data indicated that $1/A'$ is equal to the velocity of light, from which Lorentz concluded that the value of ϵ_0 must be chosen so that v becomes equal to $1/A'$. Since the quantity which one can determine by experimental means is the relative value $1 + 4\pi\epsilon$ and not the absolute value ϵ, one can assume that ϵ_0 is extremely great, an assumption that permits v to be written as $1/A'$.

In addition, this assumption obviated the difficulty of longitudinal waves. In the case where the electric vibrations lie in the plane of incidence, the theory of Helmholtz and Lorentz also must take into account a longitudinal component in the reflected and refracted rays.

46 *Ibid.*, 82 ff.

But the amplitude of longitudinal waves is proved to be of the order of $1/\epsilon$. Hence if one adopts the assumption that the constant ϵ_0, and consequently ϵ, of any body are extremely large, one can ignore the existence of longitudinal waves. Lorentz concluded that this was another reason that the electromagnetic theory was superior to the elastic-solid theory.[47] One notices, however, that, even when Helmholtz' formulation is adopted as the foundation of the theory, if account is taken only of Maxwell's case, $k = 0$, then the velocity of longitudinal waves would become infinite and these waves could be neglected without any assumption about ϵ. That in 1875 Lorentz was worrying about the appearance of longitudinal waves indicates that at this stage he did not yet feel absolutely certain about Maxwell's theory.

In spite of his uncertainty, however, Lorentz' thesis of 1875 represents a remarkable start. It consists of six chapters: the first examines the difficulties of the elastic-solid theory of light; the second is devoted to the derivation of the wave equation from Helmholtz' theory; and the third discusses the reflection and refraction of light by isotropic media. In the fourth chapter the optics of crystals is dealt with and the whole of Fresnel's theory of double refraction is derived on the basis of the electromagnetic theory. In the fifth chapter Lorentz discusses total reflection and proves that, in the case of total reflection, light penetrates slightly into the second medium. The sixth chapter deals with the optical properties of metals. He shows that the amplitude of light entering a metal rapidly attenuates because of conduction currents produced by the electromagnetic field of that light; and he concludes that the refractive index of a metal should be considered as a complex number, a result which was previously proposed by MacCullagh and Cauchy.[48] In short, Lorentz propounded most of the fundamentals of electromagnetic optics as presented in present-day textbooks. The merits of his work, however, lie not only in its first formulation of electromagnetic optics but also in its preparation for the separation of field from matter; the latter is the more important for our present subject.

In the middle of the 1870's, one could adduce, in favor of Maxwell's electromagnetic theory of light, the numerical agreement of the velocity of light with the ratio of two electric units and the validity for certain substances of the formula $n = \sqrt{K}$ which connects the specific

---

[47] Ibid., 99.
[48] Whittaker, op. cit. (note 5), 161–162.

inductive capacity with the refractive index. The former was the crucial fact which led Maxwell to propose the electromagnetic theory of light, whereas the latter needed some time before it was proved decisively by experimental data. In his *Treatise*, Maxwell could only cite Gibson's and Barkley's measurement of the inductive capacity for solid paraffin,[49] and moreover the agreement of the theoretical value of K with the measured value was not entirely satisfactory. It was a series of experimental investigations by Boltzmann[50] in 1872–1874 that brought about considerable progress here. Boltzmann obtained more precise values of the inductive capacity for various solid dielectrics by two different methods. Furthermore, he carried out precise measurements of the inductive capacity of various gases, which were in good agreement with Maxwell's prediction.

Boltzmann's work on gases also suggested to Lorentz an idea which would become part of the foundation of the theory of electrons. In chapter 3 of his thesis, Lorentz draws attention to the fact that both the refractive indices and the inductive capacities of gases are always nearly equal to unity; in other words, the velocity of light and the inductive capacity in gases and in a vacuum are nearly the same (see table).

	$\sqrt{K}$	n
Air	1.000295	1.000294
Carbonic acid	1.000473	1.000449
Hydrogen	1.000132	1.000138
Carbonic oxide	1.000345	1.000340
Nitrous oxide	1.000497	1.000503
Olefiant gas	1.000656	1.000678
Marsh gas	1.000472	1.000443

---

[49] Maxwell, *op. cit.* (note 12), 2, Part IV, Chap. XVI, §789.

[50] Ludwig Boltzmann, "Experimentelle Bestimmung der Dielektrizitätskonstante von Isolatoren," *Wiener Berichte, 67* (1873), 17–80; *Pogg. Ann., 151* (1874), 482–506, 531–572; *Wissenschaftliche Abhandlungen, 1,* 411–471. "Resultate einer Experimentaluntersuchung über das Verhalten nicht leitender Körper unter dem Einflusse elektrischer Kräfte," *Wiener Berichte, 66* (1872), 256–263; *Wiss. Abhdl., 1,* 403–410. "Experimentaluntersuchung über die elektrostatische Fernwirkung dielektrischer Körper," *Wiener Berichte, 67* (1873), 81–155; *Wiss. Abhdl., 1,* 472–536. "Über einige an meinen Versuchen über die elektrostatische Fernwirkung dielektrischer Körper anzubringende Korrektionen," *Wiener Berichte, 70* (1874), 307–341; *Wiss. Abhdl., 1,* 556–586. A summary of the last three papers is found in "Experimentaluntersuchung über das Verhalten nicht leitender Körper unter dem Einflusse elektrischer Kräfte," *Pogg. Ann., 153* (1874), 525–534; *Wiss. Abhdl., 1,* 607–615. "Experimentelle Bestimmung der Dielektrizitätskonstante einiger Gase," *Wiener Berichte, 69* (1874), 795–813: *Pogg. Ann., 155* (1875), 403–422; *Wiss. Abhdl., 1,* 537–555.

Lorentz interprets this remarkable fact as showing that electromagnetic phenomena in gases are due mainly to the ether in the intermolecular spaces, the influence of molecules being very small. With regard to solids and liquids, he supposes that the intermolecular spaces are also filled with ether, adducing as evidence the influence on optical phenomena of the motion of the earth relative to the ether. Lorentz was thus led to the view that, in order to give a complete explanation of electromagnetic phenomena in material bodies, one should primarily consider the ether and then take into account, as a small correction, the influences due to the presence of molecules: "If one wants to give to the electric motions in such a body a treatment which is satisfactory in every respect, one should take into account first the ether, and then the molecules lying within it. Then the distance, size, and form of the latter enter into consideration, from which the explanation of dispersion and the rotation of the plane of polarization will probably result." [51] In order to illustrate this way of approaching the subject, he assumed for gases

$$\epsilon = \epsilon_0 + mp,$$

where ϵ and ϵ_0 are the electric susceptibilities of the gas and the ether respectively, p is the number of molecules in a unit volume, and m is a constant. Substituting this expression into

$$n^2 = K = \frac{1 + 4\pi\epsilon}{1 + 4\pi\epsilon_0},$$

and denoting the density of gas by d, he obtained,

$$\frac{n^2 - 1}{d} = \text{const},$$

which is the well-known law of Arago and Biot. This result may be considered as confirming Lorentz' supposition "that the ether in gases has exactly the same properties as that in a vacuum," and "that in each molecule an electric moment is induced by an electromotive force X in the direction of this force, having the magnitude mX, where m is a

[51] Lorentz., op. cit. (note 30); Collected Papers, 1, 87–88. "Wil men dan een in alle opzichten voldoende behandeling der electrische bewegingen in een dergelijk lichaam geven, dan moet men vooreerst met den aether, ten tweede met de daartusschen liggende moleculen rekening houden. Daarbij komen dan afstand, grootte en vorm der laatste in aanmerking, omstandigheden, waaruit waarschijnlijk de verklaring der dispersie en der draaiing van het polarisatievlak moet voortvloeien."

constant." [52] Lorentz thus obtained the first idea of a separation of the roles of the ether and ponderable bodies and also an idea of the inner electrical structure of a molecule.

5. OPTICAL PROPERTIES OF MATTER

In a resumé at the end of his thesis Lorentz writes:

Thus the investigation of reflection and refraction leads us to the conclusion that Maxwell's hypothesis should be given priority over the old undulatory theory. Other phenomena of light also promise, with the aid of this hypothesis, greatly to augment our knowledge. Let us think of dispersion, the rotation of the plane of polarisation, and the manner in which these phenomena are related to molecular structure; further, let us think of mechanical forces that may perhaps enter into optical phenomena and the influence exerted by external forces or the motion of medium. Finally let us think of the emission and absorption of light and radiant heat.[53]

In these words we can see the germ of most of the subjects which will later be treated by the theory of electrons. Part of the program of research announced here was realized three years later in his second paper "Over het verband tusschen de voortplantingssnelheid en samenstelling der middenstoffen" ("Concerning the Relations between the Velocity of Propagation of Light and the Density and Composition of Media").[54]

The declared purpose of this work was to construct a theory of the optical properties of matter. He obtained, in fact, the so-called Lorentz-Lorenz formula and gave the first satisfactory theory of the dispersion of light. The historical importance of this article, however, is not limited to the derivation of this or that optical formula. It is more

---

[52] *Ibid.*, 88. "Dat de aether in een gas volkommen dezelfde eigenschappen heeft, als in de luchtledige ruimte . . . dat in elke molecule door een electromotorische kracht X een electrisch moment, in de richting dier kracht en met de grootte mX wordt opgewekt, waarbij m een constante is."

[53] *Ibid.*, 191–192. "Zoo brengt ons het onderzoek der terugkaatsing en breking tot de slotsom, dat aan Maxwell's hypothese de voorrang boven de vroegere undulatietheorie moet toegekend worden. Ook de andere lichtverschijnselen beloven, in verband met die hypothese beschouwd, veel bij te dragen tot vermeerdering onzer kennis. Men denke slechts aan de kleurschifting, de draaiing van het polarisatievlak en de wijze, waarop deze met de moleculaire structuur samenhangen; verder aan de mechanische krachten, die misschien bij de lichtverschijnselen kunnen optreden en aan de invloed, dien uitwendige krachten, of de beweging der middenstoffen daarop uitoefenen. Eindelijk aan de emissie en absorptie van het licht en de stralende warmte."

[54] Lorentz, *op. cit.* (note 44).

important to note that the role of the ether in optical and electromagnetic phenomena in a material body was definitely distinguished from that of the particles which constitute ponderable matter. As in 1875, Lorentz pictured material molecules as being embedded in an all-pervading ether. In contrast to Maxwell, who treated the ether and ponderable matter equally as continuous dielectrics, Lorentz regarded the ether as the sole dielectric in a proper sense; he regarded the dielectric phenomena which occur in ponderable bodies as manifestations of the interactions of the ether and molecules. But it should be remembered that Lorentz at this stage still held to the conception of action at a distance.

In his 1878 article Lorentz reconfirms the idea that the intermolecular space is filled with ether:

> That ether is actually present between the molecules is, in the case of gases, not open to doubt, since there the properties of these substances change gradually with increasing rarification into those of free ether. But even in the case of rigid bodies and fluids, difficulties would be encountered if we wished to consider the space between their particles as empty. It would then be difficult to realise why the velocity of propagation of light in these bodies is always smaller than in a vacuo and it has also appeared to me that with changes in density the index of refraction should alter far more than is the case. Lastly the influence exercised by the movement of the media on the phenomena of light also point to the presence of ether between the molecules of the bodies.[55]

But Lorentz thought that there still remained some uncertainty:

> Not only are we almost completely ignorant as to the actual structure of molecules and the electric motions which may take place within them, but the problem becomes yet more difficult from the circumstance that we must regard the space between the molecules as filled with ether, in doing which our ignorance of the manner in which the molecules are embedded in the ether provides still other difficulties.[56]

The first difficulty mentioned here, which Lorentz removed by assuming that an electric moment could appear within a molecule, is understandable to us. But the meaning of the second difficulty is difficult to gather. The clue to understanding it is to note that Lorentz regards

[55] Lorentz, *op. cit.* (note 44); *Collected Papers* 2, 23.
[56] *Ibid.*, 23.

the ether as a dielectric substance. In fact he thought that the presence of molecules could possibly alter the properties of the ether. But since he did not know if and how such an alteration occurs, he made "the very simple supposition that—except perhaps in the immediate neighbourhood of the particles—the properties of ether are the same as in a vacuum." [57] Lorentz assumed that each molecule occupied its position in a cavity whose linear dimension was small compared with the wavelength of light. Changes in the electrical states within a molecule were supposed to exert a direct influence on the state of polarization of the ether, acting across the void space between the molecule and the wall of the cavity. Such a conception is reminiscent of the way in which W. Thomson distinguished between the magnetic induction and the magnetic force.[58] This resemblance will become more remarkable when we examine Lorentz' theory of the refractive index. Anyway, it is certain that at this time Lorentz still holds to the doctrine of action at a distance and considers the ether as if it were a material body.

We shall first sketch Lorentz' derivation of the Lorentz-Lorenz formula; then we will examine his discussion of dispersion. Lorentz considers a single molecule placed in an infinite ether. If this molecule has an electric dipole moment $m(r)$ (m_x, m_y, m_z in Lorentz' notation), polarizations $P(r, t)$ (Lorentz writes ξ, η, ζ) are induced at each point of the ether. The polarizations thus induced in turn exert an electromotive force. The static part of this force is equal to that produced by a surface charge $-P_n$ on the wall of the cavity in which the molecule is placed. The other part is the contribution from a vector potential due to the variation of $m(t)$. Summing these two parts, the electromotive force acting at each point in the ether is shown to be equal to the force derived from the potentials produced by

$$\text{a moment} \qquad (\alpha + \beta)\left(\frac{8\pi}{3} + \frac{1}{\epsilon_0}\right) m$$

$$\text{a current element} \qquad (2\alpha - \beta)\frac{4}{3}\pi \frac{dm}{dt} \tag{23}$$

which are placed at the position of the molecule. The constants α and β correspond to transverse and longitudinal vibrations respectively and are determined by the condition that these expressions are equal to m

[57] *Ibid.*, 24.
[58] Cf. Whittaker, *op. cit.* (note 5), 219.

and dm/dt respectively. However, there is an electromotive force within the cavity due to the surface charge on the wall. It is given by

$$(\alpha + \beta)\frac{8\pi}{3}\frac{m}{\rho^3},\tag{24}$$

where ρ is the radius of the cavity. In these calculations, approximations are always made by regarding ρ as very small.

When many molecules are distributed in the ether, it is necessary to take into account the influences of the cavities of other molecules. In this case the electric state induced at each point of the ether by a cavity containing a molecule is equal to a state induced by

a moment $\qquad (\alpha + \beta)\left(\dfrac{8\pi}{3} + \dfrac{1}{\epsilon_0}\right)m + \dfrac{4\pi}{3}\rho^3 P'$

a current element $\qquad (2\alpha - \beta)\dfrac{8\pi}{3}\dfrac{dm}{dt} + \dfrac{4\pi}{3}\rho^3\dfrac{dP'}{dt}$

which are placed in that cavity. The last terms represent the influences of other molecules; P' (or ξ', η', ζ' in Lorentz' notation) is the polarization produced at the position of the cavity by the actions of other molecules. Now since the molecule in the cavity under consideration has been assumed to have a moment $m(t)$ and a current element dm/dt, the equalities

$$(\alpha + \beta)\left(\frac{8\pi}{3} + \frac{1}{\epsilon_0}\right)m + \frac{4\pi}{3}\rho^3 P' = m\tag{25}$$

$$(2\alpha - \beta)\frac{8\pi}{3}\frac{dm}{dt} + \frac{4\pi}{3}\rho^3\frac{dP'}{dt} = \frac{dm}{dt}\tag{26}$$

should hold. The electromotive force E' (Lorentz denotes it by X', Y', Z') exerted in the cavity by everything outside it is

$$E' = (\alpha + \beta)\frac{8\pi}{3}\frac{m}{\rho^3} + \frac{P'}{\epsilon_0} + \frac{4\pi}{3}P'.$$

The electric moment excited in the molecule is proportional to E': $m = \kappa E'$. Hence

$$\kappa\left[(\alpha + \beta)\frac{8\pi}{3}\frac{m}{\rho^3} + \frac{P'}{\epsilon_0} + \frac{4\pi}{3}P'\right] = m.\tag{27}$$

Finally, P' is calculated. The total contribution to P' from all molecules is written as an integral, the distribution of the molecular moment m being regarded as continuous. In recasting the equations, use is made of the fact that in an extended ponderable dielectric the temporal and spacial changes of m are determined by the equations

$$\Delta m = \frac{1}{V^2} \frac{\partial^2 m}{\partial t^2} \qquad \text{and} \qquad \text{div } m = 0.$$

(Note that the velocity of propagation V is not the V_0 for the free ether.) The result is

$$P' = q(m)_p, \text{ where } q \equiv \frac{4\pi}{3} p(\alpha + \beta) \frac{n^2 + 2}{n^2 - 1}, \qquad (28)$$

and where $(m)_p$ denotes the value of vector m at the position of the molecule, p is the number of molecules per unit volume, and $n = V_0/V$ represents the absolute refractive index.

Substituting this value of P' into equations (25), (26), and (27) yields three equations for the three unknown quantities α, β, and q. Solving for α, β and q and substituting into equation (28) gives, finally,

$$\frac{n^2 - 1}{(n^2 + 2)d} = k, \text{ where } k \equiv \frac{\dfrac{4\pi}{3} \rho^3 (3 + 4\pi\epsilon_0) - 4\pi\epsilon_0 \dfrac{\rho^3}{\kappa}}{m(3 + 8\pi\epsilon_0) \dfrac{\rho^3}{\kappa} - 8\pi\epsilon_0},$$

and where $d = mp$ is the density of the dielectric (m is the mass of a molecule). In this calculation, Lorentz has neglected $1/(1 + 4\pi\epsilon_0)$ and has put $4\pi\epsilon_0/(1 + 4\pi\epsilon_0) = 1$ by regarding ϵ_0 as very great. This is the Lorentz-Lorenz formula, which was found independently by L. Lorenz in 1869.[59]

Then Lorentz proceeds to the theory of dispersion. He took notice of the fact that, in deriving the Lorentz-Lorenz formula, he had ignored those terms which are of higher order in the ratios ρ/l and δ/l, where δ is the distance between adjacent molecules, and l is the

[59] Ludvig Valentin Lorenz, "Experimentale og theoretiske Undersøgelser over Legemernes Brydningsforhold," *Vid. Selsk. Skr.*, ser. 5, *8* (1869), 209; *Oeuvres scientifiques de L. Lorenz, 1*, 213–298. Cf. Mogens Pihl, *Der Physiker L. V. Lorenz* (Copenhagen, 1939), 53–74.

wavelength of light. This approximation is not valid for short wavelengths. Lorentz first thought that the correction required by this circumstance might afford a means of explaining dispersion. But the second-degree correction in δ/l and the first-degree one in ρ/l gave only a very small variation of the refractive index. Thus Lorentz was led to assume that a molecule contains a charged particle endowed with mass: "If we accept the electromagnetic theory of light, there is nothing left, in my opinion, but to look for the cause of dispersion in the molecules of the medium themselves. And we can indeed obtain formulae from which a dispersion follows if we adopt the supposition that, in such a molecule, as soon as an electric moment is excited, a certain mass is at the same time brought into motion." [60] In order to form an idea of what actually takes place in a molecule of a dispersive medium, he assumed that the molecule consists of a fixed charge $-e$ and a movable charge e whose mass is μ. Then if the movable charge undergoes a displacement x, y, z, there arises in the molecule a moment with components $m_x = ex$, $m_y = ey$, $m_z = ez$. He assumes that when the charged particle is displaced, the other parts of the molecule exert a force on it directed toward the position of equilibrium and proportional to the displacement; the components of the restoring force can therefore be written as $-gx$, $-gy$, $-gz$, and the equations of motion of the particle under a constant external electromotive force (X, Y, Z) become

$$\mu\frac{d^2x}{dt^2} = eX - gx, \mu\frac{d^2y}{dt^2} = eY - gy, \mu\frac{d^2z}{dt^2} = eZ - gz. \quad (29)$$

If one considers only a periodic motion with period T, then

$$x = a \cos\frac{2\pi}{T}(t + p), \text{ etc.}$$

Then the solutions are

$$x = \frac{e}{g - \dfrac{4\pi^2\mu}{T^2}}X, y = \frac{e}{g - \dfrac{4\pi^2\mu}{T^2}}Y, z = \frac{e}{g - \dfrac{4\pi^2\mu}{T^2}}Z.$$

Substituting these expressions in $ex = m_x = \kappa X$, we obtain the value

[60] Lorentz, *op. cit.* (note 44), 79–80.

for κ. Equations (27) and (28) together with the κ just obtained yield the dispersion relation

$$\frac{n^2 + 2}{n^2 - 1} = \frac{A - \dfrac{B}{l^2}}{\dfrac{C}{l^2} - D},$$

where A, B, C, and D are constants.

The significance of Lorentz' 1878 article is that the fundamental model of the theory of electrons (a charged harmonic oscillator within a molecule) has been firmly established, and that the supposition that the intermolecular ether retains the same properties as the vacuum has been corroborated by reasonable results from the theory. This raised the picture of a physical world consisting of an all-pervading ether and systems of charged particles. Here, despite the underlying conception of action at a distance, was the first, decisive step towards rendering the electromagnetic field independent of matter. What remained was for Lorentz to convert his standpoint from the action-at-a-distance theory to one of contiguous action.

6. MOLECULAR THEORY AND THE ELECTRODYNAMICS
OF THE CONTINENTAL SCHOOL

Before entering into the question of how and when Lorentz gave up the viewpoint of action at a distance, we shall consider some historical circumstances. Lorentz' work of 1878 naturally bore an important significance for the subsequent development of molecular theory. Lorentz himself, comparing his formulas with observed values of the refractive index for various substances, states that it would be necessary to take into account the alteration of the inner structure of molecules to explain slight discrepancies between theoretical and observed values. However, the chief importance of Lorentz' paper in relation to molecular theory lies elsewhere. From the considerations above of Lorentz' electromagnetic theory of optical phenomena, it is clear that the introduction of molecular theory into electromagnetism had an essential significance for the development of the concept of the electromagnetic field. For the decisive step in the development, the separation of the roles of the ether and of ponderable matter, was brought about by the very introduction of molecular theory.

Then the question arises as to the circumstances which induced Lorentz to introduce molecular considerations into electromagnetism. From a purely theoretical point of view, it may be said that it was necessary to consider the structure of molecules in the investigation of the optical properties of matter. But it should be stressed that the theoretical problem could, in itself, only be posed in a certain historical setting; moreover, the fact that the introduction of molecular theory, however necessary, was actually only effected in the 1870's by Lorentz requires an historical explanation.

In the first place, the rise of the molecular theoretical trends in physics in the 1860's and 1870's should be mentioned. It was under the title of "De moleculaire theorie in de natuurkunde" ("Molecular Theories in Physics") that Lorentz, in 1878, delivered his inaugural address at Leyden University.[61] He states there that "the final aim of all research must be the deduction of the innumerable natural phenomena as necessary consequences of a few simple fundamental principles,"[62] and as an illustration of this view he reviews the development of the molecular theory in physics since Gassendi. After that he outlines his program of investigating the optical properties of matter on the bases of molecular theory and the electromagnetic theory of light, suggesting that this investigation would throw light on the molecular structure of matter. At the beginning of his discourse he observes that "there will be hardly anybody nowadays, who does not know, that in the mind of physicists, a material body is a system of very small particles, so called *molecules*,"[63] which points to the fact that the molecular theory was widely accepted among physicists at that time. The cause of this wide acceptance was most likely the development of the kinetic theory of gases, a subject which had fascinated physicists in the 1860's and 1870's.

Investigations of the kinetic theory of gases, the foundations of which were laid by Clausius and Maxwell at the end of the 1850's, were extended to various, separate problems: for example, the study of heat conduction by Clausius (1862), the investigation of viscosity by

[61] Lorentz, "De moleculaire theorie in natuurkunde," *Collected Papers, 9,* 1–25; English trans., 26–49.
[62] *Ibid.,* 2. Quotation, p. 26.
[63] *Ibid.,* 3. Quotation, p. 28.

O. E. Meyer, the determination of the Loschmidt number (1865), the revised proof of Maxwell's velocity distribution, and the theories (1866) of heat conduction and viscosity based upon Maxwell's r^{-5} intermolecular force. The theoretical difficulty posed by the law of the increase of entropy (Clausius, 1865) inspired Boltzmann's H-theorem and eventually led to the emergence of statistical mechanics. Lorentz, in his inaugural address, emphasized the numerous successes of the kinetic theory as evidence for the power of the molecular theoretical approach. Moreover, he himself made a number of contributions to this field of physics: a kinetic theory of the propagation of sound (1880), in which he anticipated the relaxation of the energy partition in gases of polyatomic molecules; the derivation of van der Waals's equation of state from the virial theorem (1881); and an amendment of Boltzmann's proof of the H-theorem (1887).

Besides the background of molecular theory, the influence of the Continental school of electrodynamics must be recognized. In contrast to Maxwell's theory, where the question of the nature of electric charge was carefully set aside, the Continental school continually formulated its electrodynamic equations in terms of the movement of particles of electricity. Unlike the theory of contiguous action, the theory of action at a distance historically has a close connection with the conception of a discrete constitution of matter. The electrodynamics of the Continental school belongs to the mechanistic view of nature. The ultimate aim of the mechanistic view is to explain all physical phenomena by means of the motions of particles under the action of central forces. The notion of a charged particle, the basis of Lorentz' theory of electrons, was rooted in the development of the electrodynamics of the Continental school. A. M. Ampère, the founder of electrodynamics, concentrated on the motion of current-conductors, and not on the notions of electric fluids and particles. But the discovery of electromagnetic induction made inevitable the introduction of the electric particle. First G. T. Fechner in 1845 and then W. E. Weber in 1846 advanced a theory of electromagnetic induction assuming the electric current to consist of positively and negatively charged particles moving in opposite directions. This notion of electric current persisted in the subsequent development of Continental electrodynamics. Kirchhoff and Helmholtz should be understood literally when they called their fundamental equations of electrodynamics the "Bewegungsgleichungen der Elektrizität." The idea that the particle of electricity has a mechanical inertia is said to have been introduced by Weber and H. Lorberg.[64] Lorentz began his scientific career with an electromagnetic theory of light, conceived in the action-at-a-distance terms of Continental electrodynamics. He regarded the vibrations of light as movements of the same character as electric currents, and he always considered the motion of electricity to be represented by variations of the polarization P and not by the variations of the field intensities E and H. Thus it may be most natural to suppose that Lorentz' conception of a charged particle as the basis of electric charge and as a constituent of ponderable matter was formed under the influence of the electrodynamics of the Continental school as well as under that of the molecular theory. Lorentz' own words support this conclusion in his address to the Dutch Congress of Physics and Medicine held in April 1891, "Electriciteit en Ether" ("Electricity and Ether"). Comparing the electrodynamics of the Continental school with Maxwell's theory, he argues as follows:

> It seems to me not without interest to note that one can reach the new theory, as far at least as the form of the theory is concerned, starting from the old one with a slight modification. In the old theory one transferred a concept formed to deal with electrified conductors to the particle of imaginary electric matter. The follower of Maxwell can proceed in a similar way. One can assume that there exist small electrically charged particles, that is, particles with properties similar to those of a charged conductor, and suppose that a perceptible charge—I mean the charge of a body of perceptible dimensions—consists in an accumulation of such particles, and an electric current in their motion.[65]

Thus the rise of molecular theory and the notion of charged particles from Continental electrodynamics, by making possible the separation of the role of the molecule from that of the ether, played a significant

[64] Helmholtz, op. cit. (note 29). Wissenschaftliche Abhandlungen, 1, 552.
[65] Lorentz, "Electriciteit en Ether," Nederl. Natuuren Geneeskundig Congres, 4 April 1891. Verhandel., 3 (1891), 40; Collected Papers, 9, 89–101, esp. 100. "De opmerking schijnt mij niet zonder belang, dat men door eene kleine wijziging eene toenadering, ten minste wat den vorm betreft, van de nieuwe opvatting tot de oude kan bewerken. Men heeft vroeger op de deeltjes der denkbeeldige electrische stoffen overgedragen, wat men bij geladen geleiders had waargenomen. Iets dergelijks kan een volgeling van Maxwell doen. Men kan aannemen, dat er kleine electrisch geladen deeltjes bestaan, d.w.z. deeltjes met dergelijke eigenschappen als een geladen conductor, en onderstellen, dat eene voor ons waarneembare lading—ik bedoel de lading van een lichaam van waarneembare afmetingen—bestaat in eene opeenhooping van dergelijke deeltjes, en een electrische stroom in eene beweging daarvan."

part in the establishment of the concept of the field as an independent physical reality.

7. CONVERSION TO THE THEORY OF CONTIGUOUS ACTION

Now we shall attempt to determine when Lorentz shifted to the viewpoint of contiguous action. In 1882 he published a paper which dealt with the force between two elements of electric current from the viewpoint of the electrodynamics of the Continental school.[66] The next year he discussed the Hall effect and the rotation of the plane of polarization.[67] There is no indication that by this time he had been converted to the theory of contiguous action. These were his only two articles on electromagnetic theory in the 1880's; Lorentz undertook investigations mainly in the kinetic theory of gases and thermodynamics at this time. Hence, we cannot trace the development of Lorentz' thought in this decade; his first declaration in favor of the Faraday-Maxwell theory of contiguous action is found in his address "Electricitit en Ether" of April 1891: "Opposing the old theory of electricity stands one which Maxwell has developed following Faraday's footsteps. I believe that there are reasons to give preference to the latter conception." [68] Following this statement Lorentz describes the differences between the two theories and explains the reasons why he adopts Maxwell's in preference to the older one.

The first difference is that in the old theory the electromagnetic energy of a system of currents is considered a potential energy, whereas Maxwell's theory attributes it to motions within the medium. In the old theory this potential ought to depend on the velocities of electric particles, which, in Lorentz' opinion, seems unnatural: "Thus we are led to assume an energy which depends on the velocity of particles and nevertheless is not a kinetic energy in the ordinary sense of the word. This is a major difficulty: although it is simpler to denote energy dependent on velocity as kinetic energy, here it is not the case. In Maxwell's conception it is considered as such, and in this I find the

[66] Lorentz, "Les formules fondamentales de l'électrodynamique," *Arch. néerl., 17* (1882), 83; *Collected Papers, 2,* 120–135.
[67] Lorentz, "Le phénomène découvert par Hall et la rotation électromagnétique du plan de polarisation de la lumière," *Arch. néerl., 19* (1884), 123; *Collected Papers, 2,* 136–163.
[68] Lorentz, *op. cit.* (note 65); *Collected Papers, 9,* 93. "Tegenover de oude electriciteitsleer staat die, welke Maxwell, op het voetspoor van Faraday, ontwikkeld heeft. Ik geloof, dat er redenen bestaan, om aan de laatste opvatting de voorkeur to geven."

first reason to give his theory precedence." [69] Maxwell's conception has another advantage. The electromagnetic energy of a system consisting of two conducting wires is not equal to the sum of electromagnetic energies of the two currents individually, and, moreover, it depends on the relative position of the two wires. This may most naturally be accounted for by assuming that something else is taking part in the motion beside the electricity in the wires. In Maxwell's theory, indeed, the explanation is given by including the medium in the mutual action of conducting wires. Lorentz remarks: "it is reasonable that the way in which this medium is brought into motion may depend not only on the strength of the two currents but also on the relative position of the conductors." [70]

Secondly Lorentz adduces the novelty of Maxwell's theory that the medium "is also the seat of electrostatic energy." [71] He comments: "surely the electrostatic energy is still an energy of position, yet it is quite a different thing from what the old theory means by the same name." [72]

The third difference between the old and new theories concerns what happens in a dielectric placed between two plates of a charged parallel-plate condenser. According to Maxwell's theory, since the electric matter is considered to behave like an incompressible fluid, the same quantity of electricity supplied to the positive plate A must move from the plate into the dielectric, and from the dielectric the same quantity of electricity must enter the negative plate B. However, though the old theory supposes a similar displacement of electricity, "according to it the quantity [of electricity transferred to the dielectric] is always smaller than that which is given to plate A. The difference of these quantities gives rise to the electrostatic action of a charged condenser." [73]

[69] *Ibid.*, 93–94. "Zoo kwam men ertoe, een arbeidsvermogen aan te nemen, dat van de snelheden der deeltjes afhangt en toch geene kinetische energie in den gewonen zin van het woord is. Een overwegend bezwaar is dat nu zeker niet, maar wanneer het arbeidsvermogen toch van snelheden moet afhangen is het eenvoudiger, het als gewone kinetische energie te beschouwen. Zoo is de opvatting van Maxwell en daarin vind ik eene eerste reden, om aan zijne theorie den voorrang toe te kennen."
[70] *Ibid.*, 94. "Het is well begrijpelijk, dat de wijze, waarop dit medium in beweging gebracht wordt, kan afhangen niet alleen van de sterkte der twee stroomen, maar bovendien van den betrekkelijken stand der geleiders."
[71] *Ibid.:* ". . . ook van het electrostatische arbeidsvermogen is zij de zetel."
[72] *Ibid.*, 97. "De electrostatische energie is nog wel arbeidsvermogen van plaats, maar toch geheel iets anders dan de oude electriciteits-theorie erin zag."
[73] *Ibid.*, 99: ". . . maar volgens haar moet die hoeveelheid altijd kleiner zijn dan die, welke aan de plaat A gegeven wordt. Van het verschil hangt de electrostatische werking der geladen flesch af."

In the context of this study, the most important novelty which Lorentz recognizes in Maxwell's theory is the assumption that electromagnetic actions are transmitted with finite velocity: "another difference ... [is] that the disturbances which a particle induces in the ether and which may affect another particle do not propagate themselves instantaneously but with the velocity of light." [74]

In the old theory, too, one can assign to a medium an intervening role in electromagnetic phenomena, and thus arrive at an explanation of Hertz's experiment and the electromagnetic theory of light. But for this purpose the ratio between the quantity of electricity given to the condenser plate and the quantity transferred to the dielectric should not differ perceptibly from unity. It is difficult to make this requirement compatible with the one mentioned above that the quantity transferred to the dielectric should be smaller than the quantity supplied to the plate to give rise to the electric action of the parallel-plate condenser: "It is only through an artificial assumption that one could satisfy both requirements, and this is the second argument, to which I have already alluded, that seems to plead in favor of the new mode of conception." [75]

We have thus far shown that Lorentz' view has definitely changed by the beginning of 1891. It seems that the decisive motivation for his conversion was the work of Hertz and Poincaré. Of the reasons which Lorentz adduced in favor of Maxwell's theory, the idea that the electromagnetic energy of a system of conducting wires should be ascribed to motions in a medium had already been emphasized repeatedly by Maxwell himself. If this were the decisive reason, Lorentz would have been converted much earlier. There must then have been some new conditions revealed to him toward the end of the 1880's. First, naturally, would be Hertz's experiment on electric waves. Concerning this experiment Lorentz states: "Indeed, Hertz has to some extent stamped the experimental confirmation on Maxwell's speculation." [76] He calls Hertz's experiment "the greatest triumph that Maxwell's theory has attained." [77]

[74] *Ibid.*, 100–101: ". . . een ander onderscheid bovendien, dat hierop neerkomt, dat de veranderingen, die een deeltjes in den ether opwekt en waarvan een ander den invloed zal gevoelen, zich niet oogenblikkelijk, maar met de snelheid van het licht voortplanten."

[75] *Ibid.*, 99. "Slechts door eene gekunstelde onderstelling kan men aan beide eischen voldoen en dit is een tweede argument, ik zinspeelde er reeds op, wat mij voor de nieuwe zienswijze schijnt te pleiten."

[76] *Ibid.*, 98. "Trouwens, Hertz heeft tot op zekere hoogte op de bespiegelingen van Maxwell het zegel der experimenteele bevestiging gedrukt."

[77] *Ibid.*, 97: ". . . de grootste triomf, dien Maxwell's theorie behaald heeft."

Poincaré is cited by Lorentz in connection with the interpretation of the change of state induced within a dielectric placed between condenser plates. The difference between interpretations in the old and new theories is illustrated in Poincaré's Sorbonne lectures on Maxwell's theory, published in 1890 under the title *Électricité et optique. I. Les théories de Maxwell et la théorie électromagnétique de la lumière.*[78] In his 1891 address Lorentz refers to Poincaré only one time, without citing a published source. There he says: "Poincaré told about a physicist who claimed to have understood the whole theory of Maxwell except that he still had not grasped what an electrified sphere was." [79] Without the slightest doubt Lorentz was thinking of Poincaré's *Électricité et optique.* For, toward the end of the introduction of this treatise, Poincaré says that "one of those French scholars who have most profoundly studied Maxwell's work said to me one day: 'I comprehend all that is in his book except what an electrified ball is'." [80]

These considerations indicate that Lorentz had examined Poincaré's treatise. It may, therefore, be assumed that the decisive motivation for Lorentz' conversion to the conception of contiguous action was Hertz's experiment and Poincaré's lectures. If this is the case, it was around 1890 that Lorentz definitely changed his view from action at a distance to contiguous action.

8. HERTZ'S ELECTRODYNAMICS

Lorentz set up the concept of the electromagnetic field as an independent dynamical system in his article of 1892, "La théorie électromagnétique de Maxwell et son application aux corps mouvants," [81] which also laid the foundation of the theory of electrons. Since this work originated in Hertz's electrodynamics of 1890,[82] it is appropriate to examine some characteristic features of Hertz's theory before proceeding to Lorentz'.

[78] H. Poincaré, *Électricité et optique. I. Les théories de Maxwell et la théorie électromagnétique de la lumière* (Paris, 1890), Chap. II.
[79] Lorentz, *op. cit.* (note 65); *Collected Papers, 9,* 95. "Poincaré verhaalt van een natuurkundige, die verklaarde de geheele theorie van Maxwell verstaan te hebben, maar toch niet recht begrepen to hebben, wat nu een geelectriseerde bol was."
[80] Poincaré, *op. cit.* (note 78), xvi-xvii. "Un des savants français qui ont le plus approfondis l'oeuvre de Maxwell me disait un jour: 'Je comprends tout dans son livre, excepté ce que c'est qu'une boule électrisée'."
[81] Lorentz, "La théorie électromagnétique de Maxwell et son application aux corps mouvants," *Arch. néerl., 25* (1892), 363; *Collected Papers, 2,* 164–343.
[82] Hertz, "Ueber die Grundgleichungen der Elektrodynamik für ruhende Körper," *Wied. Ann., 40* (1890), 577; *Gesammelte Werke, 2,* 208–255; *Electric Waves,* pp. 195–240.

After the publication of Maxwell's *Treatise*, physicists continued to be confused about the meaning of the potentials which entered Maxwell's theory on the same footing as the field intensities. According to Maxwell's final formulation in his *Treatise*, the fundamental equations for the electromagnetic field in a perfect insulator are:[83]

$$B = \text{rot } A, \tag{30}$$

$$E = -\frac{\partial A}{\partial t} - \text{grad } \Psi, \tag{31}$$

$$4\pi C = \text{rot } H, \tag{32}$$

$$D = \frac{K}{4\pi} E, \tag{33}$$

$$C = \partial D/\partial t = (K/4\pi)(\partial E/\partial t), \tag{34}$$

$$\rho = \text{div } D, \tag{35}$$

$$B = \mu H, \tag{36}$$

where C is the total current and A and Ψ are vector and scalar potentials respectively. Though Maxwell wrote out these equations in components, vector notation is used here for the sake of brevity. From equations (30)–(34) and (36), Maxwell obtains

$$K\mu \frac{d}{dt}(A + \text{grad } \Psi) - \Delta A + \text{grad } J = 0, \tag{37}$$

where

$$J \equiv \text{div } A.$$

Taking the divergence of (37) gives

$$K\mu \frac{d}{dt}\left(\frac{dJ}{dt} + \Delta\Psi\right) = 0.$$

Since the second term in the bracket, $\Delta\Psi$, being equal to the volume density of free electricity, does not depend on time, it can be dropped. Then $d^2J/dt^2 = 0$, and consequently J is a linear function of t. Now in considering a periodic disturbance, Maxwell argues, such a Ψ or J may be ignored and therefore the terms in equation (37) containing Ψ or J may be dropped. Maxwell thus finally obtained the wave equation for A.[84] This procedure corresponds to adopting a Coulomb gauge, since to drop the term containing $J = \text{div } A$ is equivalent to

[83] Maxwell, *op. cit.* (note 12), 2, Part IV, Chaps. VIII–IX.
[84] *Ibid.*, Chap. XX.

assuming div $A = 0$. Since in this case Ψ expresses a static potential, a change in Ψ occurs simultaneously throughout the whole space, whereas a change in A propagates itself with finite velocity. But this seemed a very curious consequence, difficult to interpret at that time when the potentials were treated as fundamental quantities having the same rank as field intensities; also the gauge invariance was not yet recognized. And Maxwell's argument is somewhat arbitrary and artificial. Thus confusions arose over the physical meaning of potentials. The situation was aggravated especially after Hertz's experiment; Lodge reports that vigorous discussions were aroused at the annual meetings in 1888 and 1890 of the British Association.[85]

In order to clear away this confusion, it was necessary to eliminate potentials from the fundamental equations and thus reduce them to the rank of mathematical, subsidiary quantities. This task was carried out by Oliver Heaviside and Hertz. Heaviside introduced the vector notation which is employed today, defining operators such as rot and div, and thus developing vector analysis. In 1885, making use of these tools, he rewrote the system of Maxwellian equations in the form familiar today, completely eliminating the potentials.[86] Though the priority of this task belongs to Heaviside, Hertz's work is more important because of its sharp criticism of the foundation of Maxwell's formulation.

It was in his article of 1890 that Hertz launched his critique. Although he admits that Maxwell's theory may be considered perfect as regards its contents, he criticizes an inconsistency in Maxwell's conception of dielectric polarization:

> Maxwell starts with the assumption of direct actions-at-a-distance; he investigates the laws according to which hypothetical polarizations of the dielectric ether vary under the influence of such distance-forces; and he ends by asserting that these polarizations do really vary thus, but without being actually caused to do so by distance-forces. This procedure leaves behind it the unsatisfactory feeling that there must be something wrong about either the final result or the way which led to it. Another effect of this procedure is that in the formulae there

---

[85] Lodge, *op. cit.* (note 16), 133-137.
[86] Oliver Heaviside, "Electromagnetic Induction and Its Propagation"; this consists of forty-seven sections and was published section by section in numerous issues of *The Electrician* during 1885, 1886, and 1887. The sections of special interest to us are 1-9, published January–May, 1885. Heaviside, *Electrical Papers* (London, 1892), *1*, 429–560; *2*, 39–155; the sections mentioned above are in *1*, 429–476.

are retained a number of superfluous, and in a sense rudimentary, ideas which only possessed their proper significance in the older theory of direct action-at-a-distance.[87]

As examples of such superfluous and immature conceptions Hertz mentions two: the first is Maxwell's distinction between the electric force and the dielectric displacement even in the case of the free ether. Maxwell assumes that when an electric force E is exerted on the ether, a displacement D is produced proportional to E. But such a notion could, Hertz says, have meaning only when the force E stands by itself independently of the ether. This case, however, is not conceivable without admitting action at a distance, and it therefore contradicts Maxwell's fundamental assumption of contiguous action. The second thing that Hertz mentions is the predominance of potentials in the fundamental equations. In Hertz's opinion, there is no reason to replace the forces by potentials unless a mathematical advantage is thereby gained, and it does not appear that there is any such advantage. He therefore asserts that in the fundamental equations one should find only "relations between the physical magnitudes which are actually observed, and not between magnitudes which serve for calculation only." [88]

Accordingly Hertz took, as the starting point of his electromagnetic theory, the electric force and the magnetic force which were defined as forces exerted on a small electrified body and a magnetic pole respectively, and he showed that Maxwell's theory could be formulated without potentials in terms of electric and magnetic field intensities, charge, and current. This result cleared up the question as to the determining quantities of the physical state of an electromagnetic field.

The most important merit of Hertz's work with regard to the present subject is that it demolished the mechanical picture of the field. In Hertz's formulation, all problems concerning the constitution of the ether are to be put off. Having postulated the field equations, he declares: "After these equations are once found, it no longer appears expedient to deduce them (in accordance with the historical course) from conjectures as to the electric and magnetic constitution of the ether and the nature of the acting forces,—all these things being entirely unknown. Rather is it expedient to start from these equations

[87] Hertz, *op. cit.* (note 82). *Gesammelte Werke,* 2, 208–209. The quotation is from *Electric Waves,* pp. 195–196.
[88] *Ibid.,* 209; *Electric Waves,* p. 196.

in search of such further conjectures respecting the constitution of the ether." [89] Hertz's achievement would therefore have helped to clear away Maxwell's conception of the field as being borne by a material substance. In this sense Hertz's electrodynamics prepared the way for the fundamental conception of the theory of electrons, according to which the field is rendered independent of ponderable matter, and the electric charge is attributed to material particles.

The merit of Hertz's work no doubt owes much to his axiomatic treatment of the problem. He begins his theory by postulating the electric and magnetic forces E and H; he does not begin to analyze the electromagnetic field itself, the dynamical entity which stands behind the force. He says: "Starting from rest, the interior of all bodies, including the free ether, can experience disturbances which we denote as electrical, and others which we denote as magnetic. The nature of these changes of state we do not know, but only the phenomena which their presence causes. Regarding these latter as known we can, with their aid, determine the geometrical relations of the changes of state themselves." [90] This attitude certainly had the great advantage of prohibiting speculations about mechanical processes in the ether. But at the same time it inevitably imposed a limitation on Hertz's theory, especially in its treatment of the electromagnetic field in ponderable bodies. For it decreed that, once E and H have been postulated, one ought not to look further for the meaning of these quantities. Since the disturbances represented by E and H have been postulated to exist equally in ponderable matter as well as in ether, one naturally has to regard E and H within ponderable bodies as being borne, at least partially, by the ponderable matter, despite the contrary implication above of Hertz's theory. In fact, when discussing the electromagnetic equations for moving bodies, Hertz stresses that "the lines of force simply represent a symbol for special conditions of matter," [91] and that they cannot be regarded as realities independent of matter. Thus it was impossible in Hertz's conception to separate the field from matter by reducing the bearer of E and H to the ether alone.

This limitation of Hertz's theory becomes more clear in his discussion of the ether within a moving body. As the starting point of the

---

[89] *Ibid.,* 214; *Electric Waves,* p. 201.
[90] *Ibid.,* 210; *Electric Waves.* p. 197.
[91] Hertz, "Ueber die Grundgleichungen der Elektrodynamik für bewegte Körper," *Wied Ann.,* *41* (1890), 369; *Gesammelte Werke, 2,* 256–285; *Electric Waves,* pp. 241–268. The quotation is on p. 271 of *Gesammelte Werke,* and p. 255 of *Electric Waves.*

electrodynamics of moving bodies, he assumed that the ether perfectly shares in the motion of bodies. Though he admits that there are some phenomena which suggest the independence of the ether from the motion of a material body, he nevertheless declines to take them up because they do not belong to the electromagnetic phenomena which so far have been investigated seriously. He adds another reason for rejecting the hypothesis of a stationary ether: "If now we wish to adapt our theory to this view, we have to regard the electromagnetic conditions of the ether and of the tangible matter at every point in space as being in a certain sense independent of each other. Electromagnetic phenomena in bodies in motion would then belong to that class of phenomena which cannot be satisfactorily treated without the introduction of at least two directed magnitudes for the electric and two for the magnetic state." [92] It should be noted, however, that the separation of ether from ponderable bodies leads to a duplication only in the case where one does not consider the atomic constitution of matter. In fact, Lorentz could, assuming that the ether remains absolutely at rest, carry out the separation of ether and ponderable matter on the grounds of the atomic theory of matter, and thus proceed to the modern concept of the electromagnetic field. On the contrary, Hertz avoided the introduction of an atomistic viewpoint and consequently could not avoid recourse to the hypothesis of a perfectly dragged ether. He thus blocked the way to the construction of the modern concept of the field as an independent reality.

9. DYNAMICS OF THE ELECTROMAGNETIC FIELD

In Hertz's electrodynamics, since the electromagnetic field is not regarded as an independent reality, the field intensities E and H enter the equations merely as phenomenological quantities, and consequently the dynamical character of the field does not receive full expression. One of Lorentz' dissatisfactions with Hertz's theory was this lack of a dynamical formulation.

The foundation of the theory of electrons was laid by Lorentz' article of 1892, "La théorie électromagnétique de Maxwell et son application aux corps mouvants"; this work was directly motivated by Hertz's electrodynamics. In the introduction, Lorentz explains the

---

[92] *Ibid., Gesammelte Werke*, 2, 257; *Electric Waves*, p. 242.

motif of his investigation: Maxwell's equations "do not have the simplest form that one could give to them; it is even difficult to notice this clearly because of a number of auxiliary quantities which can be eliminated." [93] Though Hertz had established a system of equations of very simple form capable of accounting for all the observed phenomena, there is an essential difference between the methods of Hertz and of Maxwell, namely, Hertz "hardly concerns himself with the relation of electromagnetic actions to the laws of ordinary mechanics. He is content with a succinct and clear description, independent of all preconceived ideas as to what occurs in the electromagnetic field." [94] Lorentz, by contrast, intends to apply a dynamical method to electromagnetic theory. He also has another motivation: "In the memoir where Mr. Hertz deals with bodies in motion, he assumes that the ether contained in them moves with the bodies. Now, optical phenomena have shown that this is not always the case. I therefore would like to find out the laws governing the electric motions in bodies which traverse the ether without dragging it, and it seems to me difficult to attain this end without a theoretical idea as a guide." [95] Motivated by these considerations, Lorentz sought in his article of 1892 to bring electromagnetic theory into a form explicitly analogous to that of dynamics, and then by means of this formalism, together with the hypothesis of a stationary ether, to seek a method of treating electromagnetic phenomena in a moving body.

As a method "to bring electrical phenomena within the province of dynamics," [96] Maxwell tried, in Part IV, Chaps. V–VIII of his *Treatise,* to express electromagnetic theory in a Lagrangian form. He gave a dynamical theory of a system of circuits; this formulation is, however, restricted to circuits composed of linear conductors, and, moreover, the independent variables entering the dynamical equations represent only the current in each circuit. No independent variable representing the electromagnetic field is present. The dy-

[93] Lorentz, *op. cit.* (note 81); *Collected Papers, 2,* 168. "Elles n'ont pas la forme plus simple que l'on puisse leur donner; il est même difficile d'y voir clair, à cause d'un certain nombre de quantités auxiliaires qu'on en peut éliminer."

[94] *Ibid.,* 168. "M. Hertz ne s'occupe guère d'un rapprochement entre les actions électromagnétiques et les lois de la mécanique ordinaire. Il se contente d'une description succincte et claire, indépendante de toute idée preconçue sur ce qui se passe dans le champ électromagnétique."

[95] *Ibid.,* 168. "Dans le mémoire où M. Hertz traite des corps en mouvement, il admet que l'éther qu'ils contiennent se déplace avec eux. Or, des phénomènes optiques ont depuis longtemps démontré qu'il n'en est pas toujours ainsi. Je désirais donc connaître les lois qui régissent les mouvements électriques dans des corps qui traversent l'éther sans l'entrainer, et il me semblait difficile d'attendre ce but sans avoir pour guide une idée théorique."

[96] Maxwell, *op. cit.* (note 12), 2, Part IV, Chap. V, §553, p. 199.

namics of the electromagnetic field, therefore, are not developed in Maxwell's *Treatise*. Lorentz' attempt to develop a dynamical analogy of the electromagnetic theory was to revise and supplement Maxwell's work in this respect.

The Lagrangian equations for a system of mass points are written in the following form. The coordinates of each material point i are denoted by x_i, y_i, z_i, the mass by m_i, and the components of the forces acting on it by X_i, Y_i, Z_i; the kinetic energy of the system is written as

$$T = \frac{1}{2} \Sigma m_i(\dot{x}_i^2 + \dot{y}_i^2 + \dot{z}_i^2).$$

Then the Lagrangian equation reads

$$\delta A = \frac{d}{dt}\delta' T - \delta T, \tag{38}$$

where

$$\delta A \equiv \Sigma(X_i\delta x_i + Y_i\delta y_i + Z_i\delta z_i),$$

$$\delta' T = \Sigma\left(\frac{\partial T}{\partial \dot{x}_i}\delta x_i + \frac{\partial T}{\partial \dot{y}_i}\delta y_i + \frac{\partial T}{\partial \dot{z}_i}\delta z_i\right),$$

$$\delta T = \Sigma m_i\left(\dot{x}_i\frac{d\delta x_i}{dt} + \dot{y}_i\frac{d\delta y_i}{dt} + \dot{z}_i\frac{d\delta z_i}{dt}\right)$$

$$= \Sigma m_i(\dot{x}_i\delta\dot{x}_i + \dot{y}_i\delta\dot{y}_i + \dot{z}_i\delta\dot{z}_i).$$

In order to apply this formalism to electromagnetic theory, it is first of all necessary to find a quantity which can be interpreted as the kinetic energy of the electromagnetic field. Lorentz supposed that there exists a kind of motion in the electromagnetic field, the "electromagnetic motion," which he interpreted as the origin of Maxwell's electromagnetic energy

$$T = \frac{1}{8\pi}\int \boldsymbol{B} \cdot \boldsymbol{H} \, d\tau, \tag{39}$$

where the magnetic force $\boldsymbol{H} = (\alpha, \beta, \gamma)$ and the magnetic induction $\boldsymbol{B} = (a, b, c)$ are to be determined by the distribution of electric currents (including the displacement current). As to $\boldsymbol{H}$ and $\boldsymbol{B}$ Lorentz, following Maxwell, postulates that

$$\text{div } \boldsymbol{B} = 0, \tag{40}$$

$$\text{rot } \boldsymbol{H} = 4\pi\boldsymbol{C}, \tag{41}$$

$$\boldsymbol{B} = \mu\boldsymbol{H}. \tag{42}$$

The notation "div" and "rot" of vector analysis was not used by Lorentz; it is used here for the sake of simplicity. $C = (u, v, w)$ is the total current, and its distribution is always solenoidal. The units employed are those of the electromagnetic system. Now, introducing a vector $A = (F, G, H)$ by equations

$$\text{rot } A = B, \quad \text{and} \quad \text{div } A = 0, \tag{43}$$

the variation of T is written as

$$\delta T = \int (F \, \delta u + G \, \delta v + H \, \delta w) \, d\tau. \tag{44}$$

Corresponding to the x's in the Lagrangian equation (38), Lorentz assigns a "velocity" vector, (ξ, η, ζ), representing the electromagnetic motion at each point; he assumes that this vector is a linear function of the density of current C,

$$\xi = \int (Au + Bv + Cw) \, d\tau, \text{ etc.}$$

The virtual displacement $\delta x = \xi' \, \delta t$ caused by a virtual current (u', v', w') existing for the short time δt may be written as

$$\delta x = \int (Ae_x + Be_y + Ce_z) \, d\tau, \text{ etc.},$$

where $e_x = u' \, \delta t$, etc. In accordance with this result one may take the (e_x, e_y, e_z) as the virtual displacement of the field. Since the vector (e_x, e_y, e_z) represents, by definition, the quantities of electricity which virtually pass through unit areas perpendicular to each of the coordinate axes during unit time, it can be regarded as a variation of the electric displacement vector $D = (f, g, h)$. Comparing the foregoing expressions for $\delta' T$ and δT, it is found that $\delta' T$ is obtained formally from the expression for δT by replacing the $\delta \dot{x}$'s by δx's. Applying this procedure to the case of the electromagnetic field, one obtains, by replacing the δu etc. in equation (44) by e_x, etc.,

$$\frac{d}{dt} \delta' T = \int \left(\frac{\partial F}{\partial t} e_x + \frac{\partial G}{\partial t} e_y + \frac{\partial H}{\partial t} e_z \right) d\tau.$$

Since the result derived from the Lagrangian equation does not depend on whether the variations e_x etc. change with time or not, the e_x etc. may here be regarded as time independent. Then, since $\dfrac{d \, \delta x}{dt}$ etc. in the expression for δT may be put equal to zero,

$$\delta T = 0.$$

Denoting the components of the force acting upon an electric charge by $(-X, -Y, -Z)$, δA may be written as

$$\delta A = -\int (X e_x + Y e_y + Z e_z)\, d\tau. \tag{45}$$

Then the Lagrangian equation for the electromagnetic field is

$$\int \left[\left(X + \frac{\partial F}{\partial t} \right) e_x + \left(Y + \frac{\partial G}{\partial t} \right) e_y + \left(Z + \frac{\partial H}{\partial t} \right) e_z \right] d\tau. \tag{46}$$

According to Maxwell's picture, whenever electricity is displaced in a dielectric medium (including the ether) a restoring force is called forth. The electric displacement D is determined by the condition of equilibrium of this restoring force with the electric force E; hence, one should put $E = (X, Y, Z)$. Now the potential energy of an electric displacement may be assumed to be a homogeneous quadratic function of the displacement. However, δA is equal to the variation of potential energy with reversed sign. Hence, taking into account equation (45), one obtains

$$X = \nu_{xx} f + \nu_{xy} g + \nu_{xz} h, \text{ etc.}$$

Since (X, Y, Z) is regarded as the electric force E, this result reduces, in the case of an isotropic medium, to

$$E = \frac{1}{K} D. \tag{47}$$

In addition, by the definition of electric displacement,

$$\operatorname{div} D = 0, \tag{48}$$

and

$$C = \frac{\partial D}{\partial t}. \tag{49}$$

Now returning to equation (46), one supposes that $e = (e_x, e_y, e_z)$ vanishes everywhere except within a narrow tubular ringed region, and that the direction cosines p, q, r of e coincide with those of the tube. The cross section of the tube is denoted by ω, and the element of length by ds. Then, since $d\tau = \omega ds$, and $|e|\omega$ has the same value through the whole tube,

$$\int \left[\left(X + \frac{\partial F}{\partial t} \right) p + \left(Y + \frac{\partial G}{\partial t} \right) q + \left(Z + \frac{\partial H}{\partial t} \right) r \right] ds = 0.$$

Taking a tube of rectangular form with the sides parallel to the coordinate axes, this last equation gives

$$-\operatorname{rot} E = \frac{\partial B}{\partial t}. \tag{50}$$

10. STATIONARY ETHER AND CHARGED PARTICLES

Having formulated the dynamical analogy of the electromagnetic theory, Lorentz proceeds to his primary subject, the consideration of electromagnetic and optical phenomena in a moving body. In his formulation thus far developed, the independence of the electromagnetic field from ponderable matter is not explicitly stated. But it comes to the fore as soon as Lorentz begins to consider moving bodies upon the hypothesis of a stationary ether.

While Lorentz was inclined to Fresnel's ether from an early date, it was not until his article of 1886, "De l'influence du mouvement de la terre sur les phénomènes lumineux," [97] that he definitely decided in its favor; his preference was based on a thorough examination of two competing theories of optical phenomena in a moving body, viz., Fresnel's and Stokes's theories of aberration.

In this article Lorentz first showed that Stokes's assumption that the motion of the ether is of such a kind that the velocity distribution has a potential is incompatible with another of his assumptions that the velocity of the ether relative to the surface of the earth should vanish everywhere. Because of this conclusion he rejected Stokes's theory.[98] Lorentz then examined the possibility of a theory which is intermediate between Fresnel's and Stokes's, one based on the assumption that there is a velocity potential but that the relative velocity between the ether and the surface of the earth does not vanish. At the same time, in this modified theory, the ether is dragged by a moving body as re-

---

[97] Lorentz, "De l'influence du mouvement de la terre sur les phénomènes lumineux," *Arch. néerl.*, *21* (1887), 103 [originally *Versl. Kon. Akad. Wetensch. Amst.*, *2* (1886), 297]. *Collected Papers, 4*, 153–214.

[98] Much later M. Planck suggested that this difficulty could be avoided if one would admit that the ether was compressible. But Lorentz rejected this possibility because an extremely large ratio of compression was required in order to reconcile the theory with observation. Cf. Lorentz, "La théorie de l'aberration de Stokes dans l'hypothèse d'un éther n'ayant pas partout la même densité," *Arch. néerl.*, *7* (1902), 81 [originally *Versl. Kon. Akad. Wetensch. Amst.*, *7* (1899), 523]; *Collected Papers, 4*, 245–251.

quired by Fresnel's coefficient. Such a theory is capable of explaining the negative results of experiments by G. B. Airy, M. Hoek, F. Arago, E. Mascart, and others which were intended to disclose the influence of the earth's motion on optical phenomena. But they are equally explained by the simpler theory of Fresnel. In addition, Lorentz says, there are other facts which suggest that the ether may penetrate freely through all ponderable bodies. The example he adduces is a barometric tube: when the tube is inclined until the mercury column fills the Torricellian vacuum, the ether that has originally occupied that space must have been pushed out through the walls of the tube. Hertz also alluded to a similar argument in his article on the electrodynamics of moving bodies; he argued that the independence of the ether from tangible matter "can scarcely be avoided in view of the fact that we cannot remove the ether from any closed space." [99] Lorentz will repeat the same argument in his work of 1895; there he will adduce, as one of "the well known grounds in favor of Fresnel's theory," the "impossibility of confining the ether between solid or fluid walls"; and he will refer there to the example of the barometer.[100] It appears that such an argument was common among physicists then. Anyway, Lorentz reached the conclusion in 1886 that the hypothesis of a stationary ether should be adopted.

According to this hypothesis, matter is perfectly transparent to the ether; the ether should remain at rest even when a material body moves through it. This means that the ether has no interaction with ponderable matter insofar as the latter is conceived to be merely a mechanical substance; the ether has ceased to be a kind of mechanical substance. However, the ether is not merely the medium of light; it is also a dynamical system which bears the electromagnetic actions. Thus Lorentz' stationary ether is a non-mechanical entity which is the equivalent of an independently existing electromagnetic field.

Now, in Lorentz' view, as soon as one starts to discuss electromagnetic phenomena in a material body, one immediately encounters a grave difficulty. It is the problem of "how indeed to form a precise idea of a body which, moving through the ether and consequently

---

[99] Hertz, *op. cit.* (note 91). *Gesammelte Werke, 2,* 257; *Electric Waves,* p. 242.

[100] Lorentz, *Versuch einer Theorie der electrischen und optischen Erscheinungen in bewegten Körpern* (Leiden, 1895); *Collected Papers, 5,* 1–138. The quotation is on p. 2. "Es lassen sich zu Gunsten der Fresnel'schen Theorie verschiedene wohlbekannte Gründe anführen. Vor allem die Unmöglichkeit, den Aether zwischen feste oder flüssige Wände einzusperren."

being traversed by this medium, is at the same time the seat of an electric current or a dielectric phenomenon." [101] In short, it is the problem of how to re-establish a relation between ponderable matter and the electromagnetic field once they have been separated. Ponderable matter and the ether or the electromagnetic field are entirely independent of each other, yet in electromagnetic phenomena there are apparently interactions between them. It may therefore be said that the question raised by Lorentz is equivalent to asking what the attribute of ponderable matter is which makes it capable of interacting with the electromagnetic field. The answer to this question lies in the fundamental assumption of the theory of electrons. Lorentz says:

> In order to surmount the difficulty, I endeavored to reduce, as far as possible, all the phenomena to a single one which is simplest of all. It is nothing else than the motion of an electrified body. . . . It will be sufficient, in these applications, to admit that all ponderable bodies contain a multitude of small particles which bear positive or negative charges, and that the electric phenomena are produced by the displacement of these particles. [102]

Once the picture consisting of the stationary ether and ponderable bodies containing charged particles has been established, both the mutual independence of, and the interactions between, matter and field become intelligible at the same time. Lorentz says:

> Hence it is seen that, in the new form which I will give to it, Maxwell's theory is brought nearer to the old ideas. . . . [Weber and Clausius, however, regarded the force between electrified particles as an instantaneous action at a distance.] On the contrary, the formulas we have obtained express, on the one hand, what kinds of changes of state are provoked in the ether by the presence and the motion of electrified corpuscles. On the other hand, they make known the force with which the ether acts on any of those particles. If this force depends on the motion of other particles, it is because their motion has modified the state of the ether. Thus, the value of the force at a given instant is not determined by the velocities and the accelerations

---

[101] Lorentz, *op. cit.* (note 81), 228. "Comment, en effet, se faire une idée précise d'un corps qui, se déplaçant au sein de l'éther et traversé par conséquent par ce milieu, est en même temps le siège d'un courant électrique ou d'un phénomène diélectrique?"

[102] *Ibid.* "Pour surmonter la difficulté, autant qu'il m'était possible, j'ai cherché à ramener tous les phénomènes à un seul, la plus simple de tous, et qui n'est autre chose que le movement d'un corps électrisé. . . . Il suffira, dans ces applications, d'admettre que tous les corps pondérables contiennent une multitude de petites particules à charges positives ou négatives et que les phénomènes électriques sont produits par le déplacement de ces particules."

of the small particles at the same instant. It is rather derived from the motion which already took place.[103]

Thus we can see the modern concept of the electromagnetic field definitively settled. Though Lorentz still uses the expression "a state of the ether" instead of "a state of the electromagnetic field," his ether is no longer a dielectric medium conceived in analogy with ponderable matter; it is an independent reality which supports all electromagnetic action. Apart from the name, it is equivalent to our notion of the electromagnetic field.

In Maxwell's theory, electromagnetic phenomena are borne by conductors and dielectric media including the ether. One did not meet the problem of understanding how a material body, perfectly transparent to the ether, could at the same time be the bearer of electric currents and dielectric phenomena; instead, two different kinds of reality, field and matter, are comprehended confusingly in the single notion of a dielectric medium. Maxwell avoids the question of the nature of charge, the source of the electromagnetic field; consequently, it is impossible, on the one hand, to attribute to matter, or, more precisely, to its constituent charged particles the function of being the source of the field, and, on the other, to grasp the electromagnetic field as an independent reality distinct from matter. By adopting the hypothesis of a stationary ether and the atomistic view of matter, Lorentz both separated the electromagnetic field from the matter and made it clear that matter does function as the source of the field.

Now, having determined that electromagnetic theory is concerned with systems consisting of the stationary ether and charged particles moving through it, it becomes necessary to find equations defining the interaction of charged particles and the electromagnetic field. For this purpose Lorentz adds two assumptions.[104] First, in the interior of a charged particle, the equation

$$\operatorname{div} \boldsymbol{D} = \rho \tag{51}$$

[103] *Ibid.*, 229. "On voit donc que, dans la nouvelle forme que je vais lui donner, la théorie de Maxwell se rapproche des anciennes idées. . . . Les formules, au contraire, auxquelles nous parviendrons expriment d'une part quels changements d'état sont provoqués dans l'éther par la présence et le mouvement de corpuscules électrisés; d'autre part, elles font connaître la force avec laquelle l'éther agit sur l'une quelconque de ces particules. Si cette force dépend du mouvement des autres particules, c'est que ce mouvement a modifié l'état de l'éther; aussi la valeur de la force, à un certain moment, n'est-elle pas déterminée par les vitesses et les accélérations que les petits corps possèdent à ce même instant; elle dérive plutôt des mouvements qui ont déjà eu lieu."

[104] *Ibid., Collected Papers, 2,* 230–231.

holds, where ρ, the charge density, has definite values for each point within the particle. Second, the density of electric current $\boldsymbol{C} = (u, v, w)$ is replaced by

$$\boldsymbol{C} = \rho\boldsymbol{v} + \frac{\partial \boldsymbol{D}}{\partial t}, \tag{52}$$

where $\boldsymbol{v}$ is the velocity of each point within the particle. Lorentz says that he has borrowed this latter hypothesis from Hertz, and he recalls Rowland's experiment and the theory of electrolytes as its justification. Lorentz supposes that the particle has finite dimensions; moreover, he assumes that it is a rigid body. In calculating the electromagnetic field produced by a moving charged particle, he makes an approximation by regarding the particle as a point charge. It is much later that Liénard and Wiechert found the potential which is called after them (in 1898 and 1900 respectively). The electromagnetic field produced by an electric charge is determined by equation (51) and

$$\text{rot } \boldsymbol{H} = 4\pi\left(\rho\boldsymbol{v} + \frac{\partial \boldsymbol{D}}{\partial t}\right), \tag{53}$$

which is obtained by substituting equation (52) into equation (41). Now it is necessary to obtain the force exerted on a charged particle by the electromagnetic field with the aid of the Lagrangian equation. Fixing attention on one particle, one considers a variation δx of the coordinate x of a point within the particle. Because of condition (51), this variation produces a variation of $\boldsymbol{D} = (f, g, h)$:

$$\delta f = -\rho\,\delta x, \quad \delta g = 0, \quad \delta h = 0.$$

Denoting the force which the electromagnetic field exerts on the particle by $\boldsymbol{F} = (X, Y, Z)$, one therefore obtains the virtual work

$$\delta A = -X\,\delta x + 4\pi c^2\,\delta x \int \rho f\, d\tau,$$

where the second term of the right-hand side represents the variation of the potential energy of the system

$$2\pi c^2 \int \boldsymbol{D}^2\, d\tau.$$

If, in the previous expression for $\delta' T$, $e_x = u'\, dt$, etc. is replaced by $\rho\,\delta x + \delta f$, etc. from equation (52) and the value $\delta f = -\rho\,\delta x$ is substituted, then $\delta' T$ becomes zero. Then from equation (52), the variation of each component of C is

$$\delta u = \left\{ -\frac{\partial(\rho v_x)}{\partial x} + \left(v_x \frac{\partial \rho}{\partial x} + v_y \frac{\partial \rho}{\partial y} + v_z \frac{\partial \rho}{\partial z} \right) \right\} \delta x,$$

$$\delta v = -\frac{\partial(\rho v_y)}{\partial x} \delta x,$$

$$\delta w = -\frac{\partial(\rho v_z)}{\partial x} \delta x.$$

Substituting these values into equation (44), one obtains, by an integration by parts,

$$\delta T = \delta x \int \rho (v_y B_z - v_z B_y) \, d\tau,$$

where use has been made of the fact that for the motion of a rigid body div $v = 0$. The x-component of the force F is obtained by substituting these values of δA, $\delta' T$, and δT into the Lagrangian equation (38). The y- and z-components too are obtained in the same way. The result is the so-called Lorentz force:

$$F = 4\pi c^2 \int \rho D \, d\tau + \int \rho [v \times B] \, d\tau. \tag{54}$$

Equations (51)–(54) thus obtained, together with the earlier equation,

$$-4\pi c^2 \operatorname{rot} D = \frac{\partial B}{\partial t}, \tag{55}$$

complete the system of fundamental equations of Lorentz' electromagnetic theory.

As a first application of equations (51)–(55), Lorentz showed that results of the usual theory, e.g., Coulomb's law, the expression for the force acting on a conductor in which an electric current is flowing, and the law of electromagnetic induction, are all equally deducible from these equations. He then derives the Lorentz-Lorenz formula by the method of the effective field which acts at the position of a molecule within a dielectric body; this is the method usually adopted in present-day textbooks.[105] For this purpose Lorentz considers separately actions from inside and from outside an imaginary spherical surface around the molecule; and he introduces, for the first

[105] For example: C. Kittel, *Introduction to Solid State Physics*, 2nd ed. (New York, 1956), pp. 157–163.

time, for the force exerted by the polarization charge induced on the surface of the sphere, the so-called Lorentz local field $\dfrac{4\pi c^2}{3}$ $\boldsymbol{P}$.[106]

Lorentz next proceeds to the propagation of electromagnetic waves in a material medium conceived as an extended ether within which a number of charged particles are found. The electromagnetic waves emitted by an oscillating electric dipole are calculated and the force exerted by the field of the electromagnetic waves upon each charged particle is obtained. The self-reaction of the field of a charged particle is obtained for the first time in this calculation.[107] Making use of these results, Lorentz obtains an equation which shows that the alteration of the polarization vector at each point propagates itself in the form of a transverse wave. Then, assuming a plane monochromatic form for the transverse wave, Lorentz could determine the ratio of the velocity of its propagation to the velocity of light in a vacuum, that is, the refractive index of the dielectric substance. The result gives the dispersion and the Lorentz-Lorenz formula.

II. ELECTROMAGNETIC PHENOMENA IN A MOVING BODY

The above results are merely re-derivations of known results; though more elaborated, they could not be said to be obtained solely by Lorentz' new approach. The true merit of the theory of electrons is only revealed in the last chapter of his 1892 article where Lorentz derived the Fresnel drag coefficient. He considered the propagation of the alteration of the polarization vector at each point in a medium of refractive index v in which all charged particles share a common translational velocity p; he showed that the velocity of propagation is $\dfrac{v}{v} + \left(1 - \dfrac{1}{v^2}\right)p$, agreeing with Fresnel's hypothesis. According to Fresnel, the coefficient $\left(1 - \dfrac{1}{v^2}\right)$ represents a real dragging of the ether by a moving body. The fact that it depends on the refractive index leads to the absurd conclusion that one and the same ether is dragged by different ratios for rays of different wavelengths. This problem was examined in 1880 on the basis of Maxwell's equations by J. J. Thomson, who showed that the alteration of the velocity

---

[106] Lorentz, *op. cit.* (note 81), 256–267, esp. 262.
[107] *Ibid.*, 319.

of propagation should always be half the velocity of the moving body.[108] Lorentz too avoided the absurdity, since the coefficient in his theory does not represent a real dragging of the ether, but a secondary effect produced by a system of moving particles. Lorentz claims that "according to our theory, the value $\left[1 - \dfrac{1}{v^2} \right]$ is applicable to every sort of homogeneous light if one only understands by v the refractive index which refers to that light." [109] Three years later in his *Versuch einer Theorie der electrischen und optischen Erscheinungen in bewegten Körpern*[110] he calculated the drag coefficient to a still higher approximation,

$$\left(1 - \frac{1}{n^2} \right) - \frac{1}{n} \lambda \frac{dn}{d\lambda}, \tag{56}$$

where he now writes n in place of v.

The influences of the motion of the earth on electrical and optical phenomena are not discussed in Lorentz' first article on electron theory. The precise Michelson-Morley experiments raised a serious problem in this context; Lorentz found a solution later in the same year, 1892. When he examined the aberration theories of Fresnel and Stokes in 1886, he was aware of Michelson's 1881 experiment, which was clearly unfavorable to the hypothesis of a stationary ether. Lorentz nevertheless adopted this hypothesis because he doubted the accuracy of Michelson's experiment; he pointed out that since the true difference of times taken by each of two light beams is half the value which Michelson had expected, the accuracy of the experiment was not sufficient to draw a definite conclusion.[111] But the experiment was repeated with far more accuracy in 1887 by Michelson and Morley, leaving no room for doubt. Nevertheless Lorentz continued to hold onto the stationary ether hypothesis. His first of two reasons for this is that the contradiction in Stokes's theory seemed fatal to that theory. He wrote in 1895 that "the difficulties which this theory encounters in the explanation of aberration seem to me so serious that I could not share this opinion [that Stokes's theory was right] without trying to

[108] J. J. Thomson, "On Maxwell's Theory of Light," *Phil. Mag.*, ser. 5, 9 (1880), 284–291.
[109] Lorentz, *op. cit.* (note 81); *Collected Papers*, 2, 319. "Remarquons encore que, d'après notre théorie, la valeur (158) est applicable à chaque espèce de lumière homogène, si seulement on entend par v l'indice de réfraction qui lui est propre."
[110] Lorentz, *op. cit.* (note 100), 96–102; *Collected Papers*, 5, 95–102.
[111] Lorentz, *op. cit.* (note 97); *Collected Papers*, 4, 214.

eliminate the contradiction between Fresnel's theory and Michelson's result." [112] The second reason was the success of his theory based on the hypothesis of a stationary ether. His derivation of Fresnel's drag coefficient appears especially to have enhanced his confidence in a stationary ether, encouraging him to seek a new solution of the problem of Michelson's experiment. This may also be inferred from Lorentz' own words; in the paper in which he proposed the contraction hypothesis, he wrote:

> Fresnel's conception, on the other hand, could furnish a satisfactory explanation of all phenomena considered, if one introduced for transparent ponderable substances the "dragging coefficient," as given by Fresnel, and for which I recently derived the expression from the electromagnetic theory of light. A serious difficulty however had arisen in an interference experiment made by Michelson in order to make a decision between the two theories. . . .
>
> . . . .
>
> This experiment has been puzzling me for a long time, and in the end I have been able to think of only one means of reconciling its result with Fresnel's theory.[113]

Lorentz' contraction hypothesis was communicated to the Royal Academy of Science at Amsterdam on 26 November 1892.[114] It seems that shortly before this Lorentz had obtained a result which he used in justifying the contraction hypothesis (the deduction of this result, however, was only given in his *Versuch* of 1895[115]). He considered a body S moving with uniform velocity p. And he considered a second system S' of material particles at rest with respect to the ether; he supposed that S' has the same constitution as S, but that it is elongated in the x-direction by the ratio $\sqrt{1 - p^2/V^2} : 1$; then he

---

[112] Lorentz, *op. cit.* (note 100), 121; *Collected Papers, 5,* 120–121. "Die Schwierigkeiten, auf welche diese Theorie bei der Erklärung der Aberration stösst, scheinen mir zu gross zu sein, als dass ich dieser Meinung sein konnte, und nicht vielmehr versuchen sollte, den Widerspruch zwischen der Fresnel'schen Theorie und dem Michelson'schen Ergebniss zu beseitigen."

[113] Lorentz, "De relative beweging van de aarde en den aether," *Versl. Kon. Akad. Wetensch. Amst., 1* (1892), 74; English version, *Collected Papers, 4,* 219–223. The quotation is on p. 219 and p. 221 of the latter.

[114] The contraction hypothesis had already been proposed by FitzGerald, but it was unknown to Lorentz when the latter presented his hypothesis to the Amsterdam Academy. Cf. note 2 on p. 121 of Lorentz' *Versuch* (*Collected Papers, 5*).

[115] Lorentz, *op. cit.* (note 100), 37; *Collected Papers, 5,* 36–37. Cf. my previous paper, "Electrodynamics before the Theory of Relativity, 1890–1905," *Jap. Stud. Hist. Sci.,* No. 5 (1966), pp. 1–49, esp. pp. 15–16.

showed that the electric force E' which acts in the system S' is related to the force E acting at the corresponding point in the original system S by the following equations:

$$E_x = E_x', \; E_y = \sqrt{1 - p^2/V^2}\, E_y', \; E_z = \sqrt{1 - p^2/V^2}\, E_z'. \quad (57)$$

Now Lorentz offers as his "only one means of reconciling" the result of Michelson's experiment with Fresnel's theory "the supposition that the line joining two points of a solid body, if at first parallel to the direction of the earth's motion, does not keep the same length when it is subsequently turned through 90°."[116] Explaining that the ratio $1 : 1 - p^2/2V^2$ (an approximation to $\sqrt{1 - p^2/V^2}$) is sufficient to account for the negative result of Michelson's experiment, he asserts that this hypothesis is not inconceivable. He contends that it is not far-fetched to suppose that the molecular forces, which evidently determine the size and shape of a solid body, are propagated through the ether like electric and magnetic forces. Then the same relations as equations (57) may hold between molecular forces in a rest body and in a contracted moving body constituted of the same molecules. Hence a system of molecules, which, when at rest, is in a state of equilibrium, will maintain its state of equilibrium if the system undergoes a contraction in the ratio of $1 : 1 - p^2/2V^2$.

The same argument is found in Lorentz' *Versuch* of 1895; however, here the approximate value $1 - p^2/2V^2$ is replaced by the exact one $\sqrt{1 - p^2/V^2}$. Lorentz also discussed many other problems concerning the influence of the earth's motion.[117] He demonstrated that, to a first-order approximation in $\dfrac{P}{V}$, the earth's motion does not mainfest itself in electrostatic phenomena; further, electromagnetic induction and ponderomotive actions between an electric current and an electrified body or between two currents are all invariant to the first order. He showed that optical phenomena, too, such as reflection, refraction, and interference are not affected by the motion of the earth.

What is most remarkable from an historical point of view is the significance of the *Versuch* for the development of fundamental notions in physics. The first consequence is that the concept of the electromagnetic field as an independent physical reality has been expressed in a manner suited to its content. The notations and operators

[116] Lorentz, *op. cit.* (note 113); *Collected Papers, 4,* 221.
[117] Cf. my previous paper, *op. cit.* (note 115), 14–18.

of vector analysis are fully employed and equations (51)–(55) are postulated from the outset as the fundamental equations. All recourse to the dynamical picture which Lorentz had employed in his article of 1892 has disappeared. Everything that was transitional and inessential has been eliminated.

The second historical significance of the *Versuch* is that it provided for the first time an atomistic foundation for the macroscopic Maxwellian equations. To deal with the propagation of light in a material medium, Lorentz tried to obtain macroscopic equations for a moving body. For this purpose he defined a quantity D by

$$D = \bar{d} + M, \tag{58}$$

where d is the intensity of the electric field (denoted by D in his previous article) and M is the electric moment per unit volume. The latter is given by

$$M = \frac{1}{I} \Sigma_I \, eq, \tag{59}$$

where e is the charge of a particle, and q is its displacement from the equilibrium position. A bar over a letter indicates an average taken over a small volume I. Assuming that M is proportional to the electric force acting at the point to which all the quantities are referred, $M = \chi E$. Neglecting terms multiplied by the square of M, the following equations are obtained:[118]

$$\operatorname{div} D = 0,$$
$$\operatorname{div} \bar{H} = 0,$$
$$\operatorname{rot} \bar{H}' = 4\pi \dot{D},$$
$$\operatorname{rot} \bar{E} = -\dot{H},$$
$$\epsilon \bar{E} = 4\pi c^2 D + [p \times \bar{H}],$$
$$\bar{H}' = \bar{H} - \frac{1}{c^2} [p \times \bar{E}],$$

where $\epsilon = 1 + 4\pi c^2 \chi$. It should be noted that these equations are referred to a coordinate system moving with the body, and that only non-magnetic substances are considered here.

The third historical interest of the *Versuch* is its formulation of a partial covariance of the electromagnetic theory. Lorentz formulated this covariance as a theorem of corresponding states. He introduced

---

[118] Lorentz, *op. cit.* (note 100), 76; *Collected Papers, 5,* 75.

the following variables:

$$t' = t - \frac{1}{c^2}(xp_x + yp_y + zp_z), \tag{60}$$

and

$$4\pi c^2 \boldsymbol{D}' = 4\pi c^2 \boldsymbol{D} + [\boldsymbol{p} \times \overline{\boldsymbol{H}}]. \tag{61}$$

The variable t' defined by (60) was employed in the article of 1892, but it was not stressed there. Its important role was first recognized in the *Versuch* and was given the name of "local time" (*Ortszeit*): "the variable t' can be considered as a time measured from an instant which depends on the position of the particle in question. This variable can therefore be called the *local time* of this particle in contrast to the *general time* t."[119] If one deduces the equations for $\boldsymbol{D}'$, $\overline{\boldsymbol{H}}'$, and $\overline{\boldsymbol{E}}$, with x, y, z, and t' as the independent variables, then, neglecting terms of higher order than the second in p/V, a set of equations having the same form as the macroscopic Maxwellian equations for a stationary body is obtained. From this result Lorentz arrived at the following theorem of corresponding states (Lorentz omitted the bars over letters):

If for a system of bodies at rest a state in which

$$D_x, D_y, D_z, E_x, E_y, E_z, H_x, H_y, H_z$$

are certain functions of x, y, z, t is known, then a state in which

$$D_x', D_y', D_z', E_x, E_y, E_z, H_x', H_y', H_z'$$

are the same functions of x, y, z, and t' [i.e. $t - \frac{1}{V^2}(p_x x + p_y y + p_z z)$]

can take place in the same system when it is in motion with velocity p.[120]

This might be viewed as a remarkable first step toward the theory of relativity. Lorentz, however, was not aware until much later that his

[119] *Ibid.*, 50; *Collected Papers*, 5, 50, "... die Variable t' als Zeit betrachtet werden kann, gerechnet von einem von der Lage des betreffenden Punktes abhängigen Augenblick an. Man kan[n] daher diese Variable die *Ortszeit* dieses Punktes, im Gegensatz zu der *allgemeinen Zeit t*, nennen."

[120] *Ibid.*, 85; *Collected Papers*, 5, 84. "*Ist nämlich für ein System ruhender Körper ein Bewegungszustand bekannt, bei dem*

$$D_x, D_y, D_z, E_x, E_y, E_z, H_x, H_y, H_z \tag{69}$$

gewisse Functionen von x, y, z, und t sind, so kann in demselben System, falls es sich mit der Geschwindigkeit p verschiebt, ein Bewegungszustand bestehen, bei welchem

$$D_x', D_y', D_z', E_x, E_y, E_z, H_x', H_y', H_z' \tag{70}$$

eben dieselben Functionen von x, y, z und t' [d.h. $t - \frac{1}{V^2}(p_x x + p_y y + p_z z)$] sind."

theorem of corresponding states necessitated a modification of the concepts of space and time; consequently he had to yield to Einstein the credit for conceiving of the theory of relativity.

12. CONCLUSION

According to Maxwell's original conception, the electromagnetic field is considered as a kind of mechanical state of dielectric media, and the vacuum, too, is a dielectric medium, or "ether." This differs from our present concept of the electromagnetic field. It was Lorentz' theory of electrons of 1892 that brought about that change. In Lorentz' theory, the physical world, insofar as it is comprehended by electrodynamics, is reduced to two kinds of constituents: charged particles and the stationary ether. The electromagnetic field is regarded as a dynamical state of the stationary ether, and it is deprived of all mechanical qualities. The field, being separated from ponderable matter, has become an independent physical reality. It should be noted that in the formative stage of Lorentz' conception, the atomistic view played a decisive role.

Lorentz began his scientific career with an electromagnetic theory of reflection and refraction of light, resolving the difficulties of the elastic-solid theory of the luminiferous ether and proving the superiority of the electromagnetic theory of light. In this first work Lorentz had already taken a step toward a separation of the roles of ether and molecules. In his second work he definitely settled the fundamental picture of the theory of electrons: the relevant features of the world are an all-pervading ether and charged material particles. At this stage Lorentz shared the doctrine of action at a distance with the followers of Continental electrodynamics. After a gradual elaboration of concepts he was converted to the contiguous-action theory, and in 1892 he reached his final notion of the electromagnetic field. The influence of Continental electrodynamics over Lorentz should not be considered negative only. For there was a necessary, not contingent, relation between the introduction of the atomistic view—the notion of a charged particle—and the clarification of the concept of the electromagnetic field as an independent physical reality; and the idea of a charged particle was inherited from Continental electrodynamics.

The fecundity of Lorentz' theory of electrons need not be reemphasized. In the problem of the influence of the earth's motion on optical and electromagnetic phenomena, Lorentz' theory achieved a great success. But the theory of electrons could not evolve into the theory of relativity. The immediate reason for this is that Lorentz did not think of modifying the time-honored concepts of space and time. The limitation of the theory of electrons has, however, deeper roots. Einstein's special theory of relativity stemmed from a theoretical consideration about the electrodynamics of moving bodies. And the electrodynamics of moving bodies, as the author has pointed out,[121] was suited for the development of the theory of relativity. Since the electrodynamics of moving bodies dealt exclusively with phenomena that took place in bodies moving relative to the earth, which, in turn, was in motion relative to the ether, it had nothing to do with what had been thought of as absolute frames of reference; it was therefore more likely to suggest the idea of covariancy. But Lorentz' theory was by no means an "electrodynamics of moving bodies" in the strict sense. The electrodynamics of moving bodies took essentially a macroscopic view of matter; its aim was to consider phenomena taking place in bodies in motion. By contrast, Lorentz' theory did not admit any macroscopic bodies at all; according to its point of view there was nothing to consider beyond a system of charged particles travelling through the stationary ether. Once one has decomposed moving bodies into the stationary ether and a flow of electrons and has regarded electromagnetic processes as taking place in a stationary ether, it would be difficult, indeed impossible, to conceive of the principle of relativity. Thus it may be said that the fundamental conception of the theory of electrons, which once played an irrefutably important role in advancing physical understanding, now constituted a conceptual obstacle for conceiving of the theory of relativity. This is the basic reason why Lorentz could not propound the theory of relativity, although he had reached, in 1904, a theory which was very nearly the mathematical equivalent of it.

As to the development of Lorentz' theory of electrons after his *Versuch* of 1895 until its final formulation of 1904, the reader is referred to the author's article on electrodynamics before the theory of relativity.[122]

The author is deeply grateful to Professor Russell McCormmach for revising his English.

[121] Hirosige, "A Consideration concerning the Origins of the Theory of Relativity," *Japanese Studies in the History of Science*, No. 4 (1965), pp. 117-123.
[122] Cf. Hirosige, *op. cit.* (note 115).

The Genesis of the Bohr Atom

BY JOHN L. HEILBRON* AND THOMAS S. KUHN**

INTRODUCTION

The following pages offer a reconstruction of a momentous episode in the history of science: Niels Bohr's journey from his doctoral thesis of 1911 to the composition, some two years later, of his famous three-part paper, "On the Constitution of Atoms and Molecules." Parts of this story have been told before, most notably by Léon Rosenfeld, who has published and interpreted the most important of the relevant manuscripts.[1] Informed by its author's long acquaintance with Bohr, Professor Rosenfeld's imaginative and scholarly account will remain an essential source for students of the development of modern physics. Recent writers, however, working principally from published records to which Rosenfeld attached little weight, have suggested the need for significant modifications in his account, particularly in respect to the importance for Bohr of the work of J. W. Nicholson.[2] As a result, though

* Department of History, University of California, Berkeley, California 94720.
** Program in History and Philosophy of Science, Princeton University, Princeton, New Jersey 08540.

[1] L. Rosenfeld, "Introduction" to *On the Constitution of Atoms and Molecules* (Copenhagen, 1963), a reprinting of Bohr's three papers of 1913, hereafter cited as "Rosenfeld." See also L. Rosenfeld and E. Rüdinger, "The Decisive Years, 1911–1918," in S. Rozental, ed., *Niels Bohr: His Life and Work as seen by his Friends and Colleagues* (Amsterdam and New York, 1967), 38–73.

[2] E.g., T. Hirosige and S. Nisio, "Formation of Bohr's Theory of Atomic Constitution," *Jap. Studies Hist. Sci.*, No. 3 (1964), 6–28; J. L. Heilbron, *A History of Atomic Models from the Discovery of the Electron to the Beginnings of Quantum Mechanics,* diss. (University of California, Berkeley, 1964); R. McCormmach, "The Atomic Theory of John William Nicholson," *Arch. Hist. Exact Sci., 3* (1966), 160–184. The older literature (e.g., C. E. Behrens, "Atomic Theory from 1904 to 1913," *Am. J. Phys., 11* [1943], 60–66, "The Early Development of the Bohr Atom," *ibid.,* 135–147, and "Further Developments of Bohr's Early Atomic Theory," *ibid.,* 272–281; E. T. Whittaker, *A History of the Theories of Aether and Electricity. II. The Modern Theories, 1900–1926* [London, 1953]; and L. S. Polak, "Die Entstehung der Quantentheorie des Atoms (Das Rutherford-Borsche Atommodell)," *Sowjetische Beiträge zur Geschichte der Naturwissenschaft* [Berlin, 1960], 226–242), since its authors were necessarily ignorant of Rosenfeld's account, is not useful for a reconstruction of Bohr's *path*, though it can help to place his work in historical context.

the existing secondary literature is rich in suggestions and documentation concerning particular aspects of Bohr's route to the quantized atom, there exists as yet no treatment of the subject that is at once comprehensive and plausible. Our aim in this paper is to fill that gap, partly by a critical synthesis of suggestions in the existing literature, and partly by an elaboration of some central but previously neglected strands in Bohr's scientific development.

A cursory reading of Bohr's thesis reveals that, when he finished it early in 1911, he was already convinced that some fundamental break with classical physical theory, probably some form of Planck's quantum theory, would be necessary to resolve specific problems in the electron theory of metals. Very likely, as many have noted, that conviction was an essential prelude to Bohr's attempt, begun within a year and a half of his thesis defense, to quantize Rutherford's atom. But, if the prelude prepares the attempt, it explains neither its origin nor its nature. A concern with detailed atom models was not widespread in the years before World War I; Bohr, in any case, does not mention them in his thesis. Nor are they discussed before June 1912 in any of the many extant letters and manuscripts he wrote while a postdoctoral fellow in Cambridge and Manchester. Yet, within six weeks of the earliest surviving sign of his concern with models, Bohr had produced a quantized version of Rutherford's atom and had applied it to several problems. What suddenly turned his attention from electron theory to atom models during June 1912? Why did he then choose to develop the new, little-known Rutherford atom rather than, say, the older, more successful model proposed by J. J. Thomson? Why did he approach the quantization problem in the particular way he did, one which bore impressive fruits at once and which, a year later, began to revolutionize physics?

We are persuaded that the answer to these and to similar questions lies not in the general conviction of the need for quantum theory which Bohr drew from his thesis research, but rather in certain *specific* problems with which he busied himself until almost the end of his year in England. They helped direct his reading and uniquely prepared him to recognize the special potential of the nuclear atom. The first three sections below detail these problems, suggest how they focussed Bohr's attention on the question of atomic structure, and analyze the way they combined with other factors in the formulation of Bohr's first quantized model of the atom. That model is not, however, the one for which Bohr is known: it possessed only a single stationary state and was neither built around nor applied to the problems of atomic spectra. Sections IV and V trace the development of Bohr's radical conception of stationary states, of his concern with spectra, and of his derivations of the Balmer formula together with the first hints of a Correspondence Principle. These are Bohr's most famous contributions, and they are contained entirely in the first part of his famous trilogy, supplemented by a lecture he gave to the Danish Physical Society in December 1913. Less familiar now, but then of equal concern to Bohr, were his speculations about the stable electronic configurations of polyelectronic atoms and molecules which occupy the last two parts of the trilogy. Section VI outlines these speculations and indicates certain difficulties in Bohr's formulation which were soon to challenge those who built upon his principles.

Our reconstruction rests mainly upon the printed record and the Bohr scientific correspondence and manuscripts in the Archive for History of Quantum Physics.[3] In addition, we have been allowed to see portions of the personal papers of the Bohr family. We deeply appreciate the kindness of Professor Aage Bohr, of Mrs. Margrethe Bohr, and of Professors J. Rud Nielsen and Rosenfeld in making this personal correspondence available to us: it has helped us to understand Bohr's transition from cultivator of the electron theory of metals to developer of the nuclear atom. To the Bohr family we are grateful as well for permission to reproduce the many quotations which follow from Bohr's unpublished correspondence and manuscripts. We also wish to thank Paul Forman, Tetu Hirosige, Martin Klein, Russell McCormmach, Malcolm Parkinson and Léon Rosenfeld for their helpful comments on the final draft of our manuscript.

I. STUDIES ON THE ELECTRON THEORY OF METALS:
   COPENHAGEN, 1911

On 13 May 1911, Niels Bohr successfully defended his doctoral thesis, *Studier over metallernes elektrontheori*, before the philosophical faculty of the University of Copenhagen. The essay had grown out of his master's dissertation, completed in 1909, and so represented the fruit

---

[3] For a description, see T. S. Kuhn, J. L. Heilbron, P. Forman, and L. Allen, *Sources for History of Quantum Physics. An Inventory and Report* (Philadelphia, 1967).

of over two years of concentrated effort. It is thorough and erudite; every page displays its author's critical power, his mathematical suppleness, and his firm grasp of physical principles. His "opponents," who, perhaps, were not entirely competent to judge his work, found little to question. The "defense," which might have lasted six hours, was over in a record-making ninety minutes.[4]

The theory Bohr took as his subject traced conductivity and other metallic properties to "free" electrons, i.e., to charged particles unattached to the molecules making up the metals. Though its roots extended deep into the nineteenth century, in the work of Ampère and of Weber, its major development had occurred in the decade following the experimental isolation of the electron. Then it became one of the most exciting, promising, and popular branches of physics. Its power was increased very quickly, for the elaborate techniques of Maxwell and Boltzmann stood ready for application to the electron "gas." In 1905 H. A. Lorentz published the first systematic development of the statistical mechanics of free electrons.[5]

Lorentz' theory rested on two principles: (1) that in the absence of external fields or temperature differences the electron gas and the stationary metal molecules remain in mechanical heat equilibrium, and (2) that, whether subject or not to external forces, the molecules act isotropically on the electrons. These principles in themselves of course did not enable one to calculate; special assumptions about the interactions of the particles were also required. Here Lorentz simplified. He supposed that the electrons bounced elastically off the metal molecules like billiard balls and that the effects of collisions between electrons were negligible, assumptions which, in Bohr's words, could "scarcely hold even approximately in real metals."[6] Lorentz' principles provided the point of departure, and his special assumptions the challenge, of Bohr's thesis. "The goal I have set myself," he wrote, "is to attempt to carry out calculations for the different phenomena which are explained by the presence of free electrons in metals, in the most general manner possible consistent with the principles of Lorentz' theory."[7]

[4] N. Bohr, *Studier over metallernes elektrontheori* (Copenhagen, 1911); S. Rozental, *Niels Bohr*, 36–37.
[5] H. A. Lorentz, "The Motion of Electrons in Metallic Bodies," *Proc. Amst. Acad.*, 7:2 (1905), 438–453, 585–593, 684–691.
[6] Bohr, *Studier*, 4. Though we have profited from other translations where they exist, this and all subsequent ones are our own.
[7] *Ibid.*

Now, as Bohr observed, the first of these principles, that of heat equilibrium, implied that the electrons and the molecules interacted according to the usual laws of mechanics, that their motions always satisfied Hamilton's equations. In admitting the principle for *free* electrons, however, he emphasized that it was an assumption of very limited applicability:

> The assumption [of mechanical interaction] is not *a priori* self-evident, for one must assume that there are forces in nature of a kind completely different from the usual mechanical sort; for while on the one hand the kinetic theory of gases has obtained extraordinary results by assuming the forces between individual molecules to be mechanical, there are on the other hand many properties of bodies impossible to explain if one assumes that the forces which act within the individual molecules (which according to the ordinary view consist of systems containing a large number of "bound" electrons) are mechanical also. Several examples of this, for instance calculations of heat capacity and of the radiation law for high frequencies, are well-known; we shall encounter another later, in our discussion of magnetism.[8]

This most interesting passage, which expresses unambiguously its author's conviction of the ultimate incompetence of "the ordinary mechanics" in atomic theory, sets the tone of the thesis. When combined with different special assumptions, Lorentz' principles often gave conflicting results, none of which agreed more than approximately with experiment. By developing the electron theory in the widest generality, Bohr intended to separate the difficulties dependent on each author's special assumptions from those which were fundamental. The former he could diminish or eliminate with other special assumptions. The latter might, though Bohr did not make the hope explicit, furnish the starting point for the necessary revisions of classical mechanics.

The isotropy of molecular action independent of the presence of an external field, Lorentz' second principle, distinguished Bohr's approach from theories like J. J. Thomson's, in which the fields, by acting directly upon bound electrons, disturbed the original spatial symmetry of the molecules.[9] Bohr could have had little confidence in any theory built upon assumptions about the response of bound electrons

---

[8] *Ibid.*, 5.
[9] J. J. Thomson, *The Corpuscular Theory of Matter* (London, 1907), 86; cf. W. Sutherland, "The Electric Origin of Rigidity and Consequences," *Phil. Mag.*, *1* (1904), 417–444, esp. 423–435.

to external fields. In the one case in which he did examine such a response, magnetism, he found that ordinary mechanical principles led to absurd results, and he supposed a similar breakdown was responsible for the well-known paradoxes of blackbody radiation. The assumption of isotropy allowed him generally to disregard the conditions within molecules. It also offered a great mathematical convenience, as he could assume, for example, that the molecules acted on the electrons either continuously, with a force inversely proportional to the rth power of the distance, or in separate collisions, as in the elementary kinetic theory of gases. Bohr in fact performed most of his calculations in parallel, first using separate collisions, as had Lorentz, and then the rth power law.

Bohr's most interesting positive results concerned the famous x, the ratio between a metal's thermal and electrical conductivities, and an obscure phenomenon, the transverse temperature difference ΔT accompanying the Hall effect. Lorentz had obtained the wrong sign for ΔT and a number for x about 40 percent less than that observed. Bohr's computations based on the rth power law, however, agreed quite satisfactorily with experiment in *both* these particulars if r were set equal to three. That two such disparate phenomena should thus be brought into harmony with experience and with one another Bohr thought "remarkable." [10] Part of his enthusiasm over this coincidence may have derived, as we shall see, from a model Thomson had proposed just before this time for quite another purpose, a model which likewise required a force varying as the inverse cube.

The problem of heat radiation also claimed Bohr's careful attention. A complete account of the conduction electrons had to include both the electromagnetic waves which, according to Maxwell's electrodynamics, they emit in their violent stops and starts, and also the accelerations they suffer from any radiation to which they are exposed. In a state of steady motion the amount of energy they absorb must equal the amount they radiate, the precise quantities involved depending on the wavelength λ of the radiation and the temperature T of the steady state. Using very simple assumptions about the interactions between free electrons and metal molecules, Lorentz had derived an expression for the ratio of the coefficients of absorption, $\alpha(\lambda, T)$, to those of emission, $\epsilon(\lambda, T)$, valid for long waves.[11] This ratio was of very great

[10] Bohr, *Studier*, 57–58, 116–117.
[11] H. A. Lorentz, "On the Emission and Absorption by Metals of Rays of Heat of Great Wavelengths," *Proc. Amst. Acad.*, 5 (1903), 666–685.

interest for, as Kirchhoff had showed, it should be a universal function of λ and T, quite independent of the nature of the emitting and absorbing body. It is in fact very simply related to the energy density of blackbody radiation.

Planck, of course, had felt constrained to introduce the quantum precisely in order to obtain a formula for this energy density good for all values of λ and T. The theoretician of 1911 had thus to adopt one of two conflicting views about heat radiation. If, on the one hand, he believed that a procedure like Planck's was necessary, and that it was wholly irreconcilable with the ordinary principles of electrodynamics, it would follow from the connection between the blackbody spectrum and ϵ/α that no expression for that ratio valid for all λ and T could be deduced from the electron theory of metals. On the other hand, since Lorentz' expression agreed very nicely with experiment for long waves, our theorist might hope that, by postulating an appropriate *mechanical* interaction between electrons and metal molecules, he could extend Lorentz' computations to shorter wavelengths and explain the blackbody spectrum without recourse to Planck's questionable procedure. Thomson and other Cambridge physicists had developed the latter position.[12] Bohr, as we have seen, embraced the former.

He was led to this conviction, which was still a minority view early in 1911, on two grounds.[13] The first, and probably to him the more persuasive, derived from his own lengthy calculations confirming that Lorentz' expression for ϵ/α for long waves also followed from more general principles than those Lorentz had employed. Bohr's familiarity with the computations and his fine feel for valid approximations convinced him that the apparent success of Thomson's program rested on a flawed assumption about the dependence of absorption on frequency. This conclusion he reinforced with an appeal to the law Rayleigh and Jeans had deduced by applying the principle of energy equipartition to the electromagnetic aether conceived as a mechanical system. The Rayleigh-Jeans law, whose consequences Thomson's school, including Jeans himself, had sought to avoid,[14] agreed with Lorentz'

---

[12] J. J. Thomson, "On the Electrical Origin of the Radiation from Hot Bodies," *Phil. Mag., 14* (1907), 217–231; "On the Theory of Radiation," *ibid., 20* (1910), 238–247; J. H. Jeans, "The Motion of Electrons in Solids. I," *ibid., 17* (1909), 773–794. Jeans works out examples only for large wavelengths.

[13] Bohr, *Studier, 77,* 103–104.

[14] They argued, for example, that the flow of energy into the higher-frequency vibrations of the field takes so long that heat equilibrium is not established in the experiments on blackbody radiation. Cf. J. H. Jeans, "Temperature-Radiation and the Partition of Energy in Continuous Media," *Phil. Mag., 17* (1909), 229–254.

expression and with experiment for long waves, but failed hopelessly at the other end of the spectrum. Therefore, without applying his generalized Lorentzian procedures to a direct calculation of the *short-wave* limit of ϵ/α, Bohr concluded that all such computations must inevitably fail. In addition, he localized the difficulty: "The cause of failure is very likely this, that the electromagnetic theory does not agree with the real conditions in matter and can only give correct results if it is applied to a large number of electrons (as in ordinary bodies) or to determine the average velocity of a single electron in a comparatively long time (such as in the calculation of the motion of cathode rays), but cannot [as in short-wave radiation], be used to investigate the motion of a single electron in a short time." [15]

Closely linked to this consequence of electrodynamics, which placed the description of the motions of atomic electrons beyond the reach of the ordinary theory, was a more disconcerting failure, whose detection Bohr considered one of the chief results of his thesis. Contemporary theory traced dia- and paramagnetism to the modification of electronic trajectories occasioned by an external field. Bohr found that, if the ordinary mechanics held, neither the free nor the bound electrons contributed at all to the magnetic properties of matter, or at least not to diamagnetism. This question, because it became a *leitmotiv* of Bohr's early post-doctoral work, deserves our close attention.

In 1900 Thomson had suggested that the free electrons might account for diamagnetism, since an external magnetic field would deflect them into circular or helical orbits so oriented as to produce a moment opposed to the sense of the inducing force. [16] But, as Bohr observed, the external field, which does no work, does not alter the distribution of electrons in space or in velocity, and thus cannot create a net moment where none had previously existed. (This argument, which requires some attention to the motions of electrons close to the surface, was apparently advanced simultaneously and independently by Lorentz. [17]) Since the free electrons thus failed to perform magnetically, Bohr turned to the bound, though their consideration lay beyond the

---

[15] Bohr, *Studier*, 103.

[16] J. J. Thomson, "Indications relatives à la constitution de la matière fournies par les recherches récentes sur le passage de l'électricité à travers les gaz," *Congrès international de physique. Rapports* (Paris, 1900–01), *3*, 138–151.

[17] According to H. J. van Leeuwen, "Problèmes de la théorie électronique du magnétisme," *Journal de physique et le radium*, 2 (1921), 361–377, on 376. See J. H. van Vleck, *The Theory of Electric and Magnetic Susceptibilities* (Oxford, 1932), 101–102, for the argument.

professed scope of his thesis. Here the situation was more promising. A theory of the magnetic properties of bound electrons, which Pierre Langevin had proposed in 1905, appeared an adequate explanation of para- and diamagnetism to most physicists of the time.[18]

In Langevin's theory diamagnetism arises from alterations in the velocities of atomic electrons introduced by the external field, while paramagnetism derives from changes in orientation of the electronic orbits. Both effects occur simultaneously. The crux of Langevin's theory is the tacit assumption that, if the orbits are initially so deployed within a given molecule as to yield no average molecular angular momentum and hence no net molecular moment, a field subsequently superposed will not disarrange or reorient the orbits; it will only affect the electrons' velocities, and diamagnetism will result. If, however, the molecule originally possessed a net moment, Langevin supposed that its orientation under the field would obliterate the diamagnetic effect also present, and paramagnetism develop.

We know from Larmor's theorem that the motions of the orbiting molecular electrons under a magnetic field will be precisely the same as their motions in its absence except for a common precession about an axis parallel to it and passing through the molecule's center.[19] The magnitude of the precession is $\omega_L = eH/2mc$, where H, c, e, and m represent, respectively, the magnetic field, the speed of light, and the charge and mass of the electron; the precession is so directed as to increase the velocities of electrons whose magnetic moments oppose, and to decrease the velocities of electrons whose moments parallel the sense of the field. The accelerating force is the electric field necessarily linked with the establishment of a magnetic one; the latter, of course, does no work on an orbiting charged particle. To determine the size of the diamagnetic effect Langevin proceeded roughly as follows. Taking the z axis along H, the z component of the total angular momentum of a molecule is

$$L_z = \Sigma m\rho_i^2\dot{\phi}_i, \tag{1}$$

where ρ and ϕ form with z a cylindrical coordinate system with origin at the molecule's center. By Larmor's theorem, $\dot{\phi}_i = \dot{\phi}_{i0} + \omega_L$ and

[18] P. Langevin, "Magnétisme et la théorie des électrons," *Ann. de chimie et de physique*, 5 (1905), 70–127, esp. 73–97.
[19] The field must be applied adiabatically for this statement to be entirely true. See Van Vleck, *op. cit.* (note 17), 23.

$\rho_i = \rho_{i0}$, where the subscript $_0$ refers to the motion before the application of the field. Averaging equation (1) over time,

$$\bar{L}_z = \Sigma(\overline{m\rho_{i0}{}^2\dot\phi_{i0}}) + \Sigma\overline{m\rho_{i0}{}^2\,\omega_L} = \Sigma\overline{m\rho_{i0}{}^2\,\omega_L}, \tag{2}$$

because, by hypothesis, the molecule originally possessed no average angular momentum. Associated with $\bar{L}_z$ is a molecular magnetic moment $\bar{M}_z = -(e/2mc)\bar{L}_z$. Hence the diamagnetic susceptibility, defined as $(N/H)\bar{M}_z$, N being the number of molecules in unit volume, is

$$D \equiv (N/H)\bar{M}_z = -(e^2/4mc^2)N\Sigma\overline{\rho_{i0}{}^2}. \tag{3}$$

Equation (3) agreed approximately with experiment with acceptable values for $\Sigma\overline{\rho_{i0}{}^2}$.

To obtain an expression for the paramagnetic susceptibility P Langevin assumed that all the molecules possessed the same net moment p in the absence of the field H. In the presence of the field each molecule would have a "magnetic energy" $-pH\cos\theta_j$, where θ_j is the angle between $\mathbf{H}$ and the direction of the moment of the jth molecule.[20] If nothing opposed the orientation of the molecules, they would all line up with $\mathbf{H}$, in conflict with experimental results. Langevin, assuming that thermal agitation opposed their alignment, utilized the usual techniques of statistical mechanics to compute an average value for the paramagnetic susceptibility:

$$P = \frac{N\Sigma pH\cos\theta_j \exp\left(\dfrac{pH\cos\theta_j}{kT}\right)}{H\Sigma \exp\left(\dfrac{pH\cos\theta_j}{kT}\right)} \doteq \frac{Np^2}{3kT}. \tag{4}$$

[20] By "magnetic energy" is meant the increment ΔT in the electron's kinetic energy consequent on application of the field. Let the direction of the electron's original angular momentum $\mathbf{L}$ make an angle α with $\mathbf{H}$; $\Delta T = L\omega_L \cos\alpha = (eH/2mc)L\cos\alpha$, if we ignore a term of the order of H^2. This energy may also be computed using the convenient fiction that a circulating electron produces at great distances a magnetic field equivalent to that of a small magnet placed normal to its orbit, and possessing a moment $\mathbf{M} = -(e/2mc)\mathbf{L}$. Such a magnet would have energy $-\mathbf{H}\cdot\mathbf{M} = +(e/2mc)\mathbf{H}\cdot\mathbf{L} = (eH/2mc)L\cos\alpha = \Delta T$. This second approach is however misleading in that the magnetic field does not exert a torque on the orbit as it would on a magnet. (The accelerating force producing the Larmor precession is the electric field linked to the establishment of the magnetic.) The angle α must be altered by some extraneous influence, say a collision; then, because the magnetic flux through the orbit changes, a transient electric field is set up which alters the electron's kinetic energy by an amount precisely suitable to the new orientation of its orbit. An electron whose $\mathbf{L}$ originally lies along $\mathbf{H}$, and whose $\mathbf{M}$ therefore directly opposes it, will enjoy the greatest ΔT, or "magnetic energy." But as collisions alter the orbit's orientation, the transient electric fields decelerate the electrons until, when $\mathbf{L}$ opposes and $\mathbf{M}$ parallels $\mathbf{H}$, it reaches the "stable" position of minimum energy.

(The summations are evaluated as integrals over the solid angle $2\pi \sin \theta \, d\theta$ in the approximation $pH \ll kT$.) Of course the moments p undergo a diamagnetic change also, but in Langevin's theory one ignores the consequent alteration in P. Equation (4) was perhaps the most persuasive feature of Langevin's theory, for it agreed precisely with the temperature dependence of the paramagnetic susceptibility which Pierre Langevin had found in 1895.[21]

Unfortunately Langevin's theory is incompatible with a literal application of the principles of statistical mechanics. Bohr was the first to recognize this incompatibility, at least in respect to diamagnetism. In an extended footnote to his illuminating discussion of the failure of the free electrons to provide a metal with diamagnetic properties, he remarked that the bound electrons are likewise impotent for, if mechanical heat equilibrium is to prevail, the change of velocity introduced during the build-up of the magnetic field must quickly equalize after the field is established. If there is the "least little energy exchange" between the electrons within a molecule then, according to Bohr, Langevin's theory of diamagnetism falls to the ground. Of course, one might appeal to an agency, like the emission of radiation, which could prevent mechanical heat equilibrium from establishing itself among the bound electrons, but even so diamagnetism would remain an enigma: as Bohr observed, Thomson and Gans had on that assumption obtained only a paramagnetic effect.[22] To save Langevin's diamagnetic theory one must suppose (though Bohr does not make this consequence explicit) some non-mechanical freezing of the velocity changes derived from the field, some principle that prevents the equalization demanded by the mechanical theory of heat.

Bohr's compressed discussion of these difficulties is not complete. As noted earlier, one cannot ignore the effects of the established field because the changing magnetic flux through the electronic orbits does do work when the orbits are shifted by collisions. Such shifts, there-

---

[21] P. Curie, "Propriétés magnétiques des corps à diverses températures," *Ann. de chem. et de phys.*, 5 (1895), 289–405.

[22] Bohr, *Studier*, 103, 108; J. J. Thomson, "The Magnetic Properties of Corpuscles Describing Circular Orbits," *Phil. Mag.*, 6 (1903), 673–693; R. Gans, "Zur Elektronentheorie des Ferromagnetismus," *Gött. Nachr.* (1910), 197–273, esp. 213–230; cf. *supra*, note 20. (Incidentally Thomson, *op. cit.*, 687–688, concluded that a collection of parallel, *non-interacting* electron rings would produce no diamagnetic effect, an error arising, as Bohr, following Langevin, observed [*Studier*, 108], from neglect of the electric force accompanying the establishment of the magnetic field. Thomson's paper was nevertheless of very great importance for the theories of radiation and atomic structure.)

fore, have diamagnetic as well as paramagnetic consequences, as they change both the magnitude and the orientation of the electronic moments. In fact, these alterations just compensate one another when the angular moments of the electrons are considered statistical quantities. (Bohr may have seen, but did not describe this effect.) It is not legitimate to require, as Langevin had done, that these moments p remain sensibly constant (excepting the "small" diamagnetic effect) during the establishment of the field and the reorientation of orbits. The ordinary statistical mechanics does not restrict the values of the moments; the p under the summation sign in equation (4) should itself be averaged, and when it is D and P exactly cancel.[23]

Though his criticism of Langevin's theory is incomplete and occupies but a small portion of his thesis, Bohr clearly considered his discovery of the impotence of the ordinary mechanics in respect to magnetism one of his most important results. He refers to it in two prominent places: among the theses, or original propositions, at the end of the *Studier,* and in the Introduction, where he juxtaposes it to a recognized enigma, the ultra-violet catastrophe.[24] Indeed, Bohr thought the two difficulties intimately related, apparently because absurd results followed for radiation as well as for magnetism *if one supposed mechanical heat equilibrium to prevail* over the processes producing those phenomena. At the very least the difficulty with magnetism strengthened and confirmed Bohr's conviction that the usual mechanical laws broke down when applied to rapidly moving electrons; and, even more than the radiation problem, it isolated the breakdown in the behavior of electrons bound into atoms. We conjecture that it did much more. It left Bohr with a specific, important, and easily conceived problem, an enigma of his own creation, whose study and eventual solution promised a clue to the fundamental revisions of contemporary theory he thought inevitable. He addressed himself to this great matter from the moment he finished his thesis, if not before.

---

[23] This argument is essentially that first given by Miss Leeuwen, *op. cit.* (note 17), 373–375; cf. Van Vleck, *op. cit.* (note 17), 94–100. Miss Leeuwen did not know Bohr's thesis until after she had completed her own, of which the paper cited is an abstract.

[24] *Studier,* 5, 108, 120; cf. *supra,* 215. The magnetic difficulty also figures prominently in a resumé of his results which Bohr read the Cambridge Philosophical Society in November 1911 (*infra,* 228). Towards the end of his life he mentioned it first among examples of the fruits of his thesis and of the "enormous problems" he was then engaged with; Interviews I, 1, 5. (The roman numeral refers to the number, the arabic to the page, of the transcripts of the five interviews with Bohr conducted by the project "Sources for History of Quantum Physics" in the fall of 1962. For details see Kuhn *et al., Sources* [note 3, *supra*].)

Though the spectacular results he obtained two years later came largely from an unexpected quarter and no doubt greatly exceeded his expectations, the problem—or rather *his* problem—of magnetism had then already had an important influence on the direction of his thoughts. It had focussed his attention on the question of bound electrons, which would ultimately become, for him, the problem of atomic structure. And it had led him to consider the nature of the non-mechanical "law" which, perhaps in the form of a restriction like Planck's quantization rule, might fix the motions of the bound electrons in the manner Langevin's theory tacitly made essential to a successful derivation of the Curie law.

II. CONTINUATION OF THE ELECTRON THEORY: CAMBRIDGE AND MANCHESTER, SEPTEMBER 1911–JUNE 1912

Just as, for Bohr, the master's dissertation entailed the doctoral thesis, the latter inexorably led to a year abroad. He had two obvious alternatives: Leiden with Lorentz, or Cambridge with Thomson. He did not hesitate. His father had instilled in him a love for things English, and his wide reading had given him the greatest respect for Thomson, a pioneer of the electron theory of metals and the acknowledged world master in the design of atomic models. Cambridge also boasted Larmor and Jeans, whose work touched Bohr's in many places. "I considered first of all Cambridge as the center of physics," Bohr later said of his decision to study there, "and Thomson as a most wonderful man . . . , a genius who showed the way for everybody." [25]

In September 1911, Bohr reached his Mecca. "I found myself rejoicing this morning," he wrote his fiancée just after his arrival, "when I stood outside a shop and by chance happened to read the address 'Cambridge' over the door." [26] Thomson received him politely, and promised to read his thesis, of which he had brought a rough English translation. The promise completed Bohr's happiness. "I have just talked to J. J. Thomson," he wrote his brother, "and I explained to him as well as I could my views on radiation, magnetism, etc. You should know what it was for me to talk to such a man. He was so very kind to me; we talked about so many things; and I think he thought

---

[25] Interviews II, 6.
[26] NB to Margrethe Nørlund, 26 Sept 1911, quoted without date by Rosenfeld and Rüdinger, *op. cit.* (note 1), 39–40. Prof. Rosenfeld has kindly supplied the date.

there was something in what I said. He has promised to read my thesis, and he invited me to have dinner with him next Sunday at Trinity College, when he will talk to me about it. . . . I can't tell you how happy and thankful I am that [the translation of] my dissertation was finished and I could give it to Thomson." [27]

Unfortunately the exchange of views Bohr desired did not take place. Thomson probably never read through his enthusiastic visitor's thesis. In subsequent interviews he occasionally promised to examine a particular question, but only one, heat radiation, gave rise to any discussion.[28] That conversation came to nothing, as we may gather from Bohr's report of it to his brother, written six weeks after his arrival in Cambridge:

> [Thomson] has not yet had time to read my thesis, and I still don't know if he will agree with my criticisms. He has only chatted with me about it a few times for a couple of minutes, and that was on a single point, my criticism of his calculation of the absorption of heat rays. You may remember I remarked that in his calculation of absorption (as opposed to emission) he does not reckon the time taken up in the collisions, and that therefore he gets a value for the ratio of emission to absorption which is the wrong order of magnitude in the case of high frequencies. Thomson said that he could not see that the collision time could have so great an influence on the absorption; I tried to explain, and the following day gave him a very simple example (an example corresponding to his calculation of the emission), which showed it very clearly. Since then I've talked with him for only a moment, and that a week ago. I think he thinks my calculation is correct, but I'm not sure he doesn't believe that one can design a mechanical model to explain the law of heat radiation on the usual electromagnetic principles. . . .[29]

Bohr was never able to effect Thomson's conversion, and thus failed of one of the chief objectives of his Cambridge sojourn.[30]

[27] NB to Harald Bohr, 29 Sept 1911 (BPC). ("BPC" signifies Bohr's Personal Correspondence, for which see "Introduction," above.)
[28] NB to S. B. McLaren, 17 Dec 1911; to C. W. Oseen, 1 Dec 1911 (BSC). (The notation "BSC" refers to the Bohr Scientific Correspondence in the Archive for History of Quantum Physics, for which see Kuhn *et al., op. cit.* [note 3].)
[29] NB to Harald Bohr, 23 Oct 1911 (BPC).
[30] Jeans also remained beyond Bohr's reach, declining to discuss the thesis before it appeared in English; NB to Harald Bohr, 23 Oct 1911 (BPC). Bohr did receive some welcome confirmation of his views on heat radiation at about this time from S. B. McLaren (*ibid.;* S. B. McLaren, "The Emission and Absorption of Energy by Electrons," *Phil. Mag., 23* [1911], 66–83, and *Scientific Papers,* ed. J. Larmor [Cambridge, 1925], 22–27).

Part of his problem in communicating with Thomson must have come, as Bohr always believed, from his poor command of spoken English; Bohr could state the errors he had discovered but was unable fully to explain his arguments. Doubtlessly Thomson, unprepared for a recent graduate as thorough and erudite as Bohr, failed to recognize that conversation with his unusual student might repay a little linguistic straining. In any event he had ceased active cultivation of the electron theory of metals some time before Bohr's arrival. By 1911 his primary interest was the study of positive rays, a subject on which he set Bohr a little experimental problem that regrettably proved pointless. Thomson surely did not enjoy hearing his ancient errors rehearsed by a tenacious foreigner whose English he could hardly understand.

Even had language and divergent interests not been barriers, one doubts that the intellectual communion Bohr sought could ever have developed. Bohr's lifelong practice was to refine his ideas in lengthy conversations, which often became monologues, with informed individuals. Whether his colloquist was a full collaborator, a sounding board, or an amanuensis, he required some human contact at almost every stage of his work, even in writing. He dictated his papers, at first (as with his thesis) to his mother, then to his wife, and ultimately to a series of secretary-collaborators beginning with H. A. Kramers.[31] Thomson's method could not have been more different. He seldom solicited his students' views on scientific problems, nor did he develop his own through extended conversations with others. Bohr found that Thomson, though friendly and receptive to questions, would invariably "break off in the middle of a sentence, after a moment's conversation, when his thoughts ran on something of interest to himself."[32] Though he closely followed that part of the literature which interested him, Thomson worked very much alone, a method appropriate to the "genius that showed the way for everybody," but one fatal to anyone who lost the way, as Thomson did increasingly after 1910.

The two men also differed profoundly in their approach to physics. Already in his thesis Bohr seems to display the trait which would characterize so much of his later work. Confronting a fundamental difficulty in existing theory, he would group together problems he thought related to it and subject the complex to a slow, careful, and

---

[31] Rozental, *Niels Bohr,* 30.
[32] NB to Harald Bohr, 23 Oct 1911 (BPC).

repetitious analysis. He expected thereby to achieve a coherent new position which, expressed in consistent models, would yield results in exact quantitative agreement with experiment. Thomson's interests lay elsewhere and his expectations were more modest. To him models were mere analogies, and fundamental problems of little interest. He did not anticipate or require exact quantitative agreement between experience and calculations based upon models, nor consistency among the different pictures he employed. As Bohr said later, "Things needed not to be very correct [for Thomson], and if it resembled a little, then it was so." [33] If Thomson's imprecise, often conflicting models helped him and suggested further experiments, they fully served their purpose. [34] Their incompatibility did not disconcert him.

The design of atomic models provides a relevant example of Thomson's practice. In 1904 he suggested the familiar model called after him, in which a crowd of electrons, arranged in concentric, coplanar rings, circulated within a neutralizing sphere of positive electricity of constant charge density. [35] This picture, as he often emphasized, was doubly artificial. In real atoms the electrons must be distributed in three dimensions; he restricted them to a plane because the mathematics of a spatial distribution lay beyond his reach. The sphere of constant charge density was likewise a convenient fiction, amenable to calculation because its linear restoring force permitted mechanically stable deployments of the electrons. The artificiality of the picture notwithstanding, Thomson used it first to elucidate qualitatively the periodic table of the elements and the nature of radioactivity and then, in subsequent years, such other phenomena as chemical combination and the scattering of β rays. [36] He used other pictures concurrently. One of these, "the doublet model," which Thomson introduced in 1910 to explain the photo-effect, particularly interested Bohr and may well have influenced his later work. [37] It therefore claims our attention.

The model consisted of an electron describing a circular orbit under the influence of, and coaxial with, an electric dipole of moment

[33] Interviews II, 6–7.

[34] Cf. Russell McCormmach, "J. J. Thomson and the Structure of Light," *Br. J. Hist. Sci., 3* (1967), 362–387; Lord Rayleigh, *The Life of Sir J. J. Thomson* (Cambridge, 1942), 202.

[35] J. J. Thomson, "On the Structure of the Atom . . . ," *Phil. Mag., 7* (1904), 237–265.

[36] E.g., J. J. Thomson, *Corpuscular Theory of Matter;* "On the Scattering of Rapidly Moving Electrified Particles," *Proc. Camb. Phil. Soc., 15* (1910), 465–471.

[37] J. J. Thomson, "On the Theory of Radiation," *Phil. Mag., 20* (1910), 238–247.

P. Not all such orbits are possible: only if θ, the angle between the axis and the line drawn from the dipole's center to the electron, equals $\tan^{-1}\sqrt{2}$, will the force normal to the orbit's plane vanish. Any right section of a cone of half-angle θ with apex at the dipole's center is thus a possible orbit. The balance of centrifugal and centripetal force in the plane,

$$mv^2/r = 3\,Pe \cdot \sin^4\theta \cdot \cos\theta/r^3, \tag{5}$$

does not further define the path, as it only gives the orbital radius r in terms of the electron's velocity v. All values of kinetic energy $mv^2/2$ and of orbital frequency $\nu = v/2\pi r$ are therefore possible. But by a peculiar property of the dipole's inverse-cube force, the *ratio* $(mv^2/2)/\nu$ is independent of the orbital parameters; in fact, a simple manipulation of equation (5) yields

$$mv^2/2 = (2.76\sqrt{Pem}) \cdot \nu. \tag{6}$$

Now in his calculation of heat radiation—the calculation Bohr found faulty—Thomson traced emission to the interaction of free electrons with doublets in the metal molecules; and, by comparing his results with a well-known characteristic of the radiation, Wien's displacement law, he was able to infer a value for the doublet's moment P. That value made the expression in parentheses in equation (6) $2.1 \cdot 10^{-27}$ erg-sec, smaller than Planck's constant by a factor of three. Thomson thought this agreement extremely suggestive: with the coefficient of ν approximately equal to h, equation (6) gives Einstein's expression for the photo-effect. A photo-electron is released, according to Thomson, when light of frequency ν disrupts an unstable dipole system, sending the circulating electron on its way into the world with something like its orbital kinetic energy. From equation (6) this quantity, $mv^2/2$, is just about $h\nu$!

Bohr was interested in this representation for two reasons. First, it harmonized strikingly with the chief positive result of his thesis, the fact that the ratio of thermal to electric conductivity and the sign of the transverse temperature difference accompanying the Hall effect both agreed with experiment if the force between free electrons and molecules were assumed to vary with the inverse cube of the distance. An electric dipole produces such a force. Bohr did not miss the connection. In his thesis he had explicitly noticed the "remarkable" agreement between the ratio x, the sign of ΔT, and Thomson's dipole.[38] And at Cambridge he continued to call attention to it. Out-lining his thesis to the Cambridge Philosophical Society in November 1911, he told his auditors that his result calling for a dipole force "is perhaps of some interest, because Prof. Sir J. J. Thomson has shown that some of the optical phenomena of the metals can be explained by assuming the existence of such electrical doublets." [39]

The incorporation of Planck's constant in the dipole model would also interest Bohr, who had already decided that the quantum was implicated in the behavior of bound electrons. To be sure, Thomson's handling of h scarcely harmonized with the spirit of Bohr's thesis. For Thomson, h was a *derived* quantity, simply related to constants characterizing the electron and to a new basic parameter, the elemen-tary dipole moment P. He, like several of his contemporaries, consid-ered h shorthand for a product of mechanical quantities, and in no wise fundamental.[40] Bohr, on the contrary, very probably already re-garded h as an unanalyzable given, and perhaps also suspected that it helped determine atomic structure in a manner he aimed to un-cover.[41] If he took equation (6) seriously, he would have to read it not as a definition of h, but as a possible condition fixing the state of an atomic electron. For any force law other than the inverse-cube, equation (6) and the force balance uniquely determine both r and ω. It is consequently noteworthy that in Bohr's earliest extant treatment of atomic models, probably prepared in July 1912, he employed a

[38] Bohr, *Studier*, 34, 117.

[39] Bohr Manuscripts (BMSS) in the Archive for History of Quantum Physics.

[40] The best known of these treatments, that of A. E. Haas ("Über die elektrodynamische Bedeutung des Planckschen Strahlungsgesetzes und über eine neue Bestimmung des elektrischen Elementarquantums und der Dimensionen des Wasserstoffatoms," *Sitzb. Wien. Ak., 119:2a* [1910], 119–144; "Der Zusammenhang des Planckschen elementaren Wirkungsquantums mit den Grundgrössen der Elektronentheorie," *Jahrb. Rad. und Elek.,* 7 [1910], 261–268), sought to express h in terms of e, m, and a, the radius of the Thomson diffuse-sphere atom. The question whether to take h or a set of mechanical quantities as fundamental was by no means clear in 1911. At the Solvay Congress of that year, in response to Sommerfeld, who had said he pre-ferred "a general hypothesis about h to particular models of atoms," Lorentz remarked (*La théorie du rayonnement et les quanta. Rapports et discussions de la réunion tenue à Bruxelles du 30 octobre au 3 novembre 1911* [Paris, 1912], 124), "M. Sommerfeld does not deny that there is a relation between the magnitude of the constant h and atomic dimensions (positive Thomson spheres). One can express this in two ways: the constant h is determined by these dimensions (Haas), or else the dimensions one attributes to atoms depend on the magnitude of h. I don't see any great difference." Bohr apparently knew nothing of Haas's work at this time (Interviews I, 10).

[41] The problem of fixing atomic structure solely in terms of the fundamental constants charac-terizing the components was not new. It figures prominently, for example, in Larmor's *Aether and Matter* (Cambridge, 1900), 189–193, which Bohr read with pleasure during his stay in Cam-bridge (Interviews II, 8).

quantum condition identical in form to equation (6) and that it did then serve to fix both r and ω.[42]

Despite the disappointment of Thomson's indifference, Bohr spent the first of his two semesters at Cambridge busily, profitably, and, on balance, happily. He had arrived with letters of introduction to friends and colleagues of his former teachers. He visited Littlewood and Hardy; he dined at Thomson's house; he journeyed to Oxford to call on Dreyer, to Manchester to see Lorrain Smith, a physiologist trained by his father, and to Birmingham to talk electron theory with S. B. McLaren. He gave small entertainments in his rooms, and he joined a football club.[43] In addition, of course, Bohr took seriously his role as a student, faithfully attending the lectures of Jeans and Larmor on various aspects of electricity, and following two courses, one advanced and one elementary, given by Thomson. Bohr was very enthusiastic about both, the second because of its "most beautiful experiments," the former because in it Thomson developed his latest ideas.[44] As for his own work, Bohr spent many hours trying to produce a discharge of cathode rays with the help of positive rays, this being the little experimental task Thomson had set him. It went very slowly. Bohr did not know where the problem was to lead; the professor was not easy to consult; and the Cavendish laboratory, crowded and disordered,[45] was always trying for a new-comer.

Meanwhile Bohr worked as he could on the electron theory of metals. He labored in the old vein, criticizing the calculations of others and firing off long letters pointing out new mistakes he had encountered.[46] In addition, he tried, with Thomson's help, to interest the Cambridge Philosophical Society in printing his thesis. A fellow student was asked to revise the translation, and Bohr impatiently awaited the decision of the Society's publication committee. Articles were appearing on every side by authors unaware of his results; the longer he waited the more likely that his own contributions would be rediscovered, and the English version rendered superfluous. As it hap-

---

[42] We discuss this document at length in Section III.

[43] NB to Harald Bohr, 29 Sept and 23 Oct 1911; NB to Ellen Bohr (his mother), 31 Oct and 6 Dec 1911 (BPC). Bohr was an excellent soccer player, and Harald was an Olympic star; Rozental, *Niels Bohr*, 23–24.

[44] NB to Harald Bohr, 23 Oct 1911 (BPC).

[45] *Ibid.*; Rosenfeld and Rüdinger, *op. cit.* (note 1), 41; Interviews II, 3, 7–8.

[46] Letters to M. Reinganum, 9 Nov and 17 Dec 1911, and to C. Oseen, 1 Dec 1911 (BSC).

pened his patience was sorely tried, but to no purpose. The Society's publication committee did not act until May 1912, and they then declined to publish the thesis, at least *in toto*, because of its length.[47]

During his second Cambridge term Bohr spent less time in formal academic work. Though he continued to attend lectures, and to criticize developments of the theory of metals, he abandoned the laboratory and directed more of his time to private reading and study.[48] The problem of bound electrons doubtlessly figured prominently among his meditations, as his thesis, by its very generality, had brought him to a position from which he could advance only by making special assumptions about their motions and arrangements. We do not know precisely what directions his thought took in these first months of 1912. But we do know what he planned to ponder. Writing to his friend C. W. Oseen on 1 December 1911, Bohr concluded a very long letter devoted to special problems in electron theory with the words: "I am at the moment very enthusiastic about the quantum theory (I mean its experimental side), but I am still not sure this is not due to my ignorance. I can say the same, in a far higher degree, about my relation to the theory of magnetons. I very much look forward to trying to get all these things straight next term."[49]

The magneton to which Bohr referred had been introduced only a few months previously by Pierre Weiss as an explanation, or representation, of careful measurements he had made of the magnetic properties of certain metals.[50] The susceptibilities of these metals, whether para- or ferromagnetic, suggested the existence of a fundamental unit or atom of magnetism. A molecule, regardless of its species, would accordingly possess either no magnetic moment or one equal to an integral number of magnetons. It is evident why Bohr would have found Weiss's work of absorbing interest. If one referred the magnetic properties of matter to orbiting electrons, as Bohr did, the existence of a fundamental unit of magnetism implied that the elements of the orbits, particularly the electrons' angular momenta, were fixed. The magnetic moment $\mathbf{M}$ of a circulating electron is

[47] NB to Harald Bohr, 23 Oct 1911; Camb. Phil. Soc. to NB, 7 May 1912 (BPC).
[48] BMSS; Rosenfeld and Rüdinger, *op. cit.* (note 1); NB to Harald Bohr, 2 Feb 1912 (BPC); N. Bohr, "Note on the Electron Theory of Thermoelectric Phenomena," *Phil. Mag., 23* (1912), 984–986, a critique of a paper by O. W. Richardson. The note is dated 5 Feb 1912.
[49] BSC.
[50] P. Weiss, "Über die rationalen Verhältnisse der magnetischen Momente der Moleküle und das Magneton," *Phys. Zs., 12* (1911), 935–952.

$(-e/2mc)\mathbf{L}$, where $\mathbf{L}$ represents angular momentum; if $\mathbf{M}$ is a determinate quantity, so then is $\mathbf{L}$. Such a restriction on the electronic motions is just what Bohr sought: his thesis, one remembers, had shown that a collection of molecules which faithfully followed Boltzmann's statistics displayed no diamagnetism, and implied that the solution lay in somehow constraining the behavior of the bound electrons. The restriction implied by the magneton might resolve the diamagnetic dilemma! We do not know how far Bohr proceeded along these lines in the first months of 1912. We can, however, infer the expectations he might then have entertained from a short section on magnetism written for, but omitted from, Part II of the trilogy of 1913.[51] Bohr there wrote that if his principle of the universal constancy of angular momentum, assumed to hold without, also held with a magnetic field, then diamagnetism would indeed be a general property of matter.[52]

The connection of the magneton and the quantum, at which Bohr hints in his letter to Oseen, was straightforward. The angular momentum L of an orbiting electron is $1/\pi$ times the ratio of its kinetic energy T to its orbital frequency ν. A procedure for fixing the angular momentum therefore lay close to hand: one needed only to imitate Planck and set T/ν equal to h or a multiple thereof. Since the circulating electron possesses a magnetic moment $M = -(e/2mc)L$,

$$M = -(e/2mc)\frac{1}{\pi} \cdot \frac{T}{\nu} = -(e/2mc) \cdot (h/\pi). \qquad (7)$$

Bohr could either have produced this association for himself or have

---

[51] Many drafts of this section, amounting to over fifty pages, remain in the Bohr MSS. Prof. Rosenfeld has published the most finished of these in Rosenfeld, *op. cit.* (note 1), 75–77.

[52] Bohr here perhaps intended an argument similar to the following. As will appear below (Section VI), in the model of 1913 Bohr confined the electrons of a given atom to a plane and assigned the same value of the angular momentum, say p, to each of them. For simplicity, assume $\mathbf{H}$ acts along their common axis of rotation. By hypothesis p remains unchanged under $\mathbf{H}$ and the vector sum of the momenta is zero. The radii of the orbits must therefore change if the Larmor precession is to occur, and this makes possible a change in total kinetic energy ΔT. Consider two electrons describing the same circular orbit in opposite senses with velocity v. One will accelerate (that whose $\mathbf{p}$ parallels $\mathbf{H}$) and the other slow down by the same amount Δv. The combined ΔT for the two is $(m/2)[(v + \Delta v)^2 + (v - \Delta v)^2 - 2v^2] = m(\Delta v)^2$. A simple calculation shows that, ignoring terms of order H^2, $\Delta v = (p^2/Zmec)H$, where Z is the effective nuclear charge acting on the two electrons. Hence $\Delta T \propto +H^2$. The positive sign indicates a diamagnetic effect, but how it is to be understood in absence of a net atomic angular momentum is not evident. Perhaps it was for this reason that Bohr deleted this section. Writing Harald on 3 Aug 1913 (BPC), Bohr said: "I had thought about including some remarks about it [magnetism] in the proof to Part II: however I'm giving it up and will wait until I have had time to think more about it." (The forgoing reconstruction corresponds to Bohr's deleted section on magnetism in [1] fixing the angular momenta and [2] arguing from the positive quadratic form of the resultant energy increment.)

discovered it in the literature; it was in fact made by several people in the fall of 1911.

At the end of September, for example, in a discussion at the Karlsruhe meeting of the *Gesellschaft der deutschen Naturforscher und Ärzte* following Weiss's first major address on the magneton, Abraham, Gans, and Einstein (as reported by Weiss) all observed that the magneton should involve "the famous h." [53] Gans even supplied equation (7). In his model, electrons were attached to the ends of weightless spokes of length r, the spokes forming a rimless wheel rotating with angular velocity ω. Gans equated the kinetic energy of rotation, $\Sigma mr^2\omega^2/2$, with an integral multiple of $h(\omega/2\pi)$, "in accordance," he thought, "with Planck's theory." [54] With this assumption,

$$mr^2\omega/2 = p \cdot h/2\pi, \tag{8}$$

p being an integer; and since $M = \Sigma er^2\omega/2c$, equation (8) yields (7) as the smallest possible magneton, corresponding to $p = 1$. But this minimum magneton is unfortunately an order of magnitude too large. If one took Weiss's results seriously, however, one might consider that some submultiple of h, rather than h itself, entered into equation (6) or (7). Contemporary theory certainly did not discourage such a consideration: in the literature of the time one sometimes finds h/π or $h/2\pi$ used as the elementary quantity. In the "Rutherford Memorandum" of July 1912, which is discussed at length below, Bohr leaves the magnitude of the submultiple open.

Bohr knew Weiss's address and the following discussion at least by 1913, as both are mentioned in the deleted section on magnetism. We may be confident, however, that he knew them much earlier. They appeared in print on 15 November 1911, in a double number of the *Physikalische Zeitschrift* devoted to the Karlsruhe meeting. Exactly two weeks later Bohr wrote of his hope of straightening out the quantum and the magneton during his second term at Cambridge. It seems likely that his program reflected a reading of Weiss and Gans, and that while pursuing it he replaced the right side of Thomson's quantum relation, $mv^2/2 = h\nu$, with the less definite "$K\nu$" of the Rutherford Memorandum. Certainly questions of this sort were very much on his mind before he left Cambridge for Manchester in March 1912.

[53] *Phys. Zs., 12* (1911), 952.
[54] Gans's was not the only possible formulation of Planck's theory. See below, 244, n.85.

Manchester

Bohr may first have met Rutherford at the beginning of November 1911, in the home of Lorrain Smith, the former student of his father's who had become a professor of physiology at the University of Manchester. Rutherford would then have just returned from the Solvay Congress, which adjourned on 3 November; and its deliberations, Bohr later recalled, were the subject of his initial conversations with the man who was to influence his career so decisively.[55] In December Rutherford spoke at the annual Cavendish Dinner, a very informal affair arranged by the advanced students in the laboratory. Bohr was much taken with the talk and, if he were not already, with the man as well. Rutherford was vigorous, extroverted, modern, brilliant, and, though only fourteen years Bohr's senior, already an inspiring international success. No entirely reliable information about the subject of Rutherford's talk has survived.[56] But its substance was of secondary importance, as it was more Rutherford's personality than his message that prompted Bohr to consider the possibility of spending part of his post-graduate year at Manchester. He quickly set up a meeting with Rutherford at the home of Lorrain Smith. Tentative arrangements were probably then worked out for Bohr's removal to Manchester the following spring, arrangements confirmed after discussion with Harald, who visited him in Cambridge in January 1912.[57]

---

[55] N. Bohr, "Reminiscences of the Founder of Nuclear Science and of Some Developments Based on his Work," *Proc. Phys. Soc., 78* (1961), 1083–1115; NB to Ellen Bohr, 31 Oct 1911, announcing a planned visit to Smith at Manchester "on Friday," which was 3 November. A meeting with Rutherford later that weekend is thus just possible.

[56] Bohr ("Reminiscences," 1084) recalled that Rutherford had spoken of Wilson's newly-invented cloud chamber. Wilson's experiments would not have been novel to Rutherford's audience, however, as the cloud-chamber was developed at Cambridge. Cf. C. T. R. Wilson, "On an Expansion Apparatus for Making Visible the Tracks of Ionizing Particles in Gases and Some Results Obtained by its Use," *Proc. Roy. Soc., A, 87* (1912), 277–292.

[57] Rosenfeld and Rüdinger, *op. cit.* (note 1), 43–44; Bohr, "Reminiscences," 1083–1084, and Interviews II, 8. The chronology is not entirely secure, for Bohr's "Reminiscences," which Rosenfeld and Rüdinger follow, place the Cavendish Dinner before the Solvay Congress, and imply that the subsequent negotiations at Smith's house occurred early in November. The Dinner, however, took place in December as usual (cf., Rutherford to W. H. Bragg, 20 Dec 1911, as quoted in A. S. Eve, *Rutherford* [New York and Cambridge, 1939], 208, a reference we owe to Prof. Rosenfeld). One is left with either another Cambridge event featuring Rutherford in October, or a second visit to Smith in December. The latter is more probable: we doubt that Bohr would have seriously considered leaving Cambridge after only a month's experience of it. Moreover, Rutherford did not formally accept Bohr as a student until the end of January 1912, a puzzling delay if the negotiations occurred in early November (NB to Harald and Ellen Bohr, 28 Jan 1912 [BPC]).

This decision proved critically important for the development of modern physics, and some writers have not unnaturally interpreted it as a conscious choice between the "bitter disappointment" of Cambridge and the bright promise of Manchester, between the outdated Victorian physics of the unreachable Thomson and the bold ideas of the open and earthy Rutherford.[58] This is to read history backwards, however. If disappointed in his hopes for collaboration with Thomson, Bohr was not miserable at Cambridge. He loved the city, admired the professors, and made so many friends there it took him several days to bid them all good-bye.[59] Nor was the Cavendish passé. It led the world, for example, in studying the passage of β particles through matter, a subject which may be considered a special case of Bohr's particular interest, the electron theory of metals.[60] None of his friends could understand why he wanted to leave.[61] As for Thomson, he remained for Bohr "a tremendously great man," a teacher from whom he had learned "an enormous amount," a person he liked very much.[62] Nor was Manchester an irresistable siren. Bohr took time over his decision, stayed a second term at Cambridge, and thought seriously of spending Easter in Copenhagen.[63] In fact Bohr migrated to Manchester primarily to learn something about radioactivity, the specialty of his most recent scientific hero, Rutherford. No profound break was intended or required. It was not remarkable for a foreign student with his own fellowship to seek experience in more than one laboratory during a post-graduate year abroad. For Bohr it was not Cambridge *or* Manchester, but Cambridge *and* Manchester.

Bohr arrived in Manchester in the middle of March 1912, and immediately set about mastering what he had come to learn, the experimental side of radioactivity. He began by repeating elementary experiments on the absorption of α and β rays, introductory exercises which he took very seriously, as the careful and thorough record of them surviving among his papers testifies.[64] After working in this way

[58] E.g., Ruth Moore, *Niels Bohr. The Man, his Science and the World they Changed* (New York, 1966), 31–39; cf. Rosenfeld, *op. cit.* (note 1), xv.
[59] NB to Harald Bohr, 7 March 1912 (BPC).
[60] The chief β-ray men at the Cavendish in 1911–1912 were Thomson, Crowther, and W. Wilson (for whose work see J. L. Heilbron, "The Scattering of α and β Particles and Rutherford's Atom," *Arch. Hist. Exact Sci., 4* [1968], 247–307), C. T. R. Wilson, and R. Whiddington.
[61] NB to Margrethe Nørland, March (?) 1912, quoted by Rosenfeld and Rüdinger, *op. cit.* (note 1), 45.
[62] *Ibid.*
[63] NB to Harald Bohr, 7 March 1912 (BPC).
[64] BMSS.

for about six weeks, from 16 March to 1 May, his initiation was deemed complete, and he began an experimental investigation of his own.

Few details about this investigation, which proved no more fruitful than the positive-ray work at Cambridge, have survived. We know only that Rutherford set the problem and specified the method, that it involved radium, and that it did not arouse Bohr's enthusiasm.[65] It did, however, have the advantage of throwing Bohr into regular contact with Rutherford, who routinely visited his students and assistants in their laboratories. And, of course, it provided the occasion for the new-comer to associate with his fellows—with men like G. Hevesy and C. G. Darwin—whose work was to have a critically important effect upon his own.

The uninspiring experimental investigation, though pursued conscientiously, did not occupy all Bohr's professional time. As in Cambridge, he continued his critical examination of the electron theory and worried about printing his thesis in English. The problem of publication had become acute at the beginning of May, with the final refusal of the Cambridge Philosophical Society to print it *in extenso*. Bohr at first considered acceding to their demand to reduce it by half, but decided against doing his carefully-wrought arguments such violence.[66] Toward the end of May, and through the middle of June, his frustration reached the point that he thought seriously of bringing out the translation at his own expense.[67] In those weeks the delay in publication particularly oppressed him because he believed he had hit upon ideas which might resolve some of the difficulties the thesis had uncovered. The value of his work, as he wrote Harald, would then "be a little different from what it now seems";[68] what had before appeared a purely critical, and therefore a somewhat negative study, might now serve as the extended introduction to the positive innovations the criticism anticipated.

Two letters written Harald on successive days, 27 and 28 May,[69] permit us to discern the spirit of the innovator, if not the substance of the innovations. In the earlier letter Bohr told his brother that his new ideas would do nothing less than "answer all the chief objections which

[65] NB to Harald Bohr, 27 May 1912 (BPC).
[66] NB to Camb. Phil. Soc., 8 May 1912 (copy); to Harald Bohr, 19 May 1912 (BPC).
[67] NB to Harald Bohr, 27 May and 12 June 1912 (BPC).
[68] NB to Harald Bohr, 27 May 1912 (BPC).
[69] BPC.

can be raised (and have been raised lately) against an electron theory of the kind I have treated." He was not altogether certain, of course; Harald knew "how easily I fall into error"; and he had no opportunity in Manchester, where "no one at all is interested in such things,"[70] to test and refine his ideas in conversation. Maturer reflection, however, only slightly moderated his enthusiasm, and the second letter to Harald, on 28 May, is filled with plans for the ultimate improvement of the electron theory. His own innovations seem to explain "various difficulties of a general character," e.g., the Thomson effect and the specific heats of metals at low temperatures; a recent suggestion of Stark's offers hints towards an explanation of electrical conductivity.[71] Stark's proposal is not free from error; Bohr plans to "write a little about it," beginning his positive reforms in the old critical manner.[72] "Then," he says, "I must let a little time go by, and reach some certainty [komme lidt til Ro] about these different things. How far I can get this year I have no idea; it depends on so many outside circumstances, and also on what others find (or have found) to write about on the matter. I just feel that I must again begin to work into these subjects."

These passages are of very great interest. They reveal, we believe, that even after two and one-half months at Manchester Bohr's interests had by no means shifted to matters of common concern to Rutherford's group. In particular, he displayed no interest whatsoever in the problems of radioactivity or the new nuclear atom. The electron theory of metals, which had brought him to Cambridge, continued to be his primary commitment, a commitment which, if anything, was stronger in the last days of May 1912 than it had been at any previous time during his stay at Manchester. There lay his expertise and his most original and promising ideas.

[70] Rosenfeld, op. cit. (note 1), xvi, misleadingly prints this passage as if it referred to the attitude of Rutherford's group towards atomic models. In context it plainly refers to their attitude towards general questions in the electron theory of metals: "er det [that Bohr's new ideas clear up general difficulties in the electron theory] Tilfaeldet, vil ja vaerdien af mit Arbejde [Bohr's thesis] vaere en lidt anden, end den nu anses for. Kaere Harald, Du ved jo, hvor let jeg kan tage Fejl; og det er maaske ogsaa dumt at sige saadan noget saa tidlig; men jeg havde saadan Lyst til at kunne tale med Dig i Aften, for jeg har jo slet ingen her, der virkelig interesser sig for saadan noget."

[71] J. Stark, "Folgerungen aus einer Valenzhypothese II. Metallische Leitung der Elektrizität," Jahrb. Rad. u. Elek., 9 (1912), 188–203. Stark suggests that the valence forces constraining the electrons in a metal lattice permit them to move freely under an external force only in certain directions, so that the kinetic theory of gases is inapplicable to their motions; and that they are displaced together, almost like a lattice, along the directions permitted by the forces.

[72] A sheet of comments on and objections to Stark's ideas is preserved in the Bohr MSS.

III. ABSORPTION, ISOTOPES, AND THE FIRST BOHR ATOM:
MANCHESTER, JUNE AND JULY 1912

Two months later, however, by the time of his departure from Manchester for Copenhagen in late July, Bohr had shelved these problems indefinitely. In the event, he never returned to them, with the result that few people today realize how substantial a contribution to physics his thesis might have made if published, as Bohr had hoped, in a generally accessible language. Instead of the electron theory of metals, on which his research had focussed since at least 1909, Bohr was fully involved, by the mid-summer of 1912, with the main research problems of Rutherford's laboratory.

The earliest remaining hint of this transition is found in a letter to Harald dated 12 June 1912.[73]

It doesn't go so badly with me at the moment; a couple of days ago I had a little idea for understanding the absorption of α particles (the story is this: a young mathematician here, C. G. Darwin (grandson of the right Darwin) has just published a theory about it, and I thought that not only was it not quite correct mathematically (a rather small thing, however), but also very unsatisfactory in its basic conception), and I have worked out a little theory about it, which even if it is not much in itself, can perhaps shed a little light on some things concerning the structure of atoms.[74] I am thinking of publishing a little paper about it very soon. . . . In recent years [Professor Rutherford] has been working out a theory of atomic structure, which seems to be quite a bit more solidly based than anything we've had before. And not that my work is of the same importance or kind, yet my result does not agree so badly with his (you understand that I only mean that the basis of my little calculation can be brought into agreement with his ideas). . . .

Clearly, when this letter was written, Bohr had at last engaged a local problem, α-absorption. Furthermore, he was for the first time

[73] BPC. Rosenfeld (*op. cit.* [note 1], xvii–xviii) has reprinted the first part of this most interesting and important letter.
[74] Darwin's paper, "A Theory of the Absorption and Scattering of the α Rays," appeared in the *Phil. Mag., 23,* 901–920, in the number for June 1912. Since this issue probably did not reach Bohr before 12 June (the accession date of the British Museum's copy is 23 June), he probably read the article in manuscript or proof. But both the date of the letter to Harald and Bohr's phrase "has just published" suggest that Bohr was unaware of Darwin's work until the time of its publication. That fact illustrates the extent of Bohr's initial involvement with Rutherford's group, particularly since he and Darwin were the only mathematical physicists on the premises.

employing a structural model of the atom, Rutherford's, in his own research. Yet this was only the beginning of a transition, not a sudden conversion. The passage of charged particles through matter was a problem very closely related to the central concerns of electron theory.[75] Bohr was entirely prepared for it by past experience, and he approached it in his typical manner, the criticism of someone else's work, without for a moment setting aside the problems that had dominated his thought until June. Indeed, he returns to those problems towards the end of the letter excerpted above: "I still think that (if [my new ideas on electron theory] are right) they will perhaps be important; but I haven't time to think of publishing them in the short time I yet have here, and I have my work in the laboratory. As regards my thesis, as I said I'm making my last effort to get it published here, and if it fails I'll have to publish it myself. . . ." Initially, therefore, the absorption problem was but a minor digression. Perhaps it would not have occurred at all if Bohr had not, as he told Harald, been forced from the laboratory for a few days while awaiting the delivery of some radium.

One week later, however, the nature and depth of Bohr's involvement had entirely changed. Writing Harald on 19 June he speaks of atoms with all the enthusiasm previously reserved for electron theory, a topic he no longer mentions.

> It could be that I've perhaps found out a little bit about the structure of atoms. You must not tell anyone anything about it; otherwise I certainly could not write you this soon. If I'm right, it would not be an indication of the nature of a possibility* [the asterisk leads to a marginal note: "i.e., impossibility"] (like J. J. Thomson's theory) but perhaps a little piece of reality. It has all grown out of a little piece of information [oplysning] I obtained from the absorption of α particles (the little theory I wrote you about in my last). You understand that I still could be wrong, for it's not yet completely worked out (I believe it's not, however); nor do I believe that Rutherford thinks it's completely mad [helt vildt]; but he is the right kind of man and would

---

[75] The close parallel between electron theory and the problems of absorption and scattering would, of course, exist only for the group, still minuscule in 1912, which conceived the dimensions of the α particle to be very small compared with those of an atom. That view, which was scarcely to be found outside of Manchester, is the only consequential debt to Rutherford's model apparent in either Darwin's paper or the published version of Bohr's. Both authors show that the contribution of the nucleus to the problems which concern them is negligible and thereafter distribute electrons uniformly through the atom. For changing views of the nature of α particles, see Heilbron, "Scattering . . . and Rutherford's Atom" (note 60).

never say that he was convinced of something that was not entirely worked out. You can imagine how anxious I am to finish quickly and I've stopped going to the laboratory for a couple of days to do so (that's also a secret). This must do for a short greeting from your Niels, who would give so much to talk with you.[76]

From this time on absorption and the atom occupied more and more of Bohr's time. The calculations took far longer than his first optimistic estimate, and the laboratory work was entirely abandoned in their favor. So, as it turned out, was the electron theory of metals, for the direct pursuit of the Bohr atom had now begun. What can Bohr have learned from Darwin's paper by 12 June and what can the decisive event of the following week have been? Certainty in these matters is impossible, but we find much in the following account compelling.

Darwin's primary objective had been to investigate Rutherford's model by applying it to the computation of the velocity loss of an α particle moving through air or a thin sheet of metal. His results for the shape of the absorption curve agreed quite well with experiment, and he was also able to compute values for n, the number of electrons per atom, close to those required by Rutherford's theory. To make his computations manageable, however, Darwin had introduced two related assumptions of which Bohr was extremely critical. The first was that an α particle would not be impeded unless it actually penetrated the atom; the second that the intra-atomic forces on an electron could be neglected during the short time of its interaction with a rapidly moving α particle. The latter assumption required the former, for, as Bohr pointed out in his critique, any computation which did include the interaction between atomic electrons and non-penetrating α particles would yield an infinite result for the transferred energy unless the forces exerted by the rest of the atom on the electron were taken into account.

With his characteristically brilliant sense for legitimate approximation, Bohr rejected Darwin's first assumption at once. Outside an atom, he agreed, the *net* force on a passing particle due to the nucleus and electrons must be very nearly zero. But that net force is relevant principally to scattering computations, in which the transfer of energy from particle to atom may reasonably be neglected. Absorption computations, however, demand the consideration of energy transfer; the

[76] BPC. For a likely interpretation of the puzzling note, "i.e., impossibility," see below, 245–247. We have here reproduced the entire text of this important letter.

239

nucleus, because of its weight, scarcely contributes; and the relevant forces are those between particle and electrons alone. For purposes of absorption computations, therefore, the atom has no surface beyond which the relevant forces cancel. To avoid, without arbitrariness, an infinite product in the computations, the force binding the electron into the atom must be taken into account.[77]

Bohr's initial contribution to the problem, presumably facilitated by acquaintance with a recent paper of J. J. Thomson on ionization,[78] was to recognize that the effect of this force depended critically (and computably) on the relation between the period of the electron's motion and the collision time. The interaction between a moving charged particle and a bound electron takes the form of a pulse, and its effect on the electron is very sensitively dependent on the ratio of the pulse length to the electron's natural period. In effect, as Bohr put it in a passage that displays the preparation provided by his thesis, "the theory of the decrease of velocity of moving electrified particles on passing through matter . . . bears a great analogy to the ordinary electromagnetic theory of dispersion; the different times of vibration for the different wave-lengths considered in the theory of dispersion is here replaced by the different times of collision of particles of different velocities and at different distances from the electrons. . . ."[79] It follows that knowledge of the frequencies of the electrons in an atom should permit a computation of absorption far more accurate than Darwin's.

---

[77] These criticisms are a paraphrase from pp. 11–12 of the published paper, N. Bohr, "On the Theory of the Decrease of Velocity of Moving Electrified Particles on passing through Matter," *Phil. Mag.*, 25 (1913), 10–31, communicated by Rutherford from Manchester in August 1912. We presume that they represent the views Bohr developed in June.

[78] J. J. Thomson, "Ionization by Moving Electrified Particles," *Phil. Mag.*, 23 (1912), 449–457, a paper which Bohr cites in "Moving Electrified Particles," 11 and 12.

[79] Bohr, *op. cit.*, 13. The parallel between the absorption problem and dispersion appears clearly in Bohr's treatment. He supposes that an elastically bound electron of resonant frequency v is exposed to a perturbing force $\phi(t)$ per unit mass, with $\phi(-\infty) = 0$. The electron's equation of motion is, then, $\ddot{x} + (2\pi v)^2 x = \phi(t)$. If the electron is at rest before the force is applied, $x(-\infty) = \dot{x}(-\infty) = 0$, and the solution of the equation of motion is, for any ϕ,

$$x = \frac{1}{2\pi v} \int_{-\infty}^{t} \sin\left[2\pi v(t - z)\right] \phi(z) \, dz.$$

If $\phi(t)$ is the force exerted by a negligibly deflected α particle which passes the electron at a distance p at time $t = 0$ and with velocity v, then its component perpendicular to the particle's motion is given by $m\phi_{\perp}(t) = 2pe^2/(p^2 + v^2t^2)^{3/2}$, with m the electron's mass. A similar equation gives $\phi_{\parallel}$. For the dispersion problem, the same solution holds, but with $\phi = 0$ until some time t_0, and $\phi = (eE/m) \sin(2\pi v't + \beta)$ for $t > t_0$, where E is the amplitude of the applied radiation, v' its frequency, and β its phase.

Conversely, knowledge of the velocity loss on passage through matter should enable one "to get more information about the higher frequencies in the atoms, and from this some more information about the internal structure of the atoms." [80] Presumably this is what Bohr had in mind when he wrote Harald that "the basis of my little calculation can be brought into agreement with [Professor Rutherford's] ideas." Or, perhaps, it was the possibility of getting from absorption measurements additional information about *bound electrons*, information required for the elucidation of the two fundamental problems isolated in his thesis, that led Bohr from his criticism of Darwin to the development of an absorption theory of his own.

This far, we believe, Bohr had gone by 12 June. Reconstruction of the crucial developments of the ensuing week is necessarily more speculative, but they are likely to have taken the following form. Unlike the Thomson atom, in which the effective charge attracting an electron towards the atom's center increases with the radius of the electron's orbit, the Rutherford atom is *mechanically* unstable.[81] Whenever two or more electrons equally spaced on a ring rotate about the nucleus, there is at least one mode of oscillation produced by displacement of the electrons in the plane of the ring that will grow until the atom is ripped apart. (The atomic problem differs from that of, say, Saturn's rings because the electrons repel, while planetary particles attract, each other.) This difficulty had been discussed at length in 1904 and 1905 in connection with Nagaoka's Saturnian model (it

---

[80] *Ibid.*

[81] We italicize "mechanically" because of the persistent implication in the literature on Rutherford's atom and on Bohr's development of it that *radiative* instability was what set Rutherford's model apart. Rosenfeld even writes (*op. cit.* [note 1], xv) that Bohr, "with his dialectical turn of mind, ... greeted Rutherford's nuclear model of the atom just because its radiative instability, inescapably following from classical electrodynamics, created such an acute contradiction to chemical and physical evidence about the stability of atomic and molecular structures." There are, however, two decisive objections to giving radiative instability a special role in either the reception of Rutherford's model or in Bohr's attitude towards it, objections to which mechanical instability is immune. In the first place, radiative instability is a characteristic of any model which employs electrons in motion, as all contemporary models did. Hence radiative, unlike mechanical, instability does not distinguish Rutherford's atom from Thomson's, though the former, by supposing a one-electron atom, admitted the least stable case. Second, the problem of radiative instability was well known and seems to have caused little concern. Radiation losses rapidly decreased to manageable proportions as the number of electrons in a ring increased. Thomson was even able to make positive use of such losses in accounting for radioactive decay, and his viewpoint was for a time shared by Rutherford. (Cf., Heilbron, "Scattering ... and Rutherford's Atom," *op. cit.* [note 60], 256–257.) We are not suggesting, of course, that the problem of radiative instability was unimportant either to Bohr or to the development of quantum theory; we stress only that, unlike mechanical instability, it played no special role in Bohr's choice and initial development of Rutherford's atom.

provided one principal reason for the model's abandonment), but there is little to suggest that Bohr or anyone else at Manchester was aware of it before the fall of 1912. Neither Rutherford, nor Darwin, nor Bohr cite this earlier literature; before Bohr took up problems of electronic structure, the Manchester group had been content to treat the electrons as uniformly distributed through the atomic sphere.[82] Bohr, however, could not long have remained unaware of mechanical instability once he began to compute the displacement of electrons from their orbit by interaction with passing charged particles.

In fact, among Bohr's scientific manuscripts are some forty sheets which appear to record his first encounter with the problem of stability. They are collected in a file titled "Dispersion and Absorption of Alpha Rays," and they divide into three parts, each with its own descriptive coversheet.[83] The first deals with velocity changes caused by "an Atom consisting of an Electron which moves about a fixed Point under the Influence of a Force that varies inversely as the Square of the Distance"; the second generalizes to an arbitrary force; and the third considers the case of atoms "with several Electrons in a Ring (provisionally only 2)." The first two parts reach a conclusion, while the cover sheet of the third announces that it is "Temporarily Abandoned, since the Computation breaks down over the System's Instability, [and] cannot be continued without Applying some other Hypothesis." Unfortunately, none of these sheets is internally dated. They must,

[82] For the main early discussions of the instability of Saturnian atoms see, G. A. Schott, "On the Kinetics of a System of Particles illustrating the Line and Band Spectrum," *Phil. Mag., 8* (1904), 384–387; "A Dynamical System illustrating the Spectrum Lines . . . ," *Nature, 69* (1904), 437. On 11 March 1911 Bragg wrote Rutherford mentioning that Nagaoka had once developed a Saturnian atom and identified the article with the phrases, "Time about 5 or 6 years ago when Schott and others were on the subject: probably in the Phil. Mag." (cf., Heilbron, "Scattering . . . and Rutherford's Atom," *op. cit.* [note 60], 300 n.). Rutherford either knew the article or looked it up, for he cited it on the last page of his classic paper, "The Scattering of α and β Particles by Matter and the Structure of the Atom," *Phil. Mag., 21* (1911), 669–688. His text shows that he had not studied Schott and was unaware of instability: "[Nagaoka] showed that such a system was stable if the attractive force was large."

[83] BMSS; The coversheets are in Mrs. Bohr's hand but must have been written to her husband's dictation. The calculations appear to deal only with displacements in the plane of the orbit and could not, therefore, have taught Bohr that the atom can be made stable against perpendicular displacements. (For the importance of this distinction see 280–281, *infra.*) That these were very early computations is suggested, among other things, by the fact that they are even more closely modelled on classical dispersion theory than Bohr's published paper. For example, instead of using a pulse for the perturbing force ϕ (cf., note 79, *supra*), Bohr uses a sine wave, sometimes with an exponential damping factor. He thereby avoids integrals of the form $\int_{-\infty}^{\infty} dz \cos{(\nu z)}/(z^2 + 1)^{3/2}$, which occur in the pulse problem but not in dispersion. Since these integrals, for which Bohr developed a series solution in the published absorption paper, figure in the letter to Harald of 12 June 1912, it is almost certain that the manuscript computations were prepared before that date.

however, represent a very early stage of Bohr's concern with the absorption problem, for they embody an approach of which only mathematical traces remain in the published paper, submitted from Manchester during August. In the latter, Bohr ignores orbits entirely, presumably in order to eliminate the stability problem until prepared to confront it directly; the electrons in an atom are treated as simple harmonic oscillators, just as in classical dispersion theory.[84] We think it likely, therefore, that Bohr had completed the manuscript calculations before 12 June, the date of his first letter to Harald about Darwin's calculations and Rutherford's atom.[84a]

When that letter was written, the "other Hypothesis" demanded by the multi-electron case was not, we presume, yet at hand. We conjecture that it was just this hypothesis, together with some first fruit sufficient to account for Bohr's enthusiasm and conviction, which intervened and were discussed with Rutherford during the week prior to 19 June. The hypothesis was, of course, stabilization of the orbit by extra-mechanical fiat, through the introduction of Planck's quantum. An electron was to remain in stable orbit if and only if its kinetic energy, T, were related to its orbital frequency by the equation $T/\nu = K$, with K a constant closely related to Planck's h. About the likely first fruit of such stabilization by fiat, we reserve our conjecture until discussing the manuscript in which, together with the quantum condition, it is made explicit.

Much in Bohr's previous career leads up to this first quantization of Rutherford's atom. Convinced since the completion of his thesis that

---

[84] Though Bohr, in the published paper, never considers orbits and makes no use of his "other Hypothesis," for which see immediately below, he does twice promise a sequel which will deal with orbital dynamics in a way that only the new hypothesis would permit. In particular ("Moving Electrified Particles," *op. cit.* [note 77], 27, 23), he promises to consider "the relation between the frequencies and the dimensions of the orbits of the electrons in the interior of the atoms" and also to examine the difference between the frequencies of vibrations perpendicular and parallel to the plane of the orbit. Both of these topics are considered in Parts II and III of the trilogy published in 1913.

[84a] While this article was in press, Professor Rosenfeld kindly informed us of some of the content of letters Bohr wrote his fiancée, Margrethe Nørlund, during May, June, and July 1912. That correspondence, which has not been available to us, forces a slight alteration in the dating we had inferred from the letters to Harald: the first mention of Darwin's problem and of Bohr's interest in it must be pushed back a week, to 4 June 1912. But the correspondence with Margrethe fully confirms that the essential event in Bohr's conversion to the problem of atomic structure occurred between 12 and 19 June. A letter of 15 June mentions the discovery of a special clue that Bohr is following with alternate hope and despair, a clue which two days later (according to a letter of 17 June) he thinks has opened a prospect which might lead to something true. We should note, however, that Professor Rosenfeld, who has been so helpful in this as in other matters, does not fully share our views as to the suddenness or the lateness of the transition in Bohr's research interests.

classical mechanics must break down in the interior of the atom, he cannot have been surprised by, and may well have welcomed, mechanical instability. Both Planck's work on blackbody radiation and his own on magnetism had suggested, furthermore, that the new mechanics, whatever its form, would differ from the old in excluding a large proportion of the electron orbits permitted by classical theory. Even the quantitative formulation of Bohr's hypothesis lay close to hand in the theory of the magneton, in Thomson's model for the photoeffect, and elsewhere.[85] No wonder Bohr felt, as these elements of his previous experience coalesced about Rutherford's atom, that he had "found out a little bit about the structure of atoms . . . , perhaps a little piece of reality."

That possible "piece of reality" is what Bohr described for Rutherford in a memorandum prepared for discussion before his departure for Copenhagen.[86] Conceivably it was the basis for the discussions that occurred in the week after 12 June, for the requisite calculations are not arduous, and they contain an important error which suggests they

---

[85] *Supra*, 227, 232. There are other sources, for by 1912 *ad hoc* equations connecting the ratio of an energy to frequency with Planck's constant were becoming more usual. At the Solvay Congress in 1911 Lorentz suggested that the energy of a rotator be set equal to $nh\nu$, and Bohr could have known the procedure through the Solvay *Rapports* (*op. cit.* [note 40], 447), which were published in January 1912 (*Bibliog. de la France, 102* [1913], No. 887). Bjerrum, whom Bohr knew (though perhaps not at this time), independently applied the condition to a rotating diatomic molecule in 1912 (*Nernst-Festschrift*, 1912, 90–98), the same year in which Nicholson developed it for electron rings (*infra*, 259). In "Moving Electrified Particles," *op. cit.* [note 77], 27, Bohr uses the same device without comment to estimate the frequency of the interior electrons of oxygen. If there is a puzzle about Bohr's quantum condition, it is not his choice of the ratio of energy to frequency but the lack of an integral multiplier.

In addition to offering him possible Planck-like relations between T and ν, the accumulating contemporary literature on quantum theory—particularly the Solvay *Rapports* and perhaps also an important paper published by Poincaré in January 1912—may have been helpful to Bohr in a less specific way. Before the first Solvay Congress very few people shared the conviction Bohr had expressed in his thesis about the inevitability of a quantum-like break with classical mechanics. The Congress and Poincaré's response to it did much to spread the view that Bohr had arrived at for himself. (Cf., Russell McCormmach, "Henri Poincaré and the Quantum Theory," *Isis, 58* [1967], 37–55.) At the very least Bohr must have been reassured. Perhaps more important, Rutherford had been at the Congress and discussed it with Bohr when the two first met (Interviews II, 8; *supra*, 233). The discussions there might have reinforced Rutherford's sympathy towards the quantum, which he had already regarded favorably because of the remarkable agreement between Planck's value for the charge of the electron and the apparently anomalous value he and Geiger had obtained by counting α particles (cf., Rutherford's "Note" in the Planck-Heft of *Naturwissenschaften, 17* [1929], 483).

[86] BMSS. This crucial document (henceforth to be called the Rutherford Memorandum or the Memorandum), most of which has been reprinted by Rosenfeld (*op. cit.* [note 1], xxi–xxviii) is filed in an envelope labelled by Bohr, "First draft of the considerations contained in the paper 'On the Constitution of atoms and molecules' (written up to show these considerations to Prof. Rutherford)/(June and July 1912)." The memorandum is in English. In quoting from it, below, we have rectified Bohr's spelling and punctuation for ease in reading, but retained his characteristic turns of phrase.

were done in haste. More likely, though, the Rutherford memorandum dates from later in June or from July, for it is rich in illustrative examples which clearly display both the extent of Bohr's achievement while in Manchester and the concerns which dominated his research during the seven months after his departure.

At the start of the Memorandum Bohr points out that in Rutherford's model there can be no equilibrium configuration without motion of the electrons. "We shall therefore," he continues, "first consider the conditions of stability of a ring of *n* electrons rotating around a point-shaped positive charge of magnitude *ne*." [87] What follows without a break is almost certainly the discovery which accounts for the enthusiasm of Bohr's letter of 19 June and which rendered the Thomson model, for the first time, "an impossibility."

> By an analysis analogous to the one used by Sir J. J. Thomson, . . . it can very simply be shown that a ring [such] as the one in question possesses no stability in the ordinary mechanical sense . . . , and the question of stability may [must] therefore be treated from a quite different point of view.

> It is however immediately seen that there is an essential difference between the stability of rings containing a different number of electrons, as it can be shown that the energy of an electron in the ring (the sum of the kinetic energy and the potential energy relative to the kern [nucleus] and the other electrons) is negative if $n \leqq 7$, but positive if $n > 7$, and that therefore an electron of a ring containing more than seven electrons is able to leave the atom. It is therefore a very likely assumption that an atom consisting of a single ring cannot contain more than seven electrons. This, together with the fact that inner rings of electrons in Prof. Rutherford's atom-model will have only very little influence (and always to the worse) on the stability of outer rings* seems to be a very strong indication of a possible explanation of the periodic law of the chemical properties of the elements (the chemical properties is assumed to depend on the stability of the outermost ring, the "valency electrons") by help of the atom-model in question. . . .

> * The difference in this respect between the atom-model considered and J. J. Thomson's atom-model is very striking, and seems to make it impossible to give a satisfactory explanation of the periodic law from the last mentioned atom-model.

The ideas explicit and implicit in this striking passage mark, we believe, a critical point in Bohr's development. Bohr has discovered a second decisive contrast (the first being large-angle scattering) between

[87] *Ibid.,* xxii.

the Thomson and Rutherford models, and the comparison is for him conclusive. This discovery is no less important historically because it rests on a mistake in calculation[88] and never recurs in Bohr's later writings. What had he in mind?

In Thomson's model, as in Bohr's version of Rutherford's, electrons were arranged in rotating, concentric, coplanar rings. In both models, also, additional chemical elements were constructed by adding electrons one-by-one to the outermost ring (simultaneously adjusting the positive charge to keep the atom neutral) until the addition of one more electron would, by mutual repulsion, rip the ring apart. At this point, Thomson restabilized the ring by adding an additional electron to the interior of the atom, again adjusting the positive space-charge for neutrality; thereafter he again added electrons to the outside ring until it once more approached instability. Interior electrons were themselves arranged in concentric rings subject to the condition that each ring be as full as possible before additional electrons were placed on the ring next inside it.[89] With Rutherford's model, however, Bohr found that adding electrons *within* a ring slightly reduced stability. Once the point of instability had been reached, he therefore began a new *external* ring, adding electrons one-by-one until it too verged on instability.

Though Thomson had been able to display interesting parallels between his results and the structure of the periodic table, there were overwhelming advantages to Bohr's procedure. First, Thomson's con-

---

[88] That the argument cannot be right is indicated by its conflict with an elementary theorem which Bohr himself proves on the last page of the Memorandum: "In a system of electrical charged corpuscles which possess an axis of symmetry (common axis of rotation) the total potential energy will always be equal to the total kinetic energy multiplied by -2" (Rosenfeld, xxi n.). It follows that if T is the kinetic energy of the system, then the total energy $= T + V = T - 2T = -T \leq 0$, since kinetic energy is never negative. If the electrons are symmetrically distributed in a ring, the energy is equally divided between them, and none can have a positive total energy. No argument like Bohr's will explain the existence of periods in the periodic table. It is odd—surely evidence of haste and probably also of the need for a particular answer—that Bohr should have missed the contradiction between his key result and a theorem he himself introduces five pages after it in the Memorandum.

The computational error can be found on p. A 3b of the Memorandum (one of the two sheets which Rosenfeld, who supplied the numbering, omits). When evaluating the potential energy of one electron due to the others in the ring, Bohr sums the quantity e^2/d_s over the ring, where d_s is the distance between the Zth and the sth electron. In doing so, he forgets that e^2/d_s is the potential energy of the interaction between a *pair* of electrons and that only half of it may properly be attributed to the Zth electron. His result, which does cause the total energy to change sign between $Z = 7$ and $Z = 8$, is therefore twice too large.

[89] The *locus classicus* for these aspects of Thomson's atom and for those which follow is, J. J. Thomson, *Corpuscular Theory of Matter*, 103–120.

struction yielded periods of steadily increasing length: 5, 11, 15, 17, 21, 24; Bohr's erroneous computation gave at least roughly regular periods of length seven. Second, Thomson explained the similar properties of the elements of a given column in the periodic table by pointing to the identity (not always quite maintained) in the construction of their *interior* rings, the number of electrons in the outermost ring differing for each member of the family. In the Bohr-Rutherford model, on the other hand, all atoms in the same column had the same number of electrons in the outermost ring, the ring whose electrons, being most loosely bound, would enter most readily into chemical and optical interactions. Finally, the number of electrons required for Bohr's construction of a given element were compatible with the values of nuclear charge which Rutherford had determined from large-angle scattering. Thomson rarely committed himself to parallels between particular electronic structures and particular elements, but when he did the numbers were invariably far too high, so that oxygen, for example, received sixty-five electrons rather than the eight required by the Rutherford model.[90] Apparently it was these impressive contrasts that made Bohr so sure he was following the right path.

Having, he thought, established this much at the start of the Memorandum, Bohr turned immediately to the quantum condition.

> In the investigation of the configuration of the electrons in the atoms we immediately meet with the difficulty (connected with the mentioned instability) that a ring, if only the strength of the central charge and the number of electrons in the ring are given, can rotate with an infinitely great number of different times of rotation, according to the assumed different radii of the ring; and there seems to be nothing (on account of the instability) to allow from mechanical considerations to discriminate between the different radii and times of vibration.[91] In the further investigation we shall therefore introduce and make use of a hypothesis from which we can determinate the quantities in question. The hypothesis is: that there, for any stable

---

[90] Even in 1907 these values for n were significantly higher than those which Thomson estimated from experiments on secondary X-rays and electron scattering. By 1912 they were still further out of line, though Thomson continued to think that Rutherford's values for n were too low, perhaps by a factor of 2 (cf. Heilbron, "Scattering . . . and Rutherford's Atom," *op. cit.* [note 60]). Anyone puzzled by the popularity of Thomson's atom despite these and other difficulties should remember that most people followed its author in expecting only the crudest sort of quantitative agreement. The assumptions of coplanar distribution and of positive space charge had both been introduced solely to facilitate computation (cf., *supra*, 226).

[91] In *The Corpuscular Theory of Matter*, 158–162, Thomson had discussed a mechanical means, based upon calculations of the astronomer Sir George Darwin, of restricting the orbits.

ring (any ring occurring in the natural atoms), will be a definite ratio between the kinetic energy of any electron in the ring and the time of rotation. [Bohr clearly intends frequency, not time of rotation.] This hypothesis, for which there will be given no attempt of a mechanical foundation (as it seems hopeless*), is chosen as the only one which seems to offer a possibility of an explanation of the whole group of experimental results which gather about and seem to confirm conceptions of the mechanism of the radiation as the ones proposed by Planck and Einstein.

    * This seems to be nothing else than what was to be expected, as it seems to be rigorously proved that the mechanics is not able to explain the experimental facts in problems dealing with single atoms. In analogy to what is known for other problems it seems however to be legitimate to use the mechanics in the investigation of the behavior of a system if we only look apart from questions of stability (or of final statistical equilibria).[92]

That footnote, with its echoes both verbal and intellectual, reminds us how firmly the Rutherford Memorandum is rooted in Bohr's thesis. But he has travelled a long way from the electron theory of metals. He has, that is, chosen an atom model, justified its use, and selected a quantum condition to ensure its stability. In the Memorandum he immediately proceeds to investigate its power in application. We here examine his discussion of the hydrogen molecule, the one example for which the Memorandum provides an explicit quantitative treatment.[93]

Bohr assumes, in keeping with the requirements of Rutherford's model, that the hydrogen atom possesses only one electron, a view accepted at Manchester but not widely elsewhere. The molecule, Bohr thought, must therefore be arranged as in the figure below: two electrons, each with charge $-e$ and mass m, rotate with frequency ν at the ends of the diameter of a ring of radius a; the ring itself is centered on an axis of length $2d$ connecting the two nuclei, each with charge $+e$; the plane of the ring bisects the inter-nuclear axis perpendicularly. For equilibrium of the nuclei, their mutual repulsion must just balance the axial component of the attraction due to the two electrons. That is,

$$\frac{e^2}{(2d)^2} = \frac{2e^2d}{(a^2 + d^2)^{3/2}},$$

an equation satisfied if $a = d\sqrt{3}$. For equilibrium of the electrons, the

[92] Rosenfeld, xxiii.
[93] Rosenfeld, xxv–xxvii.

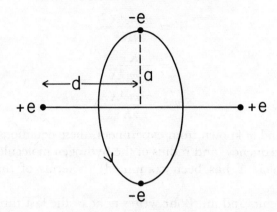

net radial force to which each is subject—i.e., the radial attraction of the two nuclei less the mutual repulsion between the electrons—must balance the centrifugal force due to the motion of the ring. The net attractive force on each electron is, therefore,

$$\frac{2e^2a}{(a^2 + d^2)^{3/2}} - \frac{e^2}{(2a)^2} = \frac{e^2}{a^2}\left\{\frac{2}{[1 + (d/a)^2]^{3/2}} - \frac{1}{4}\right\} = \frac{e^2}{a^2}X,$$

with $X = 1.049$. (Bohr discusses the problem more generally, showing first that the net radial force for any ring may be written as $(e^2/a^2)X$ and then that $X = 1$ for the hydrogen atom, 1.049 for the hydrogen molecule, 1.75 for the helium atom, etc.) For equilibrium therefore

$$\frac{e^2}{a^2}X = ma(2\pi\nu)^2, \tag{9}$$

so that the frequency is given in terms of the radius and vice versa.

This is as far as ordinary mechanics can go. To continue Bohr turns to the quantum, introducing the "*special* Hypothesis $E = K\nu$," [94] with E the kinetic energy of an electron. Equation (9) permits the kinetic and potential energy to be related, so that W, the negative of the total energy or the work required to remove an electron from the ring may be written:

$$W = -\left[\frac{1}{2}ma^2(2\pi\nu)^2 - \frac{e^2}{a}X\right] = \frac{e^2}{2a}X. \tag{10}$$

Manipulated with (9) and $W = E = K\nu$, equation (10) yields the three new equations:

[94] *Ibid.*, xxvii, italics in the original.

$$W = \frac{\pi^2 m e^4 X^2}{2K^2}, \tag{11}$$

$$a = \frac{K^2}{\pi^2 m e^2 X}, \tag{12}$$

$$\nu = \frac{\pi^2 m e^4 X^2}{2K^3}. \tag{13}$$

Assuming e and m known from experiment, these equations determine the energy, frequency, and radius of the hydrogen molecule (or of any other for which X has been computed) in terms of the universal constant K.

In the Memorandum, Bohr writes none of the last three formulas explicitly, but all of them, or equivalent forms, are implicit in his computations. (The use of [11] and [13] is illustrated below; [12] is implicated in the investigation of atomic radius.) He gives explicitly only a numerical equivalent for (11), writing that the negative total energy of a ring of n electrons is "nX^2A where A approximately is equal to $1.3 \cdot 10^{-11}$ erg." [95] That formula is then applied to the computation of the heat released when two atoms of hydrogen combine to form a molecule, yielding order of magnitude agreement with experiment. It is a remarkable computation, scarcely precedented in the literature of atomic theory.

Bohr's computation depends, however, upon the choice of K, and his Memorandum supplies neither an explicit value nor a source for this universal constant.[96] Given Bohr's reasons for adopting the "*special* Hypothesis," one would expect the value of K to equal Planck's h or one of its simple sub-multiples, but none seems reconcilable with the value Bohr attributes to A. Very likely he first tried computations using $K = h$, h/π, etc., decided they did not agree adequately with experiment, and determined, at least for the time being, to use an empirical value instead. Rosenfeld has suggested the source of the value, and his ingenious hypothesis is unlikely to be bettered.[97] Comparing $W = X^2A$ (Bohr's form for the absolute value of the energy per electron in a ring) with equation (11), yields $A = \pi^2 m e^4 / 2K^2$. Setting $A = 1.3 \cdot 10^{-11}$, as Bohr did in the Memorandum, and assuming, as Bohr did in his absorption paper, that $e = 4.65 \cdot 10^{-10}$ esu and $e/m = 5.31 \cdot 10^{17}$ esu/g, K becomes

[95] *Ibid.*
[96] Rosenfeld (*op. cit.* xxi) conjectures that a page of the Memorandum may be missing. If so, the determination of K may once have been part of Bohr's manuscript.
[97] Rosenfeld, xxx–xxxi.

$4.0 \cdot 10^{-27}$ erg-sec or approximately $0.6 \ h$.[98] This is just the value one obtains if K is computed from equation (13) with the preceding values for e and e/m, and with the frequency set at $\nu = 3.5 \cdot 10^{15}$ sec^{-1} and X at 1.049. The value of ν, as Rosenfeld points out, is the experimental figure for the resonance frequency of molecular hydrogen which Bohr employs in Part III of the trilogy. He also uses it, which makes Rosenfeld's suggestion even more plausible, in the absorption paper which he prepared at the same time as the Rutherford Memorandum.[99]

[98] Bohr, "Moving Electrified Particles," *op. cit.* (note 77), 23. Rosenfeld's calculation employs values taken from Part I of the trilogy, a slightly less likely source, but the single minor difference in value (4.7 for 4.65 in the figure for electronic charge) has no effect on the outcome of his argument.

[99] *Ibid.* Note that Bohr's value is for angular velocity and must be divided by 2π to give the figure cited above. (The values for the resonance frequencies of He and H_2 available to Bohr were determined indirectly by matching the experimental curve $\mu(\nu)$, μ being the index of refraction, to the classical formula $\mu^2 - 1 = Ne^2/\pi(\nu_0^2 - \nu^2)$. Here N is the number of dispersion electrons/cm^3 and ν_0 is the resonance frequency in question.)

More depends upon this reconstruction than is immediately apparent. Bohr's transition from the atom model of the Rutherford Memorandum to the one for which he is now known involves, among other things, equating K with $h/2$. We shall suggest in Section IV, again following Rosenfeld, that his motive for the change was simply to make his model produce the Balmer formula, including the value of Rydberg's constant. It follows that his first argument to justify the factor of ½, an argument which turns out to contain the germ of the Correspondence Principle, was in its origins a *post hoc* rationalization. One would therefore like to be quite certain that Bohr did not have other, earlier and more theoretical reasons for selecting a value of K.

One piece of evidence can be read to suggest that he did. In a typewritten letter to Hevesy dated 7 February 1913, before he had recognized the possible relevance to his model of the Balmer formula, Bohr states that his theory assumes "that the energy emitted as radiation by this binding [of an electron previously at rest relative to the nucleus] is equal to Planck's constant* multiplied by the frequency of rotation of the electron considered in its final orbit." The asterisk leads to a handwritten footnote, present in the copy sent to Hevesy but not in Bohr's carbon, which reads, "The constant entering into the calculations is not exactly equal to Planck's constant, but differs from it by a numerical factor *as was to be expected from theoretical considerations*" (BSC; Rosenfeld, xxxii–xxxiii, italics added). We now take the italicized phrase to mean only that the obvious differences between the Rutherford model and Planck's oscillators, e.g., the anharmonic structure of the former, lead one to expect a value for K different from h but otherwise unspecified. It can, however, be read to mean that Bohr's theory led him to a particular value for K, presumably $h/2$, and we initially understood it in that way.

As a result we have looked for alternate values of the physical constants current in 1912, values which might reconcile Bohr's choice of A, above, with the equation $K = h/2$. Bohr's value for e/m had, however, been standard to within a fraction of a percent since at least 1909. For Planck's constant he uses the original value, one confirmed by all subsequent measurements to within 2 percent. The electronic charge, e, was far more uncertain in 1912, but Bohr's value, $4.65 \cdot 10^{-10}$ esu, had been gospel at Manchester since the measurements of Rutherford and Geiger in 1908. Its main competitors elsewhere, Regener's value of 1909 and Millikan's of 1911, were higher, which would increase the discrepancy between K and $h/2$. To eliminate it, the electron's charge must be $4.3 \cdot 10^{-10}$ esu, and the only relatively recent measurements to give so low a value were those of Perrin, an unlikely source for Bohr, partly because the method involved was so indirect. In any case, Perrin's value of e would make $A = 1.2 \cdot 10^{-11}$ erg. We thus find Rosenfeld's reconstruction of the source of Bohr's value inescapable. (A particularly useful and convenient critical survey of the values of e and e/m current from 1897 to 1914 is included in J. S. Townsend, *Electricity in Gases* [Oxford, 1915], Chaps. VII, XII. For assistance in the search for alternate constants we are indebted to Mr. E. M. Parkinson.)

Though the hydrogen-molecule computation is the only one carried through fully in the Memorandum, the manuscript shows clearly that Bohr had already thought deeply about a considerable number of other problems. He provides, for example, models for the electronic structure of O_2, O_3, H_2O, CH_4, C_2H_2, He, and [He_2]. By comparing the energy of He with that of [He_2], he shows that the lattter, unlike H_2, cannot exist free in nature (whence the square brackets), a result that is likely to have given him particular pleasure and encouragement, for it was only with simple, single-ring atoms and molecules that he could anticipate exact results from computation. Bohr also employed his theory to estimate the dissociation temperature of H_2, and to explain why O_2 displays no infrared absorption bands and why it dissociates into two neutral atoms rather than into ions of opposite sign. Finally, he suggested that his theory could explain "the periodic law of the atomic volumes of the elements," Bragg's law relating stopping power for α rays to atomic weight, and Whiddington's law relating atomic weight to the excitation energy of characteristic X radiation.[100] Few of these results receive more than brief mention in the Memorandum. Bohr can scarcely have undertaken more than preliminary computations during late June and July, particularly since he was simultaneously deeply engaged with the absorption paper. These are the problems to which he devoted himself after his return to Copenhagen. His solutions to them provide the substance of Parts II and III of the trilogy.

The absorption paper and the quantized model are the two facets of Bohr's Manchester research which ultimately reached print under his name, but they are not the only areas in which he had made sub-

---

[100] The periodic law of atomic volume states that atomic volume (atomic weight/density), though increasing slowly with atomic weight from row to row of the periodic table, is fundamentally a periodic function dependent on position in the table. It has a recurrent high peak with the alkali metals in the first column of the table and falls rapidly to a minimum approximately half way between them. The law originated with Lothar Meyer ("Die Natur der chemischen Elemente als Funktion ihrer Atomgewichte," *Ann. d. Chem. u. Pharm.*, 1870, Suppl. 7, 354–364) and by the close of the century had become a standard topic for discussion in books on physical chemistry. For one of the many volumes from which Bohr might have known it see, W. Nernst, *Theoretical Chemistry from the Standpoint of Avogadro's Rule and Thermodynamics*, trans. R. A. Lehfeldt from 4th German ed. (London, 1904), 189–192.

Bragg's law stated that the stopping power of different metals for α rays was proportional to the square root of their atomic weights. (Cf., W. H. Bragg, "On the α Particles of Radium, and their Loss of Range in passing through various Atoms and Molecules," *Phil. Mag.*, 10 [1905], 318–340.) Whiddington's law related v, the minimum velocity an electron required to excite characteristic X rays, to the atomic weight of the target material through the equation, $v = A \cdot 10^8$ cm/sec. (Cf., A. Whiddington, "The Production of Characteristic Röntgen Radiations," *Proc. Roy. Soc. A*, 85 [1911], 323–332.) For Bohr's discussion of the first and third of these topics see Section VI, *infra*.

stantial advances before returning home. His own later recollections, supplemented, if sketchily, by letters written during 1913, show that he had also reached a clear understanding of the way in which nuclear charge, i.e., atomic number, governs the chemical properties of the elements. In the process, he had recognized, too, that radioactivity and weight must be entirely nuclear phenomena, had invented for himself the concept of isotopes, and had developed, at least in part, the radioactive displacement laws.[101] A half century after these events, Bohr suggested he had become involved with this constellation of problems during "the first weeks in Manchester," [102] and his account implies that he was much concerned with them throughout his stay. That dating, unlike the description of Bohr's achievement, seems to us extremely improbable. Such an involvement should have left traces in Bohr's rich correspondence, and there is, as we have already noted, none to be found before 12 June. We believe it far more likely that Bohr's innovations regarding the role of the nucleus, like his ideas on electronic structure, were consequences of the chain of thought initiated by Darwin's paper on absorption. Excepting recollections from the last years of Bohr's life, there is no reason to suppose that he had taken more than the most casual interest in Rutherford's model before June 1912.

Whatever their date, Bohr's nuclear innovations were important both for him and for others at Manchester. Apparently they followed quickly from a chance remark of his friend George Hevesy, who informed him that the number of known radioelements already considerably exceeded the available space in the periodic table. "Everything," Bohr later said, "then fell into line." [103] For that to have been the case, he must simultaneously have learned (or already have been aware) of one other striking feature of the surplus population, the existence of elements with different atomic weights and radioactive properties but with apparently identical chemical characteristics. Examples of such "chemically inseparable" elements were turning up

---

[101] The main contemporary testimony about this aspect of Bohr's work in Manchester is from letters: particularly, NB to Hevesy, 7 Feb 1913 (BSC); and Hevesy to Rutherford, 14 Oct 1913 (quoted in Rosenfeld and Rüdinger, *op. cit.* [note 1], 48). Fuller evidence from a date still fairly close to the event is provided in G. von Hevesy, "Bohrsche Theorie und Radioaktivität," *Naturwissenschaften, 11* (1923), 604–605. Most of the information about this episode comes, however, from two much later autobiographical sources and is correspondingly suspect as to details though not in general tenor: Bohr's "Reminiscences," *op. cit.* (note 55), 1084–1086, and Interviews I, 3–7; II, 11–16; and III, 6, 10. Only these late sources contain any implications about the point during Bohr's stay in Manchester when his ideas about the role of the nucleus emerged.

[102] Interviews II, 12–13.

[103] *Ibid.,* I, 7; II, 11–12, 16; Bohr, "Reminiscences," 1085.

frequently by 1912, and Hevesy himself had been wrestling with one pair of them, radium D and lead.[104]

Thomson's atom, which Bohr knew well, was powerless to deal with phenomena of this sort. Its chemical properties depended on all of its electrons, superficial and deep, and the total number of these atomic electrons determined the atom's weight by fixing the amount of the massive neutralizing positive charge. Bohr saw, however, that Rutherford's model permitted a quite different response to the challenges posed by Hevesy's illuminating remark. It provided the basis for a rigorous distinction between radioactive and chemical, or between nuclear and electronic, phenomena. The nuclear charge, Ze, determined the number of atomic electrons, n, and hence the atom's chemistry; while the atomic weight, A, which governed radioactivity, was essentially the mass of the nucleus, a quantity theoretically independent of n. The periodic table must, that is, be regulated by Z, not by A. Atomic weight was, as Bohr later put it, "a complete accident" [105] which might vary among atoms possessing identical chemical properties.

To this general viewpoint, Rutherford's theory of large-angle scattering added essential quantitative and qualitative detail. It required that helium have exactly two electrons and strongly suggested that hydrogen possesses only one. The experiments of Geiger and Marsden, interpreted on that theory, implied in addition that the Z's, and therefore the n's, of the metals closely approximated $A/2$. An obvious, simple, though hazardous extrapolation made the Z's of all the elements equal to their serial position in the periodic table. That novel view is one that Bohr had probably assimilated at the time he designed the atomic models in the Rutherford Memorandum and wrote his absorption paper. When, for example, he says in the latter, "According to Rutherford's theory of atoms we should expect 16 electrons in an oxygen molecule," [106] he is likely to be referring to his own interpretation of Rutherford's theory, not to a consequence that its author had yet altogether accepted.

---

[104] Bohr, "Reminiscences," 1085; G. von Hevesy, "Die Valenz der Radioelemente," *Phys. Zs., 14* (1913), 49–62.

[105] Interviews II, 12.

[106] Bohr, "Moving Electrified Particles," *op. cit.* (note 77), 26. Note, however, that in an early manuscript Rutherford, too, attributes eight electrons to the oxygen atom, presumably using simply the approximation $n = A/2$ (Heilbron, "Scattering . . . and Rutherford's Atom," *op. cit.* [note 60], 305.)

In a series of interviews held just before his death, Bohr described the source of his conviction about Rutherford's model with the words, "one knew the number of electrons in the atom, one knew the isotopes." [107] On the following day he added, "This problem of the isotopes was actually the reason that I felt we have now got some knowledge of the atom." [108] Though we do not believe that atomic number and isotopy were either the only or the very first of Bohr's reasons for taking up Rutherford's model,[109] they did provide him with essential early evidence that his research had taken the right turn, that everything would now "fall into line." As much as anything else, those discoveries must be what sustained him through the difficult and distracting year in which he attempted to work out the consequences of the quantized atomic model of which he had caught a first glimpse in June.

IV. EXCITED STATES AND SPECTRA: COPENHAGEN,
AUGUST 1912 THROUGH FEBRUARY 1913

Late in July, Bohr left Manchester for Copenhagen where he was married on 1 August. The couple spent their honeymoon in England, rather than Norway as originally planned, so that Bohr could put the

[107] Interviews I, 7.

[108] *Ibid.*, II, 16. In this connection Bohr's last comments on Moseley's work are of great interest: "And Moseley's thing, that is presented in a wrong manner, you see, because then we knew the hydrogen, we knew the helium. We knew the whole beginning [of the periodic table] . . . ." (*Ibid.*, I, 7). From Bohr's viewpoint, Moseley had confirmed an obvious extrapolation of Rutherford's model and scattering theory, and he should have said so. (See J. L. Heilbron, "The work of H. G. J. Moseley," *Isis, 57* [1966], 336–364, for Bohr's possible role in the planning of Moseley's investigation. Moseley did acknowledge that his purpose was to test the doctrine of atomic number which, however, he attributed to van den Broek.)

[109] Our conviction that Bohr's late memories of events in Manchester over-emphasize the role of his ideas about isotopy and atomic number is reinforced by our experience during the interviews from which much of the above is taken. When those interviews were conducted, the Rutherford Memorandum had not yet been discovered in Bohr's files. One of us had conjectured, however, from hints in his correspondence and in Part I of the trilogy, that Bohr had developed a detailed, non-spectroscopic, quantized version of Rutherford's atom some time before he saw the relevance of the Balmer formula. In early interviews, therefore, we repeatedly asked him for information about his work on atom models during the months before he first related the models to spectra. Bohr found such questions merely "silly" and insisted that, in the absence of the Balmer formula, he could have done no significant work on models. He consistently denied, that is, the very possibility of the sort of research which the Rutherford Memorandum ultimately documented. We conjecture that his emphasis on the role of his nuclear discoveries helped to fill the gap in memory left by the erasure of his work at Manchester (and Copenhagen, cf., *infra*) on electronic structure. Tricks of memory like this were, we should add, typical in our experience as interviewers. Bohr was by no means the only scientist unable to recall, or even to conceive as possible, participation in work which subsequent developments had removed from the corpus of proper physics.

finishing touches on the absorption paper, which he then delivered in person to Rutherford. Next they returned to Copenhagen, where Bohr undertook the duties of Assistant to the new Professor of Physics, Martin Knudsen, and also delivered a series of lectures on the foundations of thermodynamics.[110] The work on atomic and molecular models necessarily slowed down, but it was not abandoned.

In the final version of his absorption paper Bohr had promised readers a sequel, one which would treat the problems of electronic structure and orbital dynamics raised by Rutherford's atom.[111] It was to be a development of the Rutherford Memorandum, and Bohr expected in August that he could complete it quickly. As early as 4 November 1912 he wrote to Rutherford apologising for the time he was taking "to finish my paper on the atoms and send it to you."[112] Academic duties had, Bohr explained, combined with "serious trouble arising from the instability of the [atomic] systems in question" to delay the work's completion. He hoped, however, "to be able to finish the paper in a few weeks."

That estimate, too, proved excessively optimistic. Competing demands on Bohr's time continued to delay him, and he therefore asked Knudsen to relieve him of his duties, retired to the country with his wife, and "wrote a very long paper on all these things."[113] No physical trace of that manuscript remains, but its contents can be identified with assurance. Letters to Rutherford on 4 November 1912 and 31 January 1913 describe problems on which Bohr was at work, and a long letter to Hevesy on 7 February provides an extensive list of the resulting achievements.[114] The topics touched on in those letters are, in full: atomic volume and its variation with valence; the periodicity of the system of the elements; the conditions of atomic combination; excitation energies of characteristic X rays; dispersion; magnetism; and radioactivity. Those subjects are, of course, the ones on which Bohr had worked in Manchester. They are, furthermore, precisely the ones, and also the only ones, which he was to discuss in Parts II and III (including the unpublished section on magnetism) of the famous

---

[110] NB to Rutherford, 4 Nov 1912 (BSC); Rosenfeld and Rüdinger, *op. cit.* (note 1), 50–51.
[111] Cf., note 84, *supra.*
[112] BSC.
[113] Interviews III, 11; cf., *ibid.,* II, 13.
[114] BSC. The letter to Hevesy is reproduced in full by Rosenfeld, xxxii–xxxiv.

trilogy.[115] Almost certainly, therefore, those two portions of the trilogy were put together—with minor revisions to take account of what had intervened—from the "very long paper" Bohr had dictated in the country, the earlier draft vanishing through partial incorporation into the new one. As late as 7 February 1913, Bohr's research program, of which Parts II and III are the direct product, remained that of the Rutherford Memorandum.

A month later, however, that program had changed decisively, and the first fruits of the transformation were already embodied in a draft manuscript intended for publication. Part I of Bohr's trilogy was mailed to Rutherford on 6 March 1913,[116] and its subject was, for Bohr, entirely new: atomic spectra, particularly the line spectrum of hydrogen. Nothing in the correspondence or in Parts II and III suggests that he had worked on any such topic before February. In a letter to Rutherford dated 31 January 1913, he had, in fact, explicitly excluded the "calculation of frequencies corresponding to the lines of the visible spectrum" from the subject matter he took as his own. His program for model building, like that of Thomson which it closely followed, relied mainly on chemical, scarcely on optical, evidence. The complexity of spectra and the conspicuous failure of those who had tried to relate them quantitatively to models warned atom builders off the subject. Not that spectra were thought irrelevant—most physicists in 1912 would have agreed that they must directly relate to the most basic principles of atomic structure—but the evidence provided by spectra seemed inscrutable. Though Bohr, by 6 March, had proved that widespread attitude mistaken, it had been implicit in his research program until 7 February.[117] What can have happened in the interval to change his mind?

The February transformation of Bohr's research program was pre-

---

[115] N. Bohr, "On the Constitution of Atoms and Molecules [Part I, untitled]," *Phil. Mag., 26* (1913), 1–25; "Part II. Systems containing only a Single Nucleus," *ibid.,* 476–502; "Part III. Systems containing Several Nuclei," *ibid.,* 857–875. For the section on magnetism, cf., note 51, *supra.*

[116] NB to Rutherford, 6 Mar 1913 (BSC).

[117] In fact, Bohr's attitude towards the relevance of spectra may have been more negative than that of most of his contemporaries. Much later (*Interviews* I, 7) he said: "The spectra was a very difficult problem. . . . One thought that this is marvelous, but it is not possible to make progress there. Just as if you have the wing of a butterfly, then certainly it is very regular with the colors and so on, but nobody thought that one could get the basis of biology from the coloring of the wing of a butterfly."

pared during the two preceding months, beginning with his discovery, probably in December 1912, of a series of papers by J. W. Nicholson.[118] The articles of particular importance to Bohr dealt with the application of a quantized Saturnian model, much like his own, to the spectrum of the solar corona. Though they had appeared during June 1912 in the *Monthly Notices of the Royal Astronomical Society*,[119] Bohr did not encounter them until late in the year, for the journal is not one he would ordinarily have read. Perhaps someone who knew his interests called them to his attention, or he may have been led to them by the published report of Nicholson's remarks to the British Association meeting in September 1912.[120] In any case, Bohr's first known reference to Nicholson's atomic theory occurs in a Christmas card he and Mrs. Bohr sent to Harald on 23 December 1912,[121] and late in life Bohr indicated that he had been unaware of Nicholson's theory until about that time.[122] Even when he did discover it, assimilation cannot have been easy, for Bohr had known Nicholson before and been profoundly unimpressed. A letter to his close friend Oseen, written from Cambridge on 1 December 1911, mentions an "entirely preposterous" [ganske sindsvag] paper of Nicholson's on the electron theory of metals. Bohr's comments on the piece of work end with the words: "I also had a discussion with Nicholson [then at Cambridge]; he was extremely kind but I scarcely agree with him about much." [123] This was the man in whose work Bohr, a year later, recognized a severe challenge.

[118] Much of the following discussion of the effect on Bohr of his encounter with Nicholson is an elaboration of suggestions in Heilbron, *History of Atomic Structure*, 276–278, and McCormmach, "Atomic Theory of . . . Nicholson," 175–177 (both cited in note 1). Partial support for their analyses is provided by Bohr in Interviews III, 11. Asked if the notion of stationary states had first come to him only after he saw the relevance of the Balmer formula, Bohr replied: "Yes. (There you have it.) But still this is difficult because first of all the work of Nicholson is such (confusion). There I thought perhaps it is that he deals with other states. . . ." (The phrases in parentheses were unclear on the tape.)

[119] J. W. Nicholson, "The Constitution of the Solar Corona. II," *Month. Not. Roy. Astr. Soc., 72* (1912), 677–692; "The Constitution of the Solar Corona. III," *ibid.*, 729–739. They belong to a series of papers, the earliest of which appeared in November 1911. They are, however, the first articles in which Nicholson introduced Planck's quantum in discussions of an atom model and are thus the ones that would particularly have concerned Bohr.

[120] *Nature, 90* (1912), 424. This is the issue for 12 December 1912, and Bohr's first reference to Nicholson is dated 23 December.

[121] Rosenfeld, xxxvi.

[122] Interviews III, 4. Bohr here suggests that his letter of 31 January 1913 was written in the heat of first knowledge of Nicholson, but he had written on the subject to Harald five weeks before.

[123] NB to C. W. Oseen, 1 December 1911 (BSC). Nicholson's paper is "On the Number of Electrons concerned in Metallic Conduction," *Phil. Mag., 22* (1911), 245–266. Though Rutherford

To discover the nature of the challenge, consider briefly what Nicholson had done.[124] Using mechanical techniques like Bohr's, he had derived the energy of a ring of n electrons rotating about a nucleus of charge ke, and he had written the result in terms of the two related parameters, ring radius and frequency, a and v. (Nicholson considered the potential energy "of aetherial strain," not total energy, so that his formula must be divided by 2 for comparison with Bohr's.) Then, with $n = k = 5$, corresponding to "neutral protofluorine," a hypothetical element present in hot coronal gases, he had chosen the orbital frequency v so that the transverse vibration frequencies of the disturbed electrons about their equilibrium orbit would correspond to the maximum number of observed coronal lines. The determination of frequency also fixes, via the balance of centrifugal and centripetal force, the ring radius, and Nicholson could therefore compute the ratio of energy to frequency. He thus obtained,

$$mna^2(2\pi v)^2 \cdot 1/v = 154.94 \cdot 10^{-27},$$

a value which he pointed out was very nearly $25h$. The difference, he said, could easily be accounted for by uncertainties in the values of e and e/m. Performing similar computations for the singly and doubly charged ions of protofluorine gave, for the ratio of energy to frequency, $22h$ and $18h$, respectively. "These [multipliers of h]," Nicholson stated, "are the first three members of the harmonic sequence 25, 22, 18, 13, 7, 0, which would, if it continued valid, give no units to the positive nucleus alone, as would be expected."

These manipulations might have been dismissed as mere numerology if they had not resulted in impressive agreement with experiment. But they did. Nicholson's protofluorine atom accounted for fourteen previously observed but unidentified lines of the solar corona with an accuracy generally better than 4 parts in 1000. His nebulium atom, with nuclear charge $4e$, accounted for ten previously unexplained lines in nebular spectra, and his computations led, with an accuracy of one part in 10,000, to the discovery of a previously un-

---

regarded Nicholson as a promising young man, he agreed with Bohr that Nicholson could produce nonsense: "I do not know if you read Nicholson's paper and the awful hash he made of the α ray problem. I never saw so many howlers in two pages of a scientific article." Rutherford to W. H. Bragg, 23 Dec 1912 (Rutherford Correspondence, Cambridge University Library). We are indebted to Paul Forman for a copy of this letter.

[124] Cf., "Solar Corona II," *op. cit.* (note 119), 678–680, for the information cited and quoted in this paragraph.

noticed line.[125] It is no wonder that Bohr and other physicists were greatly impressed.[126]

Bohr was also troubled. His important letter of 31 January to Rutherford is primarily devoted to Nicholson's theory. "Nicholson deals, as I," he wrote, "with systems of the same constitution as your atom model; and in determining the dimensions and the energy of the systems he, as I, seeks a basis in the relation between the energy and the frequency suggested by Planck's theory of radiation." Nicholson did not, it is true, categorically identify any of his ring structures with terrestrial elements. But the four elementary ones, by compounding which Nicholson thought he could construct the periodic table, had atomic weights of the same order as hydrogen; one of them either was hydrogen or was closely related to it.[127] It was therefore disconcerting that the radii, frequencies, and energies which Nicholson computed for his rings were very different from those Bohr had found. Presumably that is what Bohr had in mind when he reported to Rutherford that Nicholson's theory "gives apparently results which are in striking disagreement with those I have obtained; and I therefore thought at first that the one or the other necessarily was altogether wrong." Unfortunately, no clearcut criterion of choice was available. Bohr's theory had greater scope,[128] and it fit better with Rutherford's scattering theory, but Nicholson had produced spectroscopic evidence of unprecedented precision. There was real cause for concern.

Bohr's concern was productive and short-lived. Though his confrontation with Nicholson did not at once lead him to embrace spectroscopic problems as his own, it did produce an important change in his understanding both of his model and of its physical basis. During December and January 1912–1913, his atom acquired excited states, and the relation between Bohr's theory and Planck's became temporarily clearer.

The first explicit sign of the change is found in a Christmas greet-

[125] Cf., J. W. Nicholson, "On the New Nebular Line at λ4353," *Month. Not. Roy. Astr. Soc.*, 72 (1912), 693, and McCormmach, "Atomic Theory of . . . Nicholson," 167–169.
[126] McCormmach, *op. cit.*, 183–184.
[127] *Ibid.*, 165–166, and Nicholson, "Solar Corona II," 682.
[128] But cf., J. W. Nicholson, "A Structural Theory of the Chemical Elements," *Phil. Mag.*, 22 (1911), 864–889, appearing in a journal that Bohr read. Furthermore, this article might well have interested him, for its aims were not unlike his own. Very likely he noticed the article when it appeared, but dismissed it in the light of his previous experience with Nicholson.

ing from the Bohrs to Harald,[129] and its nature is greatly elaborated in the letter of 31 January to Rutherford. Describing his progress, Bohr there wrote:

I am now much more clear of the foundation of my considerations, and I think that I also now better understand the relation and the difference between my calculations and, for instance, such calculations as those published in recent papers of Nicholson of [about] the spectra of stellar nebulae and the solar corona. . . .

[Nicholson's theory closely resembles mine, yet our results appear at first to be irreconcilable.] The state of the systems considered in my calculation are however—between states in conformity with the relation in question [a prescribed ratio of energy to frequency]— characterized as the one in which the systems possess the smallest possible amount of energy, *i.e.*, the one by the formation of which the greatest possible amount of energy is radiated away.

It seems therefore to me to be a reasonable hypothesis, to assume that the state of the system considered in my calculations is to be identified with that of the atoms in their permanent (natural) state. . . .

According to the hypothesis in question the states of the systems considered by Nicholson are, [on the] contrary, of a less stable character; they are states passed during the formation of the atoms, and are the states in which the energy corresponding to the lines in the spectrum characteristic for the element in question is radiated out. From this point of view systems of a state [such] as that considered by Nicholson are only present in sensible amount in places in which atoms are continually broken up and formed again; *i.e.*, in places such as excited vacuum tubes or stellar nebulae.

[129] Rosenfeld, xxxvi, lv. The remark is contained in a footnote which suggests, both in tone and condensation, that the brothers had already discussed fully both Bohr's concern and its source. The footnote reads, "P.S. Although it does not belong on a Christmas card, one of us would like to say that he thinks Nicholson's theory is not incompatible with his own. In fact his [Bohr's] calculations would be valid for the final, chemical state of the atoms, whereas Nicholson would deal with the atoms' sending out radiation, when the electrons are in the process of losing energy before they have occupied their final positions. The radiation would thus proceed by pulses (which much speaks well for) and Nicholson would be considering the atoms while their energy is still so large that they emit light in the visible spectrum. Later light is emitted in the ultraviolet, until at last all the energy which can be radiated away is lost. . . ." Cf. the following passage from Nicholson's "Solar Corona III," 730: "An atom with only two electrons . . . has a comparatively rapid rate of radiation. If it loses its energy by definite amounts, instead of in a continuous manner, it should show a series of spectrum lines corresponding to each of the stages. Moreover, its incapacity for radiating in a continuous way would secure sharpness of the lines." Russell McCormmach has called our attention to the likelihood that "Solar Corona III" was particularly important for the development of Bohr's intermediate radiation theory.

The central differences between his theory and Nicholson's were, Bohr now felt, essentially resolved. Like Planck's theory, Nicholson's permitted electron systems to possess a variety of different, though simply interrelated, values of the energy-frequency ratio. Derivation of the law of blackbody radiation demanded oscillators having the whole spectrum of energies, $W = \tau h\nu$, with τ any integer. Nicholson's ring systems were governed by an analogous law, but he investigated them only at the high energies and large values of τ to be anticipated among the particles of hot celestial gases. Bohr's interest had been in systems like Nicholson's, but he had considered them only in their lowest energy, or permanent, states. If his model were endowed with a whole series of levels, $E = \tau K\nu$, or, more generally $W = f(\tau)K\nu$, with f some function to be discovered, its permitted states should include those dealt with by Nicholson. No wonder that Bohr, having noted this deeper analogy with Planck's oscillators, felt that he was "now much clearer of the foundation" of his considerations.

The step to excited states and to a fuller use of Planck's conception of the quantized oscillator moved Bohr close to the final version of his atom.[130] The debt to Nicholson revealed in the preceding letter to Rutherford is quite sufficient to justify Jeans's remark in 1914 that,

---

[130] Bohr's study of Nicholson may also have been fruitful in two other ways. First, it may well be from Nicholson that Bohr first learned of the difference in stability between vibrations parallel and perpendicular to the plane of the orbit, a difference of which Bohr made significant use in the published paper (cf., Section VI, *infra*). Second, Bohr could have noticed that a quantitative parallel to Nicholson's theory would emerge if his constant K were set equal to $h/2$, a step which would have facilitated his recognition of the quantitative match between his equation (11) for the energy and the Balmer formula.

Besides the harmonic sequence quoted above Nicholson provides a series for the absolute potential energy per electron. The equivalent series for kinetic energy is $2\frac{1}{2}$, $2\frac{3}{4}$, 3, $3\frac{1}{4}$, $3\frac{1}{2}$, so that the energy of each permissible state differs from its predecessor by $h\nu/4$. For Nicholson, of course, each term in the series had reference to a differently ionized atom of protofluorine; the series does not yield successive energy levels of a neutral Bohr atom. Nevertheless, Nicholson had found something for which Bohr was looking, a simple numerical factor relating electron energy in a ring atom to that in one of Planck's linear oscillators. If Bohr's theory were to parallel both Planck's and Nicholson's, then his K should be some simple function of $h/4$, the Nicholson quantum unit.

Empirically Bohr's K was already very nearly two such units. In the Rutherford Memorandum it had been $0.6h$, and an equally accessible empirical source (the resonance frequency of He, also cited by Bohr in the absorption paper) would have made it $0.4h$, a fact Bohr is likely to have noted. An exploration of the quantitative fit between Bohr's theory and Nicholson's could therefore have made the equation $K = h/2$ seem very attractive. Bohr does not in fact seem to have made a final decision about the value of K at this time. Had he done so, he would presumably have specified $\frac{1}{2}$ rather than "a numerical factor" in the letter of 7 February 1913 to Hevesy (cf., note 99, *supra*). But no decision is required. The preceding exercise, if Bohr had put himself through it, would have further eased the assimilation of the Balmer formula.

whatever the ultimate fate of Nicholson's theory, "it has probably already succeeded in paving the way for the ultimate explanation of the phenomenon of the line spectrum." [131] But, as the letter to Rutherford also indicates, the largest and most radical steps to the Bohr atom were still to come. At the end of January 1913 Bohr still envisaged a radiation mechanism that was in two respects quasi-classical. First, spectral radiation must be preceded by ionization: atoms must be "broken up" before they can begin to radiate; spectral lines are produced "during the [re]formation of the atoms." Second, each optical frequency radiated corresponds to a mechanical resonance frequency of the atom or ion. Though the radiation is emitted in pulses, after (or during) which the electron ring settles into a lower energy state, the frequencies of the emitted lines are the transverse vibration frequencies of the perturbed electrons in one or another of the permitted excited states. By a mechanism which Bohr never had occasion to work out fully, an electron falling into its final orbit in the previously ionized atom causes the electrons in the high energy orbits to vibrate at their resonance frequencies, acting rather like a finger drawn across the strings of a harp.

That a spectral line of a given frequency must be produced by a charge vibrating at the same frequency was a consequence of electromagnetic theory which even Planck and Einstein had not thought to challenge. Though the nature of the atomic vibrators remained obscure, no contemporary would seriously have doubted their existence. Far more controversial was the correlation of ionization with spectral emission, but, as a natural product of research with gas discharge tubes, it too was widely held at this time. Spectra were emitted only under circumstances which, like high temperature or electric discharge, did produce ionization. The very multiplicity of lines in even the simplest spectra seemed to demand the existence of a larger number of vibrating systems than could be present in any normal atom, whatever its structure.[132] Johannes Stark, to whose influential *Prinzipien der Atomdynamik*[133] Bohr later turned to learn about spectral regulari-

---

[131] J. H. Jeans, *Report on Radiation and the Quantum Theory* (London, 1914), 50; quoted by McCormmach, "Atomic Theory of . . . Nicholson," 184.

[132] For a contemporary review of the considerable literature on this subject, one which Bohr may well have known, see, F. Horton, "On the Origin of Spectra," *Phil. Mag.*, 22 (1911), 214–219. For a fuller and more historical sketch see, Heilbron, *History of . . . Atomic Structure, op. cit.* (note 1), 176–185.

[133] J. Stark, *Prinzipien der Atomdynamik*, 3 vols. (Leipzig, 1910–1915).

ties, had argued forcefully for the association of ionization with spectra since 1902; Thomson lent his authority to that view in 1907 and developed additional arguments in its favor from that date to 1912.[134] Bohr had probably assimilated a similar position long before he applied it to his own and Nicholson's model. It was, in any case, deeply enough implanted so that he retained and used it for a time after he had recognized the relevance of the Balmer formula and revised his model to suit. Part I of the trilogy employs two different models for spectral emission: one is the now familiar process of transitions between stationary states; the other is the ionization and recombination process discussed above. The latter was, of course, incompatible with Bohr's principal innovations in Part I, and it was eliminated from all his subsequent discussions of the atom. Nevertheless, it is more than a mere out-of-date residue, overlooked as a result of hasty composition. On the contrary, it underlies Bohr's first published derivation of the Balmer formula.

Before Bohr could undertake to derive the Balmer formula, however, he had to "discover" its existence.[135] That must have occurred shortly after he wrote Hevesy on 7 February, or the manuscript which resulted could not have been sent to Rutherford so soon. Because of his concern with Nicholson, Bohr became interested in optical spectra for the first time during the early weeks of 1913.[136] On such topics H. M. Hansen, recently returned from Göttingen, was the local expert, and Bohr may well have sought him out for a reaction to his recent thoughts on radiation. In any case, it was Hansen who told him that

[134] For Stark's views, see, *ibid.* 2 (1911), 131–144. For Thomson's early statement, see, *Corpuscular Theory of Matter,* 156–160; and for an important later statement in an article Bohr used and cited, see the section on "Radiation Produced by the Recombination of Ions," 454–462, in the paper cited in note 78, *supra.*

[135] We use the word "discover" because Bohr himself repeatedly said that he had not known the Balmer formula until led to it by Hansen early in 1913 (Interviews I, 8; III, 11; Rosenfeld, xxxix–xl). But that account is unlikely to be quite right. Certainly Bohr did not take note of the Balmer formula until early in 1913; very probably he then had to look it up and did so because of Hansen's intervention. But he is likely to have seen the formula more than once before, failing to register it through disinterest. Bohr was a broadly educated physicist, and the formula was not obscure. Moreover, Bohr's teacher at Copenhagen, Christiansen, was a particular admirer of Rydberg's and had made a thorough study of his work (cf., Sister St. John Nepomucene, "Rydberg: the Man and the Constant," *Chymia, 6* (1960), 127–145, an article of which the relevance was kindly called to our attention by Professor Rosenfeld). Among other places where Bohr could easily have seen the Balmer formula, a particularly likely one is the report in *Nature* (note 120 *supra*) which may also have introduced him to Nicholson's work.

[136] Nicholson develops a formula for series spectra in "Solar Corona III." Though it was not a formula of the Rydberg type, it may well have increased Bohr's interest in the topic.

many spectroscopic lines fell into strikingly simple series, and Bohr went to look up the formulas in Stark's book. Later he said repeatedly, "As soon as I saw Balmer's formula, the whole thing was immediately clear to me." [137]

To a gifted mind like Bohr's, given also its special preparation, the Balmer formula could well have been the source of such insight. Stark gives the Balmer formula in the form $\bar{\nu} = N_0/4 - N_0/m^2$, in which $\bar{\nu}$ is the reciprocal wavelength and N_0 is a universal constant, 109675 cm^{-1}.[138] For comparison with his own formula (13) for frequency, Bohr would have multiplied N_0 by the speed of light, $3 \cdot 10^{10}$ cm/sec. The Balmer formula would then read $\nu = 3.29025 \cdot 10^{15} (1/4 - 1/m^2)$, with ν the frequency in cycles per second. In this formula the running index m, which may take any integral value greater than 2, should be related to and might be identical with the index τ which selects higher states in Bohr's radiative version of his atom. Unfortunately, τK, or $f(\tau)K$, appears cubed in the denominator of the radiative version of equation (13), and the Balmer formula calls for a square. It does, however, appear as a square in the energy formula (11), and by 1913, it was very nearly a matter of course for someone concerned with the quantum to divide an energy by h and look for the frequency that results.[139]

If Bohr's thoughts had developed in this way, everything at once would have become truly clear. Set $X = 1$ for the case of atomic hydrogen; substitute $\tau h/2$ for $f(\tau)K$ in the post-Nicholson version of equation (11), a step for which Bohr was by now well prepared; and divide once more by h to convert energy to a frequency. The result is the running term of the Balmer formula in the form $2\pi^2 me^4/h^3\tau^2$. Inserting Bohr's usual values for e and e/m, and the value $6.5 \cdot 10^{-27}$ erg-sec for h (the value Bohr uses in Part I), the constant in the Balmer formula works out to $3.1 \cdot 10^{15}$, within 7 percent of the spectroscopic value, an

---

[137] Rosenfeld, xxxix; and cf., Interviews III, 11.

[138] Stark, *op. cit.* (note 133), 2, 44–45. The formula was often written in terms of wavelength rather than wave number; in that form, which Stark also gives, its relevance to Bohr's atom is less apparent.

[139] Cf., *supra.* 244, n.85. One more example is worth recording, because it occurred at this time and was ultimately important to the development of the Bohr atom. Early in 1914 Franck and Hertz announced that they had divided what they took to be the ionization energy of mercury by h and then found the corresponding line emanating from their ionization apparatus. ("Über die Erregung der Quecksilberresonanzlinie 253.6 $\mu\mu$ durch Elektronenstösse," *Verh. deutsch. phys. Ges.* [1914], 512–517.)

extraordinarily close agreement in view of the state of the constants.[140] The first term of Balmer's formula is simply the running term with $\tau = 2$. The frequencies of the lines in the Balmer series are therefore the differences between the energies in two stationary states divided by Planck's constant. In view of the Combination Principle, the same must, as Bohr says in Part I, be true of series spectra in general. Optical and mechanical frequencies have at last been divorced, perhaps the greatest and most original of Bohr's breaks with existing tradition.

V. BOHR'S ATOM AND THE QUANTUM THEORY: FEBRUARY TO DECEMBER 1913

What Bohr had seen, shortly after 7 February 1913, was a relationship between his one-electron hydrogen model and the Balmer formula. If the latter were multiplied by h and the resulting terms interpreted as energy levels, then his post-Nicholson model, with $\tau h/2$ written for $f(\tau)K$, would yield those levels precisely. That relationship was, however, altogether *ad hoc* with respect to the determination both of energy levels and of radiated frequency. Neither the atomic, nor the spectral, nor the quantum theory of the day could justify the necessary interpretations and substitutions. Though the central step to a radiative atom model had already been taken, Bohr had still to forge a quantum theory of atom mechanics and of radiation which would permit something like a derivation of the Balmer formula. Without such a reformulation his treatment of spectra would inevitably appear—as it did, in any case, to many readers—merely an ingenious play with numbers and formulas. In the event, reformulating quantum theory to keep pace with a developing understanding of atomic spectra occupied many scientists throughout the years from 1913 to 1926. But a remarkable number of the conceptually fundamental steps were inaugurated by Bohr during 1913, mostly in February of that year. We possess, unfortunately, little explicit information about how they were made. The published papers and two letters to Rutherford provide

---

[140] Bohr demonstrates an even closer agreement in Part II, 487/39n. (Because references to the published paper will henceforth be frequent, we adopt an abridged notation for use in both footnotes and text. In "487/39n" the first number refers to the original publication in *Phil. Mag.*, 26 [1913], and the second to the reprint in Rosenfeld, *op. cit.* [note 1]. When citing Part I, only a single number is given since the pagination of the two sources is identical.) He there adopts Millikan's value for e and a recent value for e/h obtained by Warburg and his collaborators. With these constants the theoretical value of the Rydberg becomes $3.26 \cdot 10^{15}$ sec^{-1}, within one percent of the experimental value.

the only clues to the development of Bohr's thought after the first week in February.

In its essentials the draft sent to Rutherford on 6 March must have closely resembled the paper published in July, but there were a few changes. Fifteen days after mailing the first manuscript, Bohr wrote Rutherford again, enclosing a second copy in which he had "found it necessary to introduce some small alterations and additions." [141] Bohr mentions what are likely to have been the only significant changes. Both are additions: all or most of Section 4, "Absorption of Radiation," and a few paragraphs in Section 5 intended to reconcile his results with Nicholson's. Shortly after mailing that letter, Bohr travelled to Manchester to persuade Rutherford that his manuscript could not be reduced in length without grievous loss.[142] Probably further revisions were introduced during their discussions, but, given the nature of Rutherford's concerns, they are not likely to have been more than verbal. The same is true of the alterations Bohr mentions in a letter of 10 May, accompanying the return of corrected proof to Rutherford: "I have altered very little in it, and not introduced anything new. I have, however, attempted to give the main hypothesis a form which appears to be in the same time more correct and more clear." [143] Excepting the additions on absorption and on Nicholson's theory, the paper published in July was probably close in all but phraseology to the draft sent Rutherford on 6 March. The point is worth emphasizing, for the speed with which the draft was prepared may help to account for some revealing oddities in the published version.

As published, Part I contains two incompatible derivations of the energy levels in the Balmer formula, and a third is given in a talk Bohr delivered to the Danish Physical Society on 20 December 1913.[144] Examined *seriatim* the three display a fascinating developmental pattern which is at least partly autobiographical. A strong analogy between Bohr's atom and Planck's oscillator is basic to the first, muted in the second, and absent (in fact, explicitly rejected) from the third. The first depends, in addition, on the radiation-through-recombination

[141] NB to Rutherford, 21 March 1913 (BSC).

[142] Rosenfeld and Rüdinger, *op. cit.* (note 1), 54.

[143] BSC. The revision which Bohr introduced to clarify his hypothesis probably occurs on p. 7 of the published paper.

[144] An English translation of this lecture is the first essay in Bohr's *Theory of Spectra and Atomic Constitution* (Cambridge, Eng., 1922), 1–19. It had previously appeared in *Fysisk Tidsskrift, 12* (1914), 97–114. Our future references to this lecture are to the English text, but we have compared the relevant passages with the Danish original.

model which Bohr had developed when first reconciling his results with Nicholson's but which his new view of the Balmer formula might already have rendered obsolete. What emerges increasingly in the second and third derivations, gradually usurping the roles played in the first by the analogy to Planck and by the old radiation theory, is another one of Bohr's fundamental contributions to quantum theory. In a somewhat more developed form, it would later be known as the Correspondence Principle.

At the start of the first derivation Bohr restricts attention to a single electron of charge $-e$ and mass m circulating about a nucleus of charge E. If there is no energy radiated, he continues, the electron will describe stationary elliptical orbits of major axis $2a$ with a mechanical frequency ω. (In the one-electron case there is no problem of mechanical instability.) If, furthermore, W is the energy required to remove the electron from its orbit to infinity, then the preceding quantities are related by the equations:

$$\omega = \sqrt{\frac{2}{m}} \cdot \frac{W^{3/2}}{\pi e E}, \quad 2a = \frac{eE}{W}. \tag{14}$$

Maxwell's theory demands, however, that a system of this sort radiate energy and that the electron spiral rapidly into the nucleus. A paradox results, for atoms are known to possess characteristic dimensions many times larger than those of their nuclei. Planck's theory may, Bohr suggests, provide a way to resolve the dilemma:

> Now the essential point of Planck's theory of radiation is that the energy radiation from an atomic system does not take place in the continuous way assumed in the ordinary electrodynamics, but that it, on the contrary, takes place in distinctly separated emissions, the amount of energy radiated out from an atomic vibrator of frequency ν in a single emission being equal to $\tau h\nu$, where τ is an entire number, and h is a universal constant.[145]

---

145 At this point Bohr cites three recent papers by Planck (*Ann. d. Phys.*, *31* [1910], 758–768; *ibid.*, 37 [1912], 642–656; and *Verh. deutsch. phys. Ges.* [1911], 138–148). Hirosige and Nisio (*op. cit.* [note 2]) argue that Planck's revision of the quantum theory in these papers had special importance for the development of Bohr's theory of the atom. Perhaps the second and third (the first is a condensed sketch of Planck's original theory and thus irrelevant) did play a role in Bohr's localization of "the essential point of Planck's theory" in the quantization of emitted energy, and they may also have reinforced his conviction of the association between ionization and radiation. But these ideas were not novel when Planck took them up; Bohr could have found them in many other places. What was novel in the two versions of the quantum theory that Planck developed in 1911 and 1912 was the notion of continuous absorption, and this

Returning now to the simple case of an electron and a positive nucleus considered above, let us assume that the electron at the beginning of the interaction with the nucleus was at a great distance apart from the nucleus, and had no sensible velocity relative to the latter. Let us further assume that the electron after the interaction has taken place has settled down in a stationary orbit around the nucleus . . . [and] that, during the binding of the electron, a homogeneous radiation is emitted of a frequency ν, equal to half the frequency of revolution of the electron in its final orbit; then, from Planck's theory, we might expect that the amount of energy emitted by the process considered is equal to $\tau h\nu$, where h is Planck's constant and τ an entire number. If we assume that the radiation emitted is homogeneous, the second assumption concerning the frequency of the radiation suggests itself, since the frequency of revolution of the electron at the beginning of the emission is 0 (Part I, 4–5).

These assumptions permit the immediate derivation of the hydrogen energy levels. An electron initially at rest outside the atom can be bound only into an orbit satisfying the condition,

$$W_\tau = \tau h \frac{\omega_\tau}{2}, \tag{15}$$

an equation which, together with (14), yields the energy levels,

$$W_\tau = \frac{2\pi^2 m e^2 E^2}{h^2 \tau^2} = \frac{2\pi^2 m e^4}{h^2 \tau^2}, \tag{16}$$

the right-hand formula being derived from its predecessor by setting $E = e$ for the case of hydrogen, e remaining the absolute value of the electron's charge.

From equation (16) Bohr might have reached the Balmer formula in one easy step, obtaining a frequency by dividing h into the difference between two energy levels, one with $\tau = 2$. The formula which results provides, among other things, a theoretical value for Rydberg's

Bohr must necessarily have rejected. In both of Planck's revised theories vibrators could possess all classically permitted energies (quantization was restricted to the emission process), a conception incompatible with Bohr's use of the quantum to select and stabilize particular stationary states. We therefore believe that Bohr cited these papers primarily because they were up to date and that he is unlikely to have drawn anything from them he could not have taken as well, or even better, from Planck's original formulation or from other parts of the contemporary literature. When Nicholson introduced a quantum condition very like Bohr's, he did cite one of Planck's *original* papers, actually a more appropriate choice ("Solar Corona II," *op. cit.* [note 119], 677).

constant impressively within 7 percent of the one determined spectro-scopically. Bohr, however, could not take this direct route, for it had been barred by his derivation of the quantum condition, equation (15). If the radiation process is one in which a free electron at rest is bound into the τ-th energy level of a bare hydrogen nucleus, emitting in the process τ quanta all of frequency $\omega_\tau/2$, then the atom will not at the same time emit the diverse lines described by the Balmer formula. Be-fore Bohr can derive the latter, he must change his account of the mechanism of radiation. We shall shortly examine the way in which he does so, but must first detour to ask how he happened to place himself in this potentially uncomfortable position. Why, that is, did Bohr initially derive the hydrogen energy levels from a radiation mechanism incompatible with the application of his model to Bal-mer's formula? Part of the answer has already been given: Part I was prepared at white heat. But there were, in addition, two important substantive reasons for retaining the first derivation of the hydrogen levels, whatever difficulties it might later create.

In order to *derive* the Balmer formula, Bohr needed a quantum condition to determine energy levels. The required condition proves to be equation (15), $W_\tau = \tau h\omega_\tau/2$, and this equation gains whatever plausibility it possesses from its resemblance, emphasized by Bohr, to the quantum condition governing Planck's oscillator. Such an oscil-lator can emit several quanta at a time but only at a single frequency determined by its mechanical structure. Equation (15) must, by analogy, govern a process in which τ quanta are emitted, each of fre-quency $\omega_\tau/2$. To reach the Balmer formula Bohr will ultimately change this interpretation, saying instead that equation (15) repre-sents the emission of a single quantum with frequency $\tau\omega_\tau/2$. That in-terpretation, however, at once destroys the analogy between the Bohr and Planck radiators, and some other justification for (15) is there-fore required. In fact, Bohr found none. In the later portions of Part I and for some time after its publication, equation (15) ceased to be a quantum condition and became instead a derived formula. In its der-ivation, the Balmer formula itself became a premise, and it could not therefore be deduced from first principles. One function, then, of this first derivation of energy levels was to provide a quasi-derivation of the Balmer formula, an objective which Bohr was forced to renounce after the transition to a more appropriate radiation mechanism.

Nevertheless, Bohr would, we presume, have abandoned the first derivation if its conflict with the radiation process required to produce the Balmer formula had been as clear to him as it is in retrospect. The published text of Part I, however, demonstrates that he had not seen the full depth of the difficulty when he submitted revised proof in May. On the one hand, his discussion of the Balmer formula, of Rydberg's series, and, above all, of absorption spectra, shows a complete and subtle conceptual command of the process in which a single quantum of homogeneous radiation is emitted or absorbed in conjunction with the transition of an electron between two specified energy levels. On the other hand Bohr refers repeatedly to "the assumption used in this paper that the emission of line-spectra is due to the re-formation of atoms after one or more of the lightly bound electrons are removed" (Part I, 18). In deriving the Rydberg-Ritz formula, for example, he writes: "Let us assume that the spectrum in question corresponds to the radiation emitted during the binding of an electron" (Part I, 12). And in explaining why the Pickering series is not found in gas discharge tubes filled with helium, he states that "the condition for the appearance of the [Pickering] spectrum is . . . that helium atoms are present in a state in which they have lost both electrons" (Part I, 11).[146] Clearly all these passages refer to the process of spectral emission-through-recombination that Bohr had outlined in the letter of 31 January 1913, the result of assimilating Nicholson's theory to his own. When Part I was prepared he had not yet quite seen that he could not retain both that process and a theory of radiation through transitions. That is another reason why Bohr, who needed it to reach the Balmer formula, could retain his first derivation of energy levels. It fitted closely with an older view of the radiation process which he had not yet altogether abandoned.

One last aspect of Bohr's initial quantum condition, equation (15), requires discussion, namely, the source of the factor of ½. Following Rosenfeld, we suggest that, well prepared as Bohr was to find his old K close to $h/2$, his definitive choice of the value was probably determined by the need to match the constant in his theoretical formula to the one deduced from experiment. In that case, his remark that the frequency $\omega_\tau/2$ is an average of the electron's mechanical frequencies, o and ω_τ, in its initial and final states, appears an *ad hoc* ration-

---

[146] The Pickering series is described by a formula just like Balmer's but with half-integers rather than integers in the denominators of the energy levels. Before Bohr attributed it to ionized helium, it had been observed only in stellar spectra and in gas tubes containing a mixture of hydrogen and helium. Bohr's success in demonstrating that it derived from helium rather than hydrogen was one of the most persuasive early arguments for his theory. (Cf., Rosenfeld and Rüdinger, *op. cit.* [note 1], 59–60.)

alization, designed to preserve the parallelism between Bohr's radiator and Planck's.[147] Probably that was its origin, but it quickly acquired a far greater significance. In Bohr's second and third derivations of the hydrogen energy levels, this passing remark about an average of mechanical frequencies has, we shall soon discover, been transformed to a first, but already powerful, formulation of the Correspondence Principle.

How then does Bohr manage the transition from his first derivation of the hydrogen levels to the apparently incompatible Balmer formula? He interrupts his argument to discuss the work of Nicholson

---

[147] Hirosige and Nisio (note 2) argue that Bohr's idea of averaging frequencies was suggested by Planck's averaging of oscillator energies in the later formulations of his theory. But Planck's technique of averaging *actual* oscillator energies is entirely straightforward and can scarcely have been a source of novel insights. Besides, Bohr's argument for frequency would parallel Planck's for energy only if the electron, during binding, had radiated all frequencies between o and ω_r. He would then be determining the average of the *actual* frequencies radiated, and the factor of ½ would present no problems. It did present problems to contemporaries, particularly at Göttingen (Interviews I, 5), and no parallel to Planck's modified theory was, to our knowledge, educed in its defense.

A passage in Bohr's lecture to the Danish Physical Society (*op. cit.* [note 144], 13–14) can be read as supplying still another explanation of the factor of ½.

> By introducing the expression, which has been found for R [Rydberg's constant], we get for the nth state $W_n = \frac{1}{2}nh\omega_n$. This equation is entirely analogous to Planck's assumption concerning the energy of a resonator. W in our system is readily shown to be equal to the average value of the kinetic energy of the electron during a single revolution. The energy of a resonator was shown by Planck you may remember to be always equal to $nh\nu$. Further the average value of the kinetic energy of Planck's resonator is equal to its potential energy, so that the average value of the kinetic energy of the resonator, according to Planck, is equal to $\frac{1}{2}nh\omega$. This analogy suggests another manner of presenting the theory, and *it was just in this way that I was originally led into these considerations.* When we consider how differently the equation is employed here and in Planck's theory it appears to me misleading to use this analogy as a foundation, and in the account I have given I have tried to free myself as much as possible from it. (Italics added.)

The italicized clause seems to say that Bohr adopted the factor of ½ in equation (15) because he recognized early in his research that the parallel to Planck should be drawn through kinetic energy rather than total energy and the average kinetic energy of Planck's oscillator was $nh \cdot \nu/2$. It need, however, mean only that Bohr was "originally led into these considerations" by a general analogy to Planck's theory, the particular parallel being incompletely specified.

Strong arguments favor the latter interpretation. First, footnotes 99 and 130 provide evidence that Bohr had not settled on the factor of ½ before 7 February 1913 (or had at least found no theoretical justification for it). Second, if Bohr had developed the preceding derivation of (15) by the spring of 1913, he would surely have mentioned it in Part I, perhaps as an alternate to the less plausible averaging of mechanical frequencies. (He did, in fact, use it in a subsequent recapitulation of his theory: "On the Quantum Theory of Radiation and the Structure of the Atom," *Phil. Mag., 30* [1915], 394–415, esp. 396.) Finally, the argument that the quantization of kinetic rather than total energy requires, at least in the case of a rotating dipole, setting $T = nh\nu/2$ was first published during 1913 by Paul Ehrenfest ("Bemerkung betreffs der spezifischen Wärme zweiatomiger Gase," *Verh. deutsch. phys. Ges.* [1913], 451–457; "A Mechanical Theorem of Boltzmann and its Relation to the Theory of Energy Quanta," *Proc. Amsterdam Acad., 16* [1913], 591–597). Bohr may well have heard of Ehrenfest's argument before the lecture, or at least before its publication. Possibly the paragraph above was intended to record his reasons for ignoring it in discussing the atom.

on which his own has so far been partially dependent.[148] In the discussion he isolates and rejects an implausible assumption that he and Nicholson have to this point shared. The way to a new radiation mechanism and to the Balmer formula is then open to him.

Bohr first briefly describes Nicholson's model, points out that its agreement with observation provides strong arguments in its favor, and then notes that "serious objections . . . may be raised against the theory" (Part I, 6–7). Minor objections include the theory's failure to yield formulas like Balmer's and Rydberg's as well as the problem of mechanical instability to which Bohr promises to return. The central criticism, however, is of a different sort, one "intimately connected with the problem of the homogeneity of the radiation emitted":

> In Nicholson's calculations the frequency of lines in a line-spectrum is identified with the frequency of vibration of a mechanical system in a distinctly indicated state of equilibrium. As a relation from Planck's theory is used, we might expect that the radiation is sent out in quanta; but systems like those considered, in which the frequency is a function of energy, cannot emit a finite amount of a homogeneous radiation; for, as soon as the emission of radiation is started, the energy and also the frequency of the system are altered (Part I, 7).

In part Bohr is saying that one must, as he already has, abandon the connection between mechanical and radiated frequency. The latter, he immediately points out, is to be computed from Planck's theory, the emitted energy being divided by h. But Bohr seems also to be saying—or else his next step is incomprehensible—that during a nonclassical transition between stationary states only a single quantum may be emitted. The frequencies of successive quanta would otherwise be different, and the radiation would not be homogeneous.

That point Bohr makes explicit only five pages later, at the start of his second derivation of the energy levels, but he has meanwhile used it. Immediately after the preceding critique of Nicholson, he determines the frequency emitted in a transition from state τ_1 to τ_2 by the equation $W_{\tau_2} - W_{\tau_1} = h\nu$. The Balmer formula, including the theoretical value of Rydberg's constant, follows at once. The latter provides strong evidence for Bohr's theory, and he immediately reinforces it by attributing the Pickering series to ionized helium and by

---

[148] This discussion of Nicholson is *not* the one that Bohr added to his manuscript after its first submission (*infra*, 281). We shall examine that addition later.

a persuasive argument explaining why the identical multiplicative constant appears in the formulas of all spectral series (Part I, 10–12). Only after this powerful evidence has been educed does Bohr put the foundations of his theory in order with a new derivation of the hydrogen energy levels.

He prefaces his second derivation with a paragraph rejecting the assumption which was fundamental to the first:

> We have assumed that the different stationary states correspond to an emission of a different number of energy-quanta. Considering systems in which the frequency is a function of the energy, this assumption, however, may be regarded as implausible; for as soon as one quantum is sent out the frequency is altered. We shall now see that we can leave [out] the assumption used and still retain . . . [equation (15)] and thereby the formal analogy with Planck's theory (Part I, 12).

The alternative interpretation of (15) makes the energy radiated when an electron is bound from rest consist of a single quantum with frequency $\tau\omega_\tau/2$ rather than of τ quanta with frequency $\omega_\tau/2$. One sees why Bohr now describes the analogy with Planck's theory as "formal."

The derivation itself is very different from its predecessor. Bohr first assumes that the quantum condition determining energy levels must take the form $W_\tau = f(\tau)h\omega_\tau$, with f an unknown function. By a process exactly parallel to that which took him from equation (15) to (16), he finds,

$$W_\tau = \frac{\pi^2 m e^4}{2h^2 f^2(\tau)} \quad \text{and} \quad \omega_\tau = \frac{\pi^2 m e^4}{2h^3 f^3(\tau)}. \tag{17}$$

No formula like Balmer's will result, he points out, unless $f(\tau) = c\tau$ with c a constant to be determined.

It is in the determination of c that the Correspondence Principle first emerges clearly in Bohr's work. Setting $f(\tau) = c\tau$, he applies equation (17) to the determination of the optical frequency emitted in a transition between neighboring states, from $\tau = N$ to $\tau = N - 1$. The radiated frequency is,

$$\nu = \frac{\pi^2 m e^2 E^2}{2c^2 h^3}\left\{\frac{1}{(N-1)^2} - \frac{1}{N^2}\right\} = \frac{\pi^2 m e^2 E^2}{2c^2 h^3} \cdot \frac{2N-1}{N^2(N-1)^2}, \tag{18}$$

and the corresponding mechanical frequencies are,

$$\omega_N = \frac{\pi^2 m e^2 E^2}{2c^3 h^3 N^3} \quad \text{and} \quad \omega_{N-1} = \frac{\pi^2 m e^2 E^2}{2c^3 h^3 (N-1)^3}.$$

"If N is great," Bohr continues, "the ratio between the frequency before and after the emission will be very nearly equal to 1, and according to the ordinary electrodynamics we should therefore expect that the ratio between the frequency of radiation and the frequency of revolution also is very nearly equal to 1. This condition will only be satisfied if $c = \frac{1}{2}$" (Part I, 13). Thus, equation (15), the Balmer levels, and the Balmer formula are produced again.

Bohr also notes briefly a further, and historically even more pregnant, parallel between classical electrodynamics and the emerging quantum theory. If one considers a transition between states N and $N - n$, with n small compared to N, one again finds $c = \frac{1}{2}$ provided that $v = n\omega_N$. (The second factor on the right side of equation [18] becomes in this case $[2nN - n^2]/N^2[N - n]^2$.) This occurrence of an n-th harmonic of the orbital frequency ω_N provides Bohr with a further analogy between classical and quantum results. He points out that "with the ordinary electro-dynamics . . . an electron rotating round a nucleus in an elliptical orbit will emit radiation which according to Fourier's theorem can be resolved in homogeneous components, the frequencies of which are $n\omega$, if ω is the frequency of revolution of the electron" (Part I, 14). That is the interpretation which Bohr would later invert to arrive at his consequential selection rules: an atom will emit all and only those frequencies which correspond to Fourier components of its permissible classical orbits.

However fragmentary their initial formulations, these insights were of decisive importance for the subsequent development of the old quantum theory.[149] From Bohr's first paper on "Atoms and Molecules" through Heisenberg's first publication on matrix mechanics, many of the considerable triumphs of the quantum were associated with the discovery of new and more precise ways of employing classical mechanics and electrodynamics to determine the appropriate quantum formulation of special problems. More than any particular correspondences, that program of exploring ways of relating classical and quantum computations was the core of the Correspondence Principle. As it appears in Part I of Bohr's trilogy, the Principle is in embryo,

---

[149] For a perceptive study of the development of Bohr's conception of the Correspondence Principle, see K. M. Meyer-Abich, *Korrespondenz, Individualität, und Komplementarität*, Boethius, Texte und Abhandlungen zur Geschichte der Exakten Wissenschaften (Wiesbaden, 1965). In an appendix Meyer-Abich also provides a useful schematic sketch of the structure of what we have here called Bohr's three derivations of the hydrogen energy levels. His monograph is, however, mainly concerned to discover what Bohr took Correspondence to be. A comprehensive account of the role of the Correspondence Principle in the technical development of the old quantum theory remains to be written.

and its parentage is still clear. Planck's blackbody-radiation formula was known to reduce to the classical Rayleigh-Jeans law in the low-frequency limit, and Bohr anticipates that radiation from his atom will approach the classical limit in the same way. But Bohr, unlike Planck, employs the classical result to determine the appropriate quantum mechanical treatment of his atom. Besides, as Bohr develops the limiting case, it is apparent that classical and quantum computations coincide only in their results: *the models and the mechanisms of radiation remain distinct.* Planck's oscillator, in contrast, could with relative ease be viewed as itself behaving classically in the low frequency limit. Both as a heuristic tool and as a mark of the chasm separating classical from quantum physics, the Correspondence Principle thus transcended the limiting principle that had emerged when h was allowed to approach zero in the blackbody distribution law. Used with imagination and skill, as it was by Bohr and his Copenhagen colleagues during the years before 1926, it proved consequential in ways that its parent principle was not.

Shortly before Christmas, 1913, Bohr presented a revised version of this second derivation in a lecture to the Danish Physical Society. By that time radiation had for him become entirely a transition process, and he no longer referred at all to radiation during the binding of an electron initially at rest. As a result, his third derivation of 1913 caps the line of development inaugurated by its predecessors. In Bohr's lecture, the essence of Planck's contribution is reduced to the fundamental assumption,

$$hv = W_1 - W_2, \qquad (19)$$

and he emphasizes that "it is possible to derive Planck's law of radiation from this assumption alone," without reference to a Planck resonator or to any mechanical frequency of oscillation.[150] A comparison of (19) with the Balmer formula yields $W = Rh/\tau^2$ for the permitted energy levels of the hydrogen atom, where R is the Rydberg constant (more accurately, the usual Rydberg constant divided by the velocity of light, since spectral formulas are generally written for wave number rather than, as here, for frequency). By reference to (14) the mechanical frequencies of the orbital electrons are found to be,

$$\omega_\tau{}^2 = \frac{2R^3h^3}{\pi^2me^4\tau^6}. \qquad (20)$$

[150] *Op. cit.* (note 144), 11.

Bohr now compares that mechanical frequency with the frequency ν radiated in a transition from the state τ to the state $\tau - 1$. For large values of τ,

$$\nu = R\left\{\frac{1}{(\tau - 1)^2} - \frac{1}{\tau^2}\right\} = \frac{2R}{\tau^3}. \tag{21}$$

The Correspondence Principle demands that the mechanical frequency (20) and the radiated frequency (21) be equal for large τ, a condition which will be fulfilled only if,

$$R = \frac{2\pi^2 me^4}{h^3}.$$

Once again Bohr has derived a theoretical value for the constant in the Balmer formula.

That, however, is all that Bohr now derives. The very clarity of his third derivation highlights an element that has been gradually lost with the rejection of the detailed analogy to a Planck oscillator. That analogy provided the first derivation with an explicit quantum condition, $W_\tau = \tau h \omega_\tau / 2$, and that condition permitted a derivation, if an imperfect one, of the Balmer formula as well as of the Rydberg constant. In the second derivation a looser analogy to Planck permitted only the more general quantum condition, $W_\tau = f(\tau) h \omega_\tau$; reference to the Balmer formula was required to determine the form of $f(\tau)$; and the formula itself was only in part derived. By the time of the third derivation Bohr was convinced that it was "misleading to use this analogy [to the Planck oscillator] as a foundation," [151] and he had therefore to proceed without any quantum condition at all. He took the Balmer formula, interpreted from the start as a statement about energy levels, as his point of departure, and could deduce only the value of the multiplicative constant, the Rydberg coefficient. Before the Balmer formula could be derived again, Bohr would need a new quantum condition or at least a new justification for the old one.

Not until 1915 was a satisfactory new formulation developed, and then it was not Bohr who supplied it. He had, however, already reinterpreted his own initial quantum condition in ways that provided two sorts of significant guidance to his successors. In both Bohr's first and second derivations of the hydrogen energy levels, the quantity W_τ, when it appeared in a quantum condition like (15), stood for the

[151] *Ibid.*, 14.

"amount of energy emitted" during the binding of an electron into the τ-th stationary state (Part I, 5, 12–13). W_τ was also, of course, the negative of the energy of the electron in that state, but the quantum condition was, in the first instance, a condition on emitted energy rather than on the mechanical variables determining the τ-th orbit. Bohr, however, was necessarily aware of the possibility of the latter interpretation, which in any case was not new. A quantum limitation on the values of mechanical variables was the condition he had sought in order to resolve the problem of magnetism, and he had used just such a condition in developing his atom model until February 1913, the time at which the radiation of line spectra first came to concern him. In Part I of the trilogy, though he first developed a condition on emitted radiation, he made explicit the possibility of an alternate mechanical interpretation, and he put the point in an especially consequential way, one that he noted had also been developed by Nicholson:[152]

> While there obviously can be no question of a mechanical foundation of the calculations given in this paper, it is, however, possible to give a very simple interpretation of the result of the calculation [of stationary states] by help of symbols taken from the ordinary mechanics. Denoting the angular momentum of the electron round the nucleus by M, we have immediately for a circular orbit $\pi M = T/\omega$, where ω is the frequency of revolution and T the kinetic energy of the electron; for a circular orbit $T = W \ldots$ and from [15] we consequently get
>
> $$M = \tau M_0$$
>
> where
>
> $$M_0 = \frac{h}{2\pi} = 1.04 \times 10^{-27}.$$

> If we therefore assume that the orbit of the electron in the stationary state is circular, the result of the calculation can be expressed by the simple condition: that the angular momentum of the electron round the nucleus in a stationary state of the system is equal to an entire multiple of a universal value, independent of the charge on the nucleus (Part I, 15).

[152] Like Bohr, Nicholson had initially set the ratio of energy to frequency equal to a multiple of Planck's constant ("Solar Corona," *op. cit.* [note 119], 679). He immediately noted, however, that that ratio was proportional to angular momentum and that the introduction of the quantum might therefore "mean that the angular momentum of an atom can only rise or fall by discrete amounts when electrons leave or return. It is readily seen that this view presents less difficulty to the mind than the more usual interpretation, which is believed to involve an atomic constitution of energy itself."

From the time this passage was written, Bohr consistently deployed quantum conditions that restricted mechanical variables rather than emitted energy. At the end of Part I of the trilogy, for example, he announced the hypothesis which he would use when treating multi-electron systems in Parts II and III (Part I, 24–25; both the quotation marks and italics are Bohr's):

> *"In any molecular system consisting of positive nuclei and electrons in which the nuclei are at rest relative to each other and the electrons move in circular orbits, the angular momentum of every electron round the center of its orbit will in the pemanent state of the system be equal to h/2π, where h is Planck's constant."*

That condition on the mechanical variable, angular momentum, was, Bohr thought, applicable only to atoms in their permanent state. (His derivation of the angular momentum condition from [15] was restricted to circular orbits in which the electrons' kinetic energy is constant. In excited states, where orbits might be elliptical [Part I, 21–22], no angular-momentum formulation could be assumed to apply.[153]) When dealing with radiation problems, Bohr therefore continued to employ the more general quantum condition (15), but after Part I of the trilogy he consistently read it as a restriction on the values of mechanical variables. In his lecture to the Physical Society late in 1913 W is always the energy of an orbital electron, never an emitted energy. The analogy to Planck (which Bohr mentions to illustrate his point but then rejects as not fundamental) is to the mechanical energy possessed by an oscillator, not to the energy it emits.[154] The same approach is used in Bohr's subsequent publications on radiation problems up to the development of the phase-integral quantum conditions in 1915.[155]

Those new quantum conditions, which proved vital to the further development of the old quantum theory, were like Bohr's in two important respects. They limited, not emitted energy, but the permitted values of mechanical variables. More significant, they were momentum conditions, a generalization of the form that Bohr had developed for the circular orbits of the permament state. Clearly their emergence in

---

[153] For elliptical orbits, of course, the average value of kinetic energy equals W, and the angular momentum can be equated to the ratio of average energy to frequency. But there is a considerable conceptual difference between quantizing a constant of the motion, as Planck had done, and quantizing the average value of a variable.

[154] Cf., note 147, *supra*.

[155] Cf., N. Bohr, "On the Effect of Electric and Magnetic Fields on Spectral Lines," *Phil. Mag.*, 27 (1914), 506–524; "On the Quantum Theory of Radiation and the Structure of the Atom," *ibid.*, 30 (1915), 394–415.

1915 depended in major ways on the work Bohr had published two years before. The men who developed the phase-integral conditions were all quite consciously seeking a quantum formulation that could cover the two quite different problems—one-dimensional linear oscillator and two-dimensional planetary system—to which the quantum had previously been applied so successfully.[156] All of them, that is, forged their conditions in an attempt to reconcile Planck and Bohr, discovering and making explicit what the two had in common. Sommerfeld, who prepared the most influential version of the phase-integral conditions, actually credited Bohr with having quantized angular momentum in his derivation of the Balmer formula, a historical slip which is nonetheless revealing of Bohr's role in the evolution of the new mode of quantization.[157]

Bohr's reinterpretation of his quantum condition was also implicated in another aspect of the development of the old quantum theory, the treatment of dispersion. When he dealt with rings containing more than a single electron, Bohr had, for the first time in print, to confront the problem of mechanical instability. His solution was, in its essentials, the same quantum fiat offered in the Rutherford Memorandum: "The stability of a ring of electrons rotating round a nucleus is [in the atom's permament state] secured through the above condition of the constancy of the angular momentum" (Part I, 23). In discussing this version of the stability condition, however, Bohr made explicit a feature of classical orbital mechanics of which he had been unaware at the time the Rutherford Memorandum was written. Citing Nicholson's calculations, Bohr pointed out that it is only for displacements in the plane of the ring that the atom is incurably unstable. Electrons displaced perpendicular to the orbital plane will, unless the ring is already overloaded, vibrate parallel to the atomic axis until their vibrational energy is dissipated through radiation.[158]

---

[156] The complex history of the evolution of generally applicable quantum conditions demands separate treatment, which one of us (T. S. K.) is undertaking. The generalizations above are based on the work of Sommerfeld, Wilson, and Ishiwara, who are generally commonly credited with independent inventions of the phase-integral conditions during 1915 (cf., Max Jammer, *The Conceptual Development of Quantum Mechanics* [New York, 1966], 91–93). A fuller account of the sources of these new conditions would also emphasize the role of Planck's phase-space reformulation of quantum statistics, e.g., at the Solvay Congress in 1911 (cf., note 85, *supra*).

[157] A. Sommerfeld, "Zur Theorie der Balmerschen Serie," *Sitzungsb. Bayer. Akad. zu München,* 1915, 425–458, esp. 428, 431. That Bohr derived the Balmer formula by quantizing angular momentum is now a recurrent myth. On the functions of such myths, cf., Thomas S. Kuhn, *The Structure of Scientific Revolutions* (Chicago, 1962), 138–139.

[158] Bohr probably learned this important distinction between parallel and perpendicular displacements from Nicholson, whom he mentions when he first introduces it. There are many other

The orbital plane is also, however, the locus of the electron displacements or transitions which permit atoms to form themselves by electron capture. Those phenomena are just the ones which could not, Bohr had found, be treated classically. It was they which demanded the introduction of a quantum condition. That condition, viewed as a restriction on mechanical variables, need apply, however, only to displacements parallel to the orbital plane. Displacements perpendicular to the orbit, since they presented none of the quantum paradoxes, could remain the preserve of classical theory. For Bohr that distinction between the two types of displacement was the clue to an explanation of the striking success of Nicholson's theory. In a passage presumably added to Part I in the new draft mailed to Rutherford on 21 March, he suggested that Nicholson had not been dealing with a case of genuine emission but rather with scattering of the sun's light by the widely spaced atoms in the solar corona. Electrons vibrating perpendicular to the plane of their ring would at once emit the energy they absorbed, and there need be no change in the ring's size or mechanical frequency. Unlike spectral emission, dispersion need not be a quantum phenomenon (Part I, 23–24; cf., Part II, 482/34).

Except for the restriction to perpendicular vibrations, that explanation of dispersion is the same one Bohr had used in the Rutherford Memorandum when relating orbital frequency to the measured frequencies of anomalous dispersion. In Parts II and III of the trilogy he used it again to compute theoretical dispersion frequencies for comparison with experiment, and his procedure was soon taken up by other physicists.[159] Not until a few years later did Bohr or anyone else recognize that a classical treatment of dispersion could not possibly be right. On such a treatment the frequencies of anomalous dispersion must necessarily occur at the mechanical resonance frequencies of the electrons in the rings. Experimentally, however, they were found at the same frequencies as lines in the emission spectrum, and the latter, Bohr had just shown, occurred at frequencies different from those dic-

---

places he could have learned it (cf., note 82, *supra*) but he mentions none of them when discussing stability problems. He could also have worked it out for himself, but he had certainly not done so by August 1912 and probably not by the following November (cf., note 161 *infra*).

[159] P. Debye, "Die Konstitution des Wasserstoff-Moleküls," *Sitzungsb. d. Bayer. Akad. zu München*, 1915, 1–26; A. Sommerfeld, "Die allgemeine Dispersionsformel nach dem Bohrschen Modell," in *Arbeiten aus den Gebieten der Physik, Mathematik, Chemie—Festschrift Julius Elster und Hans Geitel* (Braunschweig, 1915), 549–584; and "Die Drudesche Dispersionstheorie vom Standpunkte des Bohrschen Modelles und die Konstitution von H_2, O_2, und N_2," *Ann. d. Phys.*, 53 (1917), 497–550.

tated by classical mechanics. Dispersion is necessarily a quantum phenomenon if emission is, a fact that Bohr had recognized by 1916.[160]

That being the case, it is particularly interesting that in Parts II and III Bohr had already begun to develop a quantum theory of dispersion side by side with the classical theory. In his absorption paper he had noted that, if dispersion electrons were treated as linear oscillators, the number of dispersion electrons in helium proved to be 1.2 rather than 2, a conflict with Rutherford's theory. In a footnote he promised to deal with the source of the discrepancy in a sequel, and on 4 November 1912 he reported to Rutherford that he had somewhat improved the agreement by substituting an inverse-square force for the elastic type he had used before.[161] But the problem had not been solved, for in Part II (489–490/41–42) Bohr found that the transverse vibration frequency $\nu_\perp$ of his two-electron model was $20.3 \cdot 10^{15}$ sec^{-1}, more than three times the experimental figure, $5.9 \cdot 10^{15}$. To remove the discrepancy he computed a frequency $\nu_\parallel$ which, he thought, might plausibly correspond to a resonance vibration in the plane of the orbit. Since, classically, such a vibration would rip the atom apart, Bohr set $\nu_\parallel = I/h$, I being the energy needed to remove one of the helium electrons from the atom. The result, $\nu_\parallel = 6.6 \cdot 10^{15}$ sec^{-1}, agreed reasonably

---

[160] N. Bohr, "Die Anwendung der Quantentheorie auf periodische Systeme," in *Abhandlungen über Atombau aus den Jahren 1913–16,* trans. H. Stintzing (Braunschweig, 1921), 123–151; cf., iv–v, 138–139. The article had reached corrected proof for the *Phil. Mag.* for April 1916, but was withdrawn when Bohr saw the paper (note 157, *supra*) Sommerfeld had delivered to the Munich Academy.

By about 1920, as the gap between quantum theory and Newtonian mechanics widened, classical stability considerations ceased to seem a part of quantum physics, and physicists tended to forget that they had ever made creative use of them. (Bohr, for example, talked in the interviews [I, 7] as though his approach to the quantized atom had always been incompatible with a quasiclassical treatment of dispersion.) We suspect that this shift in perspective is one source of the myth (cf., note 91, *supra*) that radiative, rather than mechanical, instability was a special characteristic of Rutherford's atom and played a major role in Bohr's development of it. Even Bohr (Interviews II, 13) spoke of radiative instability as central to his earliest work on Rutherford's model. Apparently he remembered wrestling with stability considerations but misplaced what, in 1912 and 1913, had been the most relevant sort of stability.

[161] Bohr, "Moving Electrified Particles," *op. cit.* (note 77), 23 n. The extent of the discrepancy is not, however, indicated there but in the trilogy, Part II, 490/42. The letter of 4 November 1912 to Rutherford is in BSC.

Bohr cannot have known about the distinction between parallel and perpendicular displacements when he wrote the footnote to the absorption paper, for he suggests that the difficulty in the computation of the number of electrons in helium may have to do with the difference in the frequencies of vibrations of electrons displaced parallel and perpendicular to the atomic axis. He could have discovered it by the time he wrote Rutherford, for he says that he has been delayed by difficulties due to the instability of the two-electron system, but has made progress. If, however, he had learned that parallel displacements were unsalvageable, perpendicular not, he would likely have said so.

well with experiment. Dispersion in helium was, he concluded, primarily due to vibrations in the orbital plane, and was thus governed by the quantum. Perpendicular displacements, he later suggested (Part II, 865/63), occurred in helium at a frequency too high to make a significant contribution. In Part III (864/62) Bohr performed a similar set of calculations, classical and quantum, for the dispersion frequency of the hydrogen molecule. The fact that the frequencies obtained for parallel and perpendicular displacement were almost identical explained, he thought, why classical theory had given so much better results when applied to H_2 than to He.

These computations were, of course, but primitive first steps towards a quantum theory of dispersion, and it was a long time before the next ones were taken. In the event, dispersion turned out to present one of the central difficulties that undermined the old quantum theory and provided clues towards a new one.[162] But, at least conceptually, those next steps, as taken in Copenhagen, were closely linked to these first efforts of Bohr's. Like any revolutionary contribution to science, his "Constitution of Atoms and Molecules" provided a program for research as well as a concrete research achievement.

VI. THE PRINCIPLES OF ATOMIC STRUCTURE

Parts II and III of "On the Constitution of Atoms and Molecules," published in the *Philosophical Magazine* for September and November 1913, present Bohr's solution to the problem of the principles of atomic structure, the question that had led him to Part I. Though published later, these portions of the trilogy were prior to the first: they are the elaborations of the matter and method of the Rutherford Memorandum on which Bohr labored so vigorously after his return to Copenhagen in the fall of 1912. The chief aim of Part II is to assign definite ring configurations to the various chemical atoms; that of Part III is to urge Bohr's novel view of molecular binding by a girdle of electrons circulating about the axis formed by the united nuclei. Part II also touches on X radiation and radioactivity, making public for the first time that qualitative picture of the nuclear atom, regulated by the principles of isotopy and atomic number, which still survives in introductory physics courses. But most of Parts II and III are given up to

---

[162] T. S. Kuhn, "The Crisis of the Old Quantum Theory, 1922–1925," to appear.

material much less familiar: elegant mechanical considerations about the behavior of interacting rings of electrons, dexterous applications of stability conditions to radically unstable systems, and arguments almost numerological about the structure of the elements.

Part II derived little benefit from the exact results of its predecessor, a fact which should occasion no surprise. Not only was the second part essentially completed before the first, but the primary concern of the latter, the excited states of one-electron systems, barely brushed that of the former, the normal configurations of poly-electronic atoms. In only one respect did the treatment of the hydrogen spectrum enable Bohr, in Part II, to go beyond the formulations of the Rutherford Memorandum. In place of the indefinite quantity K representing the ratio of kinetic energy to orbital frequency for each electron in its ground state, he could substitute the exact law of the universal constancy of the angular momentum: "in the permanent state of an atom, the angular momentum of every electron around the center of its orbit is equal to the universal value $h/2\pi$" (Part II, 477/29). But even this improvement was largely formal.

In Part II, as in the Memorandum, the electrons are arranged in coaxial, coplanar rings, a distribution Bohr justified on the slippery ground of stability. He supposed a ring system stable if its electrons, which of course all satisfy the principle of the constancy of the angular momentum, are so distributed that their total energy is less than that of any neighboring configuration satisfying the same principle of angular momentum (Part II, 477/29). This condition, as he neatly showed, insures the ground state stability against *all* displacements in the plane of the rings (Part II, 480/32). As for displacements normal to the rings, Bohr subjected them to the ordinary mechanics (Part II, 481–482/33–34), a refinement of the procedure in the Memorandum most likely adopted during his study of Nicholson's papers. These conditions make possible the investigation of the stability of certain three-dimensional configurations, for example of coaxial but not coplanar rings, or of mutually inclined rings passing through the nucleus. Bohr implied that he had examined some of these possibilities, and that the outcome vindicated the earlier restriction of the electrons to a single plane. "Calculation indicates," he said, "that only in the case of systems containing a great number of electrons will the plane of the rings separate; in the case of systems containing a moderate number of electrons, all the rings will be situated in a single plane through the nucleus" (Part II, 483/35). Bohr adopted a two-dimensional model not, as had Thomson, as an artificial alternative to a mathematically intractable spatial atom, but as a supposed consequence of his principles of atomic structure.

In his assignment of ring arrangements Bohr was of course guided by the doctrine of atomic number, which immediately specified the *total* number of non-nuclear electrons associated with each chemical atom. He had then to decide how many rings a given element required, and how its electrons were distributed among them. The stability conditions and the principle of angular momentum are insufficient to resolve the problem, as they are compatible with various distributions, and regrettably tend to favor those which conflict with the chemical evidence. The most helpful condition is that on the transverse displacements, for it does provide a relation between n, the number of electrons in a single-ring atom, and Ne, the nuclear charge just necessary to retain them against such displacements. The most interesting feature of this relation is that $N < n$ for $n \leqq 7$, and that $N > n$ for $n > 7$ (Part II, 482/34). Hence the largest possible neutral, single-ring atom contains only seven electrons, precisely the result Bohr had obtained in the Memorandum by an entirely different, erroneous procedure.[163] For the rest, the calculation yields the information that N increases rapidly with n; a central charge of ten, for example, is necessary to bind a ring of eight, and one of seventy-two to bind a ring of sixteen. Bohr concluded that the innermost ring of the atoms of the lighter elements contains small numbers of electrons (Part II, 482/34).

To fix the ring populations precisely, however, Bohr was obliged to use his intuition more often than his principles. He invoked the chemical properties of the elements, and, more directly, the well-known "curve of atomic volumes," which, as Bohr had recognized in the Memorandum, agreed well with his principles and gave an important clue to the number of rings each atom possessed. The curve, one recalls, is periodic, the "atomic radius" decreasing regularly from the alkalis to the inert gases, and increasing abruptly from the latter to the former.[164] According to Bohr's principles, the jump is associated with the beginning of a new ring, and the decrease with the addition

---

[163] *Supra*, 245–246.
[164] Rosenfeld, xxiii, and cf., note 100, *supra*.

of further electrons to it.[165] The place where the jump occurs cannot, however, be determined by those principles; and thus Bohr's assignment of ring numbers remained phenomenological, a direct translation into the terms of his model of the empirical curve of atomic volumes.[166]

The structure of hydrogen of course presented no new problems. Helium's two electrons Bohr assigned to a common ring, an arrangement which coincidentally yielded an ionization potential in close agreement with contemporary measurements (Part II, 488–490/ 40–42).[167] The third element in the periodic table, lithium, presented a severe problem. Since the energy of its three electrons is least when they occupy a common ring, Bohr's principles required that lithium possess the structure 3(3), where the notation $N(n_1, n_2, \ldots)$ gives the ring distribution of the Nth element, n_1 referring to the innermost ring; while the atomic-volume curve and the chemical properties of that element, which suggested that the lithium atom held one electron very loosely, pointed to the structure 3(2, 1). Bohr chose the double-ring arrangement (Part II, 490–492/42–44). Similarly, he selected 4(2, 2) for beryllium, although the single-ring system 4(4) has the lower energy. Continuing in this way, always guided by chemical data and atomic size, he arrived at the following suggestions for the electronic configurations of the lighter atoms (Part II, 49/497):

1(1)	5(2,3)	9(4,4,1)	13(8,2,3)	17(8,4,4,1)	21(8,8,2,3)
2(2)	6(2,4)	10(8,2)	14(8,2,4)	18(8,8,2)	22(8,8,2,4)
3(2,1)	7(4,3)	11(8,2,1)	15(8,4,3)	19(8,8,2,1)	23(8,8,4,3)
4(2,2)	8(4,2,2)	12(8,2,2)	16(8,4,2,2)	20(8,8,2,2)	24(8,8,4,2,2).

Particularly noteworthy is the confluence of inner rings between elements 9 and 10, and between 17 and 18, giving neon the structure

[165] Consider a neutral atom with nuclear charge Ne and an external ring of n electrons, and imagine that each internal ring acts on an electron outside it as if it (the internal ring) formed a continuous ribbon of current. The effective charge acting on an external electron is then $(N - s_n - \alpha)e$, where s_n represents the effect of the mutual repulsion of the outer-ring electrons and α, a complicated function, expresses the effect of the inner "continuously charged" rings. The radius of the outer ring, according to Bohr's theory, is $a_0/(N - s_n - \alpha)$, a_0 being the radius of the hydrogen atom in its ground state. As $(N + 1 - s_{n+1} - \alpha)$ is always greater than $(N - s_n - \alpha)$, the atom becomes smaller as electrons are added to its outermost ring.

[166] In the Rutherford Memorandum (Rosenfeld, xxiii), Bohr claimed that his quantization rule "explained" the atomic-volume curve. This is true in the sense of the previous note, but not in respect to predicting the periodicity, the most characteristic aspect of the curve.

[167] "Coincidentally" because the "ionization potential," as measured by Franck and Hertz, was in fact the first excitation potential. Cf., Heilbron, *History of . . . Atomic Structure, op. cit.* (note 1), 313–319.

10(8,2), precisely the reverse of that later accepted, and argon the assignment 18(8,8,2). The importance of the number eight in the electronic arrangements of the light elements is of course obvious from the periodicity of Mendeleev's table. Its connection with the inner rings, however, required a special argument, which we recapitulate as an illustration of Bohr's adroit exploitation of the ordinary mechanics in the service of his unmechanical model.

Imagine that the inner of two concentric and equally populated rings is slowly moved normally to their common plane by appropriate external forces. Electrostatic repulsion between the electrons pushes the outer ring out of the plane in the opposite direction. The inner ring expands, since the centripetal force on it decreases; the outer contracts, for the inverse reason. An equilibrium position might then be reached in which the rings attain the same size, with the electrons in one situated just opposite the intervals between the electrons of the other. If now the extraneous forces relax, the rings will coalesce in their original plane, providing, of course, that the central charge is large enough to protect the combined ring against disruptive oscillations perpendicular to its plane. Bohr concluded that there is a marked tendency for two adjacent rings to combine when each contains the same number of electrons. The formation of inner rings containing two, four, eight . . . electrons is thus likely, larger numbers being favored as the central charge increases. At some point in the periodic table two inner rings of four electrons should flow together into one ring of eight. Bohr set this point at neon because of the periodicity of the chemical elements; and neon, happily, is the first element whose central charge is large enough to bind eight electrons into a single ring stable against displacements normal to itself. As for the confluence of the two eight-rings of argon, it cannot occur until late in the periodic table, for, as we have said, a ring of sixteen requires a central charge of seventy-two. Bohr did not discuss the higher elements in detail. What he did say, however, proved prescient, for he observed that the properties of the iron group and of the rare earths suggested that the members of these families differed among themselves only in the arrangement of their inner electrons (Part II, 493–496/45–48).

The last few pages of Part II record Bohr's explanation of radioactivity and his elucidation of Whiddington's law (498–502/50–54). The latter provided a most fitting finale. It had been on Bohr's mind from the days of the Rutherford Memorandum, where it figured, along with the periodic law of atomic volumes, among the experimental facts marshalled to support the relation $T = Kv$. Then he had supplied no details. Now, having fixed the value of K, he was able to use Whiddington's intriguing relation to bring the principle of angular momentum, deduced in Part I from a consideration of the hydrogen spectrum, into semi-quantitative agreement with the relation $Z \doteq A/2$, the chief result of Rutherford's scattering theory and a main attraction of the nuclear atom. He had only to assume, as many physicists did in 1913, that the prerequisite for characteristic radiation was the removal of an electron from the innermost electron ring.[168] Whiddington had found that the velocity w just necessary to excite characteristic radiation in an element of atomic weight A is $w = A \cdot 10^8$ cm/sec; the orbital velocity v of one of the innermost electrons, assuming it feels the full force of the nuclear charge Ze, is

$$v = (2\pi e^2/h)Z = 2.1 \cdot 10^8 Z \text{ cm/sec.}$$

Now the energy required to remove an electron describing a circular orbit under an inverse-square force is the electron's kinetic energy. Hence one would expect v to equal w. This indeed follows from the preceding relations and the Rutherfordian approximation, $Z \doteq A/2$.

Part III develops the molecular theory of the Memorandum with the help of the stability considerations elaborated in Part II. The models are those Bohr had invented in the summer of 1912: collinear nuclei or positive ions held together by a ring of electrons each regulated by the principle $T = Kv$.[169] The Memorandum supported this picture with several impressive qualitative arguments, e.g., that a molecule so joined would dissociate into neutral atoms, in conformity with current experiments on oxygen; that two atoms of hydrogen could, while two of helium could not, remain together in such structures;[170] and that symmetric diatomic molecules, like H_2 or O_2, built on Bohr's plan, would show no infra-red absorption bands corresponding to vibrations of the nuclei along their axis.[171]

Part III improves upon the Memorandum via an obvious gener-

---

[168] See, e.g., J. J. Thomson, "Ionization by Moving Electrified Particles," *Phil. Mag.*, *23* (1912), 449–457, and Heilbron, "Moseley" (note 108, *supra*), 345.

[169] Rosenfeld, xxiv–xxviii; *supra*, 248–249.

[170] *Ibid.*, xxvi. The condition is that the energy of the molecule be less than the sum of the energies of its separated atoms.

[171] *Ibid.*, xxv: "The absence of absorption bands in the ultra-red for H_2 and O_2 follows . . . from the symmetrical condition of the two kerns (the same ratio of charge to mass)."

alization of the earlier quantum condition, namely, the requirement that each of the bonding electrons possess an angular momentum $h/2\pi$ about the molecular axis (Part III, 858/56). Then, Bohr showed, the condition for mechanical stability of the electronic vibrations perpendicular to the bonding ring restricts the number of bonding electrons to two or three, and the charges of the ions to unity (Part III, 861/59). This result admirably supported the molecular model for hydrogen, on which Bohr based most of his discussion (pp. 863–871/61–69). He described the formation of molecules much as he had the confluence of electron rings in Part II. He produced the consequential argument about disperson which we noticed earlier.[172] And he computed, as in the Memorandum, the heat of dissociation of a mole of hydrogen, using $K = h/2$ in place of the earlier $K = .6h$. The outcome was not as satisfactory as Bohr deserved; his result, $6.10 \cdot 10^4$ cal, was less than half the best experimental value, which was later found to be far too large.[173]

Parts II and III of "On the Constitution of Atoms and Molecules" evidently differ from Part I in achievement as well as in content. The theory of the Balmer formula was quantitative and, whether or not one approved of the principles from which Bohr deduced it, the demonstrations did follow from the assumptions. The unique and extremely suggestive specifications of atomic and molecular models, on the other hand, gave largely qualitative results, despite the mathematical ingenuity with which Bohr laboriously pursued their consequences, while the details of the structures did not follow from, and often conflicted with the principles deduced in Part I. Contemporaries were not unaware of this difference. Nicholson, for example, though he thought Bohr's theory of hydrogen and ionized helium "very attractive," regarded as wholly unjustified the extension to higher atoms. He even took the trouble to prove that, assuming Bohr's principles and the validity of the ordinary mechanics in the stationary states, the concentric ring model for lithium, $3(2,1)$, was impossible.[174]

The successful extension of the approach of Part I to the problems of Parts II and III required a decade, and enlisted the collaboration

---

[172] *Supra*, 282. The quantum computation obtains ν from the difference in energy between H_2^+ and the system $H^+ + H$.

[173] Rosenfeld, xxvii, xlvii, 61.

[174] For bibliography see Heilbron, "Moseley," *op. cit.* (note 108), 361, n. 81.

of many physicists. The universal principle of angular momentum, assigning the *same* quantum number, unity, to each electron in the ground state *regardless of its ring,* proved fallacious. New quantum numbers appeared. The concentric, coplanar rings burst asunder, scattering their electrons throughout the atomic volume.[175] Meanwhile Bohr was honing the Correspondence Principle, the most promising product of Part I, into a powerful instrument for probing the atom. In 1922 he returned to the problem of Parts II and III, bringing his sharpened Principle to bear on the new spatial models. Once again he built up the atoms with a deft combination of intuition and deduction, and once again he gave his readers the impression that his models followed directly from his principles.[176] This time, however, his profound insight penetrated nearer the atom's heart. Others quickly provided a narrower discrimination of the atomic subshells, culminating, in 1924, in the Pauli Principle, the closest solution the old quantum theory gave to the problem with which Bohr had begun his momentous journey to the quantized atom.[177]

---

[175] Cf., J. L. Heilbron, "The Kossel-Sommerfeld Theory and the Ring Atom," *Isis, 58* (1967), 451–485.

[176] N. Bohr, "Der Bau der Atome und die physikalischen und chemischen Eigenschaften der Elemente," *Zs. Phys., 9* (1922), 1–67.

[177] E. C. Stoner, "The Distribution of Electrons Among Atomic Levels," *Phil. Mag., 48* (1924), 719–736; W. Pauli, "Über den Zusammenhang des Abschlusses der Elektronengruppen im Atom mit der Komplexstruktur der Specktren," *Zs. Phys., 31* (1925), 765–783, esp. 773–776.

Why Was It Schrödinger Who Developed de Broglie's Ideas?

BY V. V. RAMAN* AND PAUL FORMAN**

I. INTRODUCTION

In 1923–1924 Louis de Broglie published a half dozen short articles on various aspects of his general conception that there was a very nearly complete parallel between matter (electrons, protons, etc.) and radiant energy, considered as quanta or "atoms of light."[1] The most prominent and important feature of this parallel was the attribution of a "phase wave" to every material particle, the frequency of the wave being given by $mc^2 = E = h\nu$, the wave length by $p = h/\lambda$, and the group velocity by $v = p/m$, where m is the mass of the particle and p its momentum. One by-product of this view was the interpretation of the prevailing quantum conditions, $\oint p_i \, dq_i = n_i h$, as expressing the conditions for resonance of the phase wave along an electronic orbit.[2] De Broglie gave a coherent exposition of his ideas in *Recherches sur la théorie des quanta* (Théses, Paris), published in November 1924. One year later Erwin Schrödinger began to apply himself seriously to the problem of developing the concept of phase waves presented in de Broglie's thesis.[3] After two months work Schrödinger

* Department of Physics, Rochester Institute of Technology, Rochester, New York 14623.
** Department of History, University of Rochester, Rochester, New York 14627.

[1] Items 16–18, 21–24 in the "Bibliographie Générale (dressée par l'auteur)" appended to *Louis de Broglie—physicien et penseur* (Paris, 1953) and·also to Louis de Broglie, *Recherches sur la théorie des quanta. réédition du texte de 1924* (Paris, 1963). Cf. (7) and (16), *infra*.

[2] L. de Broglie, "Ondes et quanta," *Comptes rendus, 172* (1923), 507–510. Accounts of de Broglie's work are given by Wm. T. Scott, *Erwin Schrödinger, An Introduction to His Writings* (Amherst, Mass., 1967), pp. 43–46; Max Jammer, *The Conceptual Development of Quantum Mechanics* (New York, 1966), pp. 243–247; Edmund T. Whittaker, *A History of the Theories of Aether and Electricity* (London, 1953), 2, 214–217.

[3] Schrödinger to A. Landé, 16 November 1925; quoted and discussed in Section VI below.

found the wave equation which bears his name and began the series of papers on "Quantisierung als Eigenwertproblem" which constituted wave mechanics.[4] The question which inevitably arises is: why did Schrödinger, of all theoretical physicists, take up de Broglie's ideas and develop them into a wave mechanics.[5] In what follows we offer some tentative answers to this question, first by arguing that Schrödinger was among the small fraction of quantum theorists who would not have been *a priori* ill-disposed towards de Broglie, and second by showing that Schrödinger previously had had certain ideas strikingly similar to those advanced by de Broglie.

II. ACCESS

In approaching the question "why Schrödinger?" we must first consider whether de Broglie's writings were uniquely accessible to Schrödinger, and whether Schrödinger had a unique sort of exposure to de Broglie's ideas. In some degree both were actually the case. In particular, the French scientific publications of this period were *not* widely available in Germany, but were fully accessible to Schrödinger

---

[4] Schrödinger, "Quantisierung als Eigenwertproblem (Erste Mitteilung)," *Annalen der Physik, 79* (Mar. 1926), 361–376, received 27 Jan. 1926. The four *Mitteilungen* and Schrödinger's paper "Über das Verhältnis der Heisenberg-Born-Jordanschen Quantenmechanik zu der meinen" are reprinted in Armin Hermann, ed., *Dokumente der Naturwissenschaft, Abteilung Physik, 3* (Stuttgart, 1963).

There is a well known anecdote due to Dirac (see Jammer, *C D Q M*, pp. 257–258; Scott, *Erwin Schrödinger*, pp. 50–51) that i) the first wave equation which Schrödinger found was the relativistic Klein-Gordon equation. When this equation, applied to the hydrogen atom, did not yield the familiar results, ii) Schrödinger abandoned his "method." "Only after some months did he return to it" and consider the non-relativistic approximation which he then published. There is good reason to credit i): Schrödinger, in his "Erste Mitteilung" (p. 372), explicitly stated the results he had obtained by treating the hydrogen atom with the Klein-Gordon equation (cf. "Quantisierung als Eigenwertproblem [Vierte Mitteilung]," *Ann. d. Phys., 81* [1926], 132); Schrödinger's notebooks in sections 5 and 6 of Sources for History of Quantum Physics microfilm nr. 40 (see the *Inventory and Report* published by the American Philosophical Society, Philadelphia, 1967) show him searching for a relativistic wave equation before he had found the non-relativistic equation; D. M. Dennison recalls (interview by Sources for History of Quantum Physics, 30 Jan. 1964) that in the fall of 1926 Schrödinger showed him the carbon copy of a paper on the relativistic wave equation which Schrödinger said had been written before the "Erste Mitteilung." There does not, however, appear to be the least reason to credit ii), i.e., the assertion that after carrying through a treatment of the hydrogen atom with the relativistic wave equation Schrödinger abandoned this general line of investigation for some considerable time (months).

[5] We are acquainted with only one attempt to raise and answer this question: Martin J. Klein, "Einstein and Wave-Particle Duality," *The Natural Philosopher, 3* (1964), 3–49, especially sections I and V. Klein's suggestions have been the stimuli for many of the considerations presented in this paper.

in Zurich.[6] When compared with the entire population of German language theorists—and they were certainly the largest group, and the group most involved with atomic physics—Schrödinger had unusually good opportunities for early acquaintance with de Broglie's ideas. Yet the fact remains that Schrödinger did not take up de Broglie's ideas until November 1925. By that time de Broglie's work, or at least its existence and its general tendency, had become rather generally known in Germany, and elsewhere. This came about in part through de Broglie's article in the *Philosophical Magazine*, but principally through Einstein's citations and advocacy.[7] Not merely at Zurich, but also at Cal Tech and probably a number of other institutions, an exposition of de Broglie's thesis was assigned as the topic for a colloquium talk.[8] Thus accidents of access or exposure are not likely to have been significant factors in Schrödinger's development of de Broglie's ideas.

III. DE BROGLIE'S REPUTATION

The case is quite different, however, when we ask how Schrödinger, in contrast to the great majority of theorists in atomic and quantum physics, would have been predisposed towards de Broglie.

[6] On the availability of foreign scientific publications in post World War I Germany: P. Forman, *Environment and Practice of Atomic Physics in Weimar Germany* (Ph.D. dissertation, Berkeley, 1967; Ann Arbor: University Microfilms, 1968), pp. 276–288. There are also anecdotes alleging transmission of de Broglie's thesis to Schrödinger (e.g. Jammer, *C D Q M*, p. 258). A. F. Ioffe (see Jammer, *C D Q M*, pp. 248–249) recalled that Langevin talked about de Broglie's ideas at the fourth Solvay Congress, April 1924. Both Einstein and Schrödinger attended this Congress.
[7] L. de Broglie, "A Tentative Theory of Light Quanta," *Phil. Mag., 47* (Feb. 1924), 446–458. A. Einstein, "Quantentheorie des einatomigen idealen Gases. 2. Abhandlung," *Preuss. Akad. d. Wiss., Sitzungsber.* (9 Feb. 1925), 3–14. Pauli (see Klein, *op. cit.* [note 5], 38) recalled that "Einstein proposed a search for interference and diffraction phenomena with molecular beams" in discussions at the *Naturforscherversammlung* in Innsbruck in September 1924. Schrödinger attended this congress (A. Sommerfeld to Schrödinger, 12 November 1924. Sources for History of Quantum Physics microfilm nr. 34, section 2). In June 1924 the *Physikalische Berichte, 5* (1924), 782–783, carried half-page abstracts by A. Smekal of de Broglie's "Waves and Quanta," *Nature, 112* (1923), 540, and "Quanta de lumière, diffraction et interférences," *Comptes rendus, 177* (1923), 548–550. The former paper, a letter to the editors of *Nature*, dated 12 September, summarizing his recent results, is not included in the "bibliographie générale" prepared by de Broglie. In October 1925 the *Physikalische Berichte, 6* (1925), 1250–1251 carried Smekal's 1-½ page abstract of de Broglie's thesis, "Recherches sur la théorie des quanta," as published in *Annales de physique, 3* (Feb. 1925), 22–128.
[8] Debye claims to have, so to speak, assigned de Broglie's thesis to Schrödinger as the topic of a colloquium talk (Jammer, *C D Q M*, p. 257; interview by Sources for History of Quantum Physics, 3 May 1962, p. 14). Epstein asserts that he assigned it to two different students to report on in his seminar (interview by S. H. Q. P., 26 May 1962, pp. 10–11).

We consider two aspects of this question: first the repute in which de Broglie would have been held on account of his previous work in atomic physics, and second the alignments which had already formed over the character of the as yet undiscovered quantum mechanics. De Broglie's reputation in Copenhagen must have been very poor. In 1921 he had been enthusiastic about Bohr's correspondence principle—at a time when it had not yet been widely appreciated or employed. But de Broglie had the temerity to give it an interpretation and application differing from Bohr's,[9] and indeed rather arbitrary ones. Kramers' response that de Broglie's usage "seems not to be consistent with the way in which the quantum theory at present is applied to atomic problems,"[10] for all its apparent mildness, will be recognized as a Copenhagen euphemism for "nonsense." In the following three years there accumulated a long list of disagreements between de Broglie and the Copenhageners over "the way in which the quantum theory at present is applied to atomic problems." There was the question whether or not every degree of freedom of every electron in an atom is entitled to a quantum number ("le système de Smekal").[11] De Broglie: yes; Bohr, and virtually all other theoretical atomic physicists: no. There was the question of the number of X-ray and optical energy levels associated with each value of the principal quantum number, and also the number of electrons with given quantum numbers in the atom of each element.[12] Here de Broglie, having teamed up with A. Dauvillier, entered into direct competition and conflict with the Copenhagen team of Bohr and D. Coster.[13] Finally, there was the very sensitive question of element 72. Urbain, who believed this elusive element to be a rare earth, had named it celtium, and in 1922 Dauvillier claimed to have identified its X-ray lines in a mixture of rare earths. But in 1923 Coster and Hevesy,

[9] L. de Broglie, "Sur la théorie de l'absorption des rayons X par la matière et le principe de correspondance," *Comptes rendus, 173* (1921), 1456–1458; "Rayons X et equilibre thermodynamique," *Journ. de phys. et le radium, 3* (1922), 33–45.

[10] H. A. Kramers, "On the Theory of X-Ray Absorption and of the Continuous X-Ray Spectrum," *Phil. Mag., 46* (1923), 836–870; on p. 840.

[11] L. de Broglie and A. Dauvillier, "Le système spectral des rayons Röntgen et la structure de l'atome," *Journ. de Phys., 5* (Jan. 1924), 1–19; on p. 3. De Broglie reasserted his view in *Recherches sur la théorie des quanta* (Nov. 1924; reissued 1963), pp. 51–52.

[12] L. de Broglie and A. Dauvillier, "Sur le système spectral des rayons Roentgen," *Comptes rendus, 175* (Oct. 1922), 685–688; "Sur les analogies de structure entre les séries optiques et les séries de Röntgen," *ibid.* (Nov. 1922), 755–756; *op. cit.* (note 11).

[13] N. Bohr and D. Coster, "Röntgenspektren und periodisches System der Elemente," *Zeits. für Physik, 12* (Jan. 1923), 342–374.

working in Copenhagen and guided by the Bohr-Coster theory, found the X-ray spectrum and isolated the element from zirconium ores. "For the new element we propose the name Hafnium (hafniae = Copenhagen)." [14] Dauvillier contested their claim to the discovery of element 72, and de Broglie associated himself with Dauvillier's refusal to yield the name "celtium." [15]

As this last matter suggests, the Copenhageners would not have regarded their disagreements with de Broglie as honest differences of scientific opinion. Although by the end of 1923 it was evident that de Broglie and Dauvillier's energy level scheme was wrong, they continued to advance it vigorously, along with assertions of their priority.[16] Thus in Copenhagen—and in Göttingen, where atomic physics was pursued in the Copenhagen spirit—de Broglie would certainly have had the reputation of a renegade, if not exactly a crank, who stuck obstinately to his own ill-conceived theories.[17]

Lying somewhat off the Copenhagen-Göttingen axis was the other important school of theoretical atomic physics, that of Sommerfeld in Munich. Although the number of points of friction between de Broglie and the Munich school were fewer, they were aggravated by the French-German animosities of these years. In a summarizing article, with Dauvillier, published in January 1924, de Broglie spoke rather scornfully of Sommerfeld's procedure in applying the quantum conditions to electronic orbits, and of his introduction of "inner" and "fun-

---

[14] D. Coster and G. Hevesy, "On the Missing Element of Atomic Number 72," *Nature, 111* (20 Jan. 1923), 79; "On the New Element Hafnium," *ibid.* (10 Feb. 1923), 182.

[15] G. Urbain and A. Dauvillier, "On the Element of Atomic Number 72," *Nature, 111* (17 Feb. 1923), 218; L. de Broglie and A. Dauvillier, *op. cit.* (note 11), 17. For Background: J. L. Heilbron, "The Work of H. G. J. Moseley," *Isis, 57* (1966), 336–364; for bibliography on the controversy over the discovery and the name: M. E. Weeks, *Discovery of the Elements,* 3rd ed. (Easton, Pa., 1935), pp. 322–324. "Nevertheless the impression which Coster and v. Hevesy's discovery made in Germany is not to be compared with the agitation which it produced in the news sheets of Denmark, France, and England. For quite apart from the fact that in Germany at present the columns of the newspapers are so crammed with the most ungratifying political reports that, naturally, scientific interests have to step entirely into the background, there was also the circumstance that the question of the discovery of element 72 gave rise to a national *Prioritätsstreit* in the aforementioned countries." (F. Paneth, "Über das Element 72 [Hafnium]," *Ergebnisse der exakten Naturwissenschaften, 2* [1923], 163–176; on pp. 168–169.)

[16] L. de Broglie and A. Dauvillier, *op. cit.* (note 11), 14. Also, L. de Broglie and A. Dauvillier, [letter to the editors], *Phil. Mag., 49* (April 1925), 752–753. This "priority claim" is not included in the "bibliographie générale" prepared by de Broglie.

[17] It must be admitted, however, that one of the early attempts to employ de Broglie's ideas was made by W. Elsasser, "Bemerkungen zur Quantenmechanik freier Elektronen," *Naturwiss., 13* (August 1925), 711, then a student at Göttingen, and his efforts received a measure of encouragement from Franck and Born. (W. Elsasser, interview by J. L. Heilbron for Sources for History of Quantum Physics, 29 May 1962.)

damental" quantum numbers to account for the optical and X-ray energy levels, respectively. "In reality," de Broglie charged, "this system has been adopted for the purpose of safeguarding the explanation of the L doublet given by Sommerfeld." [18] Then after some condescending remarks on Sommerfeld's and Heisenberg's treatment of the optical doublets, de Broglie and Dauvillier concluded: "One must, moreover, remark that Sommerfeld's school is tending gradually to reconcile itself to the viewpoint developed here, as a result of the increasing precision of the experimental data." [19] If "Sommerfeld's school" was not infuriated by these sneers, it was because they knew that, in fact, experiment had shown the de Broglie-Dauvillier scheme to be out of the question, and they could dismiss the authors as stubborn chauvinists.

Thus among the central European physicists deeply involved in the problems of theoretical spectroscopy—and this was indeed the great majority of those seriously concerned with the quantum theory —de Broglie must have had a very bad reputation. These theoretical spectroscopists were not likely to invest much effort in trying to make sense of and extend de Broglie's recent phantasies about phase waves associated with material particles. As Klein has emphasized, one of the first theorists to see real merit in de Broglie's work was Einstein, who had never allowed himself to become involved in the problems of theoretical spectroscopy. If, as de Broglie testified, it required Einstein's backing to obtain a sympathetic hearing for his ideas, that was at least as much on account of de Broglie's personal reputation as on account of any characteristics of the ideas themselves.[20]

While matrix mechanics was developed by those concerned with theoretical spectroscopy, Klein argued that the other quantum mechanics, the wave mechanics, was a product of a wholly distinct tradition and group of theorists concerned with quantum statistics, "and the

---

[18] Op. cit. (note 11), 3. That is, Sommerfeld interpreted the prominent doublet in the L series of the X-ray spectra as a greatly enlarged version of the relativistic fine structure of the hydrogen spectrum. P. Forman, "The Doublet Riddle and Atomic Physics circa 1924," Isis, 59 (1968), 156–174.

[19] Op. cit. (note 11), 6.

[20] L. de Broglie, New Perspectives in Physics (New York, 1962), p. 140; quoted by Klein, op. cit. (note 5), 38. In the five years 1920–1924 fourteen papers authored or co-authored by de Broglie had been abstracted in the Physikalische Berichte, so that it seems fairly certain that he was not without a reputation among the Central European theoretical physicists at the time his notion of a wave phenomenon associated with material particles began to be widely known.

presiding genius and principal guide was not Bohr, but Einstein."[21] Yet it seems that the man who was to develop de Broglie's ideas into a wave mechanics—which was, after all, in the first instance, a spectroscopic theory—would have had to have, as de Broglie himself had, one foot in each of these two camps. The reproduction of the quantum conditions for the hydrogen atom was for de Broglie, as the energy levels would be for Schrödinger, "the best justification that we are able to give of our manner of attacking the problem of quanta."[22] And yet, de Broglie's discussion of the dynamics of the atom, the interpretation of the quantum conditions, $\oint p_i \, dq_i = n_i h$, as the conditions for a stationary phase wave regime, occupied only a very small fraction of his dissertation. Thus the theoretical spectroscopist, who looked forward to the quantum mechanics as the quantum theory of the atom, would again have found little to sustain his interest.

Schrödinger, a marginal man, a loner, a member of no school, had interests as broad as, and largely congruent with, de Broglie's. In contrast with other leading statistical mechanicians such as Einstein, Planck, Laue, Ehrenfest, and R. H. Fowler, Schrödinger was not merely abreast of, but made important contributions to, theoretical spectroscopy.[23] Yet, as Klein properly emphasizes, Schrödinger was not fully committed to the field and kept aloof from its internal conflicts. This detachment received clear expression in Schrödinger's review, in the summer of 1925, of volume VI of the *Handbuch der Radiologie: Die Theorien der Radiologie.* The review contains no word of criticism: each author wrote "what lay closest to his heart" and the volume thus reads like a "Glaubensbekenntnis," and as such "will attain lasting classical value."[24] At the same time Schrödinger was more acutely aware than most theoretical spectroscopists of the logical gaps in the theories employed,[25] while still sharing their general view that the difficulties in the physical interpretation of the X-ray energy levels show "ein tiefes Nichtverstehen der Systemmechanik."[26] Schrödinger,

[21] Klein, *op. cit.* (note 5), 4.

[22] L. de Broglie, *Recherches* (Nov. 1924; reissued 1963), *op. cit.* (note 1), 53.

[23] Schrödinger, "Versuch zur modelmässigen Deutung des Terms der scharfen Nebenserien," *Zeits. für Phys., 4* (1921), 347–354; "Die Wasserstoffahnlichen Spektren vom Standpunkte der Polarisierbarkeit des Atomrumpfes," *Ann. d. Phys., 72* (June 1925), 43–70.

[24] Schrödinger, *Naturwiss., 13* (14 Aug. 1925), 710–711.

[25] Schrödinger, *op. cit.* (note 23), 43.

[26] Schrödinger, notes for lectures on the "Theorie der Spektren," Winter Semester 1925–1926, p. 79 (Sources for History of Quantum Physics microfilm nr. 40, section 4). Cf. Forman, "Doublet Riddle," *op. cit.* (note 18).

then, of all theoretical spectroscopists, was one of those least likely to
have been strongly prejudiced against de Broglie, and thus most likely
to give serious consideration to the problem of developing de Broglie's
ideas into an atomic mechanics.

IV. PROGRAMMATIC PREJUDICES

There is yet another, more "intellectual," aspect to the question of
predispositions towards de Broglie, namely, prior judgments about de
Broglie's "manner of attacking the problem of quanta" deriving from
prejudices, commitments, and alignments on the question of the central problem and probable character of the sought-for quantum mechanics. This is an exceedingly large and complex problem, and the
following remarks are both summary and tentative, indeed speculative.

Is it merely a coincidence that de Broglie, Einstein, and Schrödinger—the three principal figures on the path to wave mechanics
—also became the most prominent opponents of a probabilistic and
indeterministic interpretation of that theory? Not at all. The well-
known conflict between Bohr and Einstein after 1926 is but the pro-
jection of an alignment which had earlier formed over the question of
the likely, if not desired, character of the new mechanics. Einstein had
been clear all along about his own position and the tendencies
he found distasteful,[27] but it was only about 1923 that this developed
into a sharp antithesis between him and Bohr.[28]

Left:	1. light quanta	Right:
	2. description in space-time	
Bohr, *anti*	3. description by differential eqs.	Einstein, *pro*
	4. complete continuity and causality	

[27] Einstein to M. Born, 27 Jan. 1920, extract quoted in Born, "Physics and Relativity,"
Helvetica Physica Acta, Supplementum, 4, (1956), 257; quoted at somewhat greater length in Born,
Physik im Wandel meiner Zeit, 4th ed. (Braunschweig, 1966), p. 293.
[28] Einstein, "Bietet die Feldtheorie Möglichkeiten für die Lösung des Quantenproblems?"
Preuss. Akad. d. Wiss., Sitzungsber. (1923), 359–364; Einstein to Born, 24 April 1924, as quoted in
Physik im Wandel, op. cit. (note 27), 294. Bohr, ". . . The Fundamental Postulates of the Quantum
Theory," *Camb. Phil. Soc., Proc., Suppl.* (1924), dated Nov. 1922, on p. 35; Bohr, H. A. Kramers,
and J. C. Slater, "The Quantum Theory of Radiation," *Phil. Mag., 47* (1924), 785–802,
on p. 790. (The first of these papers by Bohr appeared in German in the *Zeits. für Phys.* early in
Feb. 1923, and the second in late May 1924.) On the development of Bohr's views: K. M.
Meyer-Abich, *Korrespondenz, Individualität und Komplementarität* (Wiesbaden, 1965). Although it

After the discovery of the quantum mechanics Bohr came to accept light quanta, while "description by differential equations," which had been thought to be equivalent to "complete continuity and causality," was found not to have any physico-philosophical bearing. However, the other two bones of contention, "space-time description" and "complete causality," continued to be discussed under the rubrics "complementarity" and "indeterminacy."

We cannot here attempt to analyze the stages in the formation of this alignment or the grounds upon which these four issues were regarded as intimately linked. We may note, however, that partisans of the left, in their effort to eliminate light quanta entirely, came, by way of virtual oscillators, to the matrix mechanics; partisans of the right, in their effort to justify light quanta, came, by way of a radical parallelism between light quanta and material particles, to the wave mechanics. For the purposes of the question in hand—"Why Schrödinger?"—the relevant fact is that among the leading European physicists actively concerned with the quantum theory, Einstein's faction was a very small minority. The theoretical spectroscopists, who were by far the largest group, tended to side with Bohr, not so much because of his leadership as because of the current difficulties with atomic models.[29] More than that, Bohr's position was in the air and had been, as Bohr himself said when putting it forward, "advocated from various sides" for some years past.[30] The question of the

---

might seem that the acceptance of light quanta is a more radical position than the rejection of that concept (and so Planck, and also Bohr, represented the situation), in this period the opposite was actually the case, at least in Germany. The light quantum was relatively *anschaulich*, a distinction at the crux of the left-right opposition in German physics; in practice the real threat to the lofty structure of classical physical theory came from those who sought to circumvent light quanta rather than from those who made use of that concept.

[29] Forman, "Doublet Riddle," *op. cit.* (note 18).

[30] Bohr, Kramers, Slater, *op. cit.* (note 28), 790. Bohr there cites O. W. Richardson, *Electron Theory of Matter*, 2nd ed. (Cambridge, 1916), p. 507 as, "perhaps for the first time," clearly expressing doubts about the possibility of a "causal description in space and time." Serious doubts whether natural processes at the atomic level could be described by means of differential equations had been raised circa 1911 in connection with the problem of deriving Planck's blackbody radiation formula. See: R. McCormmach, "Henri Poincaré and the Quantum Theory," *Isis, 58* (1967), 37–55. It had, of course, been clear from the fact of the atomistic structure of matter that even if Hamiltonian mechanics were universally valid the differential equations which so successfully described macroscopic processes—such as propagation of light, heat conduction, and hydrodynamics—were only "limiting" or "abbreviated" forms of the true equations, which were necessarily equations of finite differences. See, for example, the remarks by Boltzmann (1899) and Schrödinger (1914) quoted by Jammer, *C D Q M*, p. 256. Cf. Felix Klein: "Now it is again the molecular theory which, of course in a form appropriate to the time, advances more into the foreground; the differential equations of physical problems appear to be scarcely more than abbreviated forms of difference equations." (W. Lexis, ed., *Das Unterrichtswesen im Deutschen Reich. 1. Die Universitäten* [Berlin, 1904], 248.)

validity of the law of causality was a favorite topic for popular scientific lectures in Germany at this time, and public sentiment was evidently not in its favor.[31] Indeed, some of the most prescient statements came from quasi-cranks: "the possibility of describing a process in space and time with arbitrary accuracy (by coordinates) is in principle limited by Planck's constant h."[32]

In aligning himself with Einstein on the right-hand side of the fence de Broglie had, *ipso facto,* drastically limited the audience for his ideas. The bulk of the quantum theorists would simply have been unsympathetic to that "manner of attacking the problem of quanta." Schrödinger, again, was different—although his views were anything but uncomplicated. In the years 1922–1924 Schrödinger seems to have maintained the same sort of nonpartisan attitude which he had preserved in respect to the disputes among the several schools of theoretical spectroscopy. He had, on the one hand, expressed a belief that the "serious intrinsic incoherences" in the electrodynamics of the atom would only be eliminated "once we have discarded our rooted predilection for absolute causality."[33] And when Bohr, Kramers, and Slater proposed a theory which circumvented the light-quantum interpretation of the Compton effect by abandoning causality and the conservation laws, Schrödinger published an enthusiastic exposition of their theory and its consequences.[34] On the other hand, Schrödinger

---

[31] Born, Nernst, Planck, Schrödinger, among others, gave lectures which explicitly called attention to this issue in the title. Many popular essays on contemporary physics included some discussion of it. Planck regarded the law of causality (in an exceedingly generalized sense, to be sure) as the *sine qua non* of natural science and complained bitterly that "precisely in our time, which indeed plumes itself so highly for its progressiveness, the belief in miracles in the most various forms—occultism, spiritualism, theosophy, and however all the numerous shadings may be called —penetrates wide circles of the educated and uneducated more mischievously than ever." (*Kausalgesetz und Willensfreiheit; öffenlicher Vortrag gehalten in der preussischen Akademie der Wissenschaften am 17. Februar 1923* [Berlin, 1923], p. 44; see also Forman, *op. cit.* [note 6], 11–24.) Most other physicists discussing the validity of the law of causality would, however, have agreed with Walther Nernst that "the previously generally accepted conception of the principle of causality as an absolutely rigorous law of nature laced the intellect in Spanish boots, and it is surely at present the obligation of scientific research to loosen these fetters sufficiently so that the free stride of philosophical thought is no longer hindered." ("Zum Gültigkeitsbereich der Naturgesetze," *Naturwiss., 10* [1922], 489–495; on p. 495.)

[32] H. A. Senftleben, "Zur Grundlegung der 'Quantentheorie'," *Zeits. für Phys.,* 22 (Mar. 1924), 127–156; on p. 131.

[33] Schrödinger, "What is a Law of Nature," *Science, Theory, and Man* (New York, 1957), p. 146. Schrödinger's inaugural lecture at the University of Zurich, of which this is a translation, was not published until 1929. It is alleged to have been delivered on 9 Dec. 1922, but as Schrödinger took up his duties at Zurich in the fall of 1921, Dec. 1921 seems more probable.

[34] Schrödinger, "Bohrs neue Strahlungshypothese und der Energiesatz," *Naturwiss., 12* (5 Sept. 1924), 720–724. Cf. Jammer, *C D Q M,* p. 184; Scott, *Erwin Schrödinger,* pp. 49–50.

was quite prepared to take light quanta seriously, and had himself shown in 1922 that the classical Doppler shift of the frequency of radiation from a moving Bohr atom followed immediately from the application of the laws of conservation of energy and momentum to light quanta.[35]

Yet in late 1924 or early 1925 Schrödinger came down off the fence, onto the right-hand side. Schrödinger's close personal relations with Willy Wien suggest that the stimulus to this shift is likely to have been as much personal and political as scientific, and this supposition is supported by the fact that after 1924 Schrödinger made no use of the *Zeitschrift für Physik*, the organ of the left wing of German physics, although it would have been the most appropriate vehicle for his papers on the wave mechanics.[36] Thus by 1925 Schrödinger would appear to have become one of the few members of the Einstein-de Broglie faction.

In concluding this discussion of "programmatic prejudices" it is necessary to consider briefly how de Broglie's ideas, apart from any stigma deriving from their position in the left-right alignment, would have been regarded by contemporary theoretical physicists generally, and by Schrödinger in particular. It is perhaps only a slight exaggeration to suggest that, disregarding de Broglie's personal reputation, the ideas put forward in his thesis would have appeared to be those of a cross between a crank and a reactionary. After all, not since W. H. Bragg's "neutral pair" theory of X-rays, which had been abandoned

---

[35] Schrödinger, "Dopplerprinzip und Bohrsche Frequenzbedingung," *Physikalische Zeits.*, 23 (1922), 301–303. Scott, *Erwin Schrödinger*, p. 48, gives but a hurried glance to this paper which was quite famous at the time. Cf. A. Sommerfeld, *Atombau und Spektrallinien*, 4th ed. (Braunschweig, 1924), p. 53.

[36] Schrödinger quite consciously allowed extra-scientific considerations great weight in the formation of his scientific opinions; as he wrote Wien on 25 August 1926: "Bohr's standpoint, that a spatio-temporal description is impossible, I reject *a limine*. Physics consists not merely of atomic research, science not merely of physics, and life not merely of science. The purpose of atomic research is to fit our experiences from this field into the rest of our thought; but all the rest of our thought, in so far as it has to do with the external world, moves in space and time." W. Wien, *Aus dem Leben und Wirken eines Physikers* ... (Leipzig, 1930), p. 74. This line of argument, which was basic to the contemporary opposition to the theory of relativity, suggests a strange rapport between Schrödinger and the right wing of German physics. The division in scientific opinion between those with and those without enthusiasm for either relativity or the quantum theory, or for both, was also expressed as a political division on national as well as professional issues, and was reflected in the editorial policies and lists of contributors to the various physical journals and handbooks. See P. Forman, *op. cit.* (note 6), 55–58, 161–205; cf. Siegfried Grundmann, "Der Deutsche Imperialismus, Einstein und die Relativitätstheorie (1914–1933)," *Relativitätstheorie und Weltanschauung* (Berlin-Ost, 1967), pp. 155–285.

a dozen years before, had a respectable physicist proposed "atoms of light" with non-zero rest mass, "possessing the same symmetry as an electric dipole," and moving with a velocity slightly less than c.[37] And if de Broglie's attempt to make light quanta material particles was simply mistaken, his efforts to attach a wave phenomenon to every material particle might also have been dismissed, not because that notion was so very novel, but because it sounded all too familiar. The late nineteenth- and early twentieth-century physicists were acquainted with several attempts to frame "wave" theories of gravitational attraction—by Challis, Bjerknes, Leahy, A. Korn—in which the mass of a particle determined the properties of the wave-phenomenon of which it was the source.[38] The recent attempt by the elder Brillouin to "represent the essential properties of the Bohr atom" by phase relations between the orbiting electron and the waves it produces in the hypothetical elastic medium through which it moves[39] would certainly not have helped his compatriot de Broglie win serious consideration of his own "resonance" explanation of those orbits. Once again, it is surely no accident that Einstein, who, unlike his colleagues, was very patient and conscientious in answering crank mail,[40] was among the first to see real promise in de Broglie's ideas.

If, as seems likely, Schrödinger, along with his fellow theorists, was repelled by these latter aspects and associations of de Broglie's ideas, he would nonetheless have been attracted far more than his fellows by certain other aspects. First of all, the attempt to combine atom mechanics and statistical mechanics would have been highly congenial. Secondly, the attempt to use relativity as a guide and as a source of formal apparatus would have attracted Schrödinger. Thirdly, the stress on Hamilton's analogy—i.e., the action integral, $\delta C = \delta \int 2T \, dt = 0$, may be regarded as determining the trajectory of a particle or the ray of a wave—would have struck a responsive chord. In certain respects Schrödinger had himself carried this analogy even farther than de Broglie in unpublished investigations of "Tensoranalytische Me-

[37] L. de Broglie, *Recherches* (1924; reissued, 1963), p. 65.

[38] J. D. North, *The Measure of the Universe* (Oxford, 1965), pp. 32–34; M. Jammer, *Concepts of Mass* (Cambridge, Mass., 1961; New York, 1964), p. 141.

[39] Jammer, *C D Q M*, p. 242. Marcel Brillouin, "Actions méchaniques à hérédité discontinue par propagation; essai de théorie dynamique de l'atom à quanta," *Comptes rendus, 168* (1919), 1318–1320, left it undecided whether the waves are emitted by the particle "by its own vibrations or as a result of its motion in the medium."

[40] P. Frank, *Einstein, His Life and Times* (New York, 1947), p. 117.

chanik" which he had made between 1918 and 1922.[41] In the first of three research notebooks, he developed analytical mechanics in tensor notation, laying stress upon Hamilton's analogy in a more suggestive way than de Broglie had; he set the velocity v of a particle proportional to the gradient of the action function C, which forms a family of surfaces in configuration space:

$$v \propto \frac{\delta C}{\delta s} = \sqrt{2(E - V)}, \quad \int_{c_0}^{c_1} dC = C_1 - C_0 = \int \sqrt{2(E - V)} \, ds.$$

The third notebook, entitled "Tensoranalytische Mechanik und Optik inhomogener Medien," with a section on *"Analogien zur Optik. Huyghensche Prinzip und Hamilton'sche partialle Differentialglng* [sic]," shows clearly that Schrödinger had examined closely the analogy to which de Broglie now attributed a fundamental significance. Finally, the second notebook explored connections with Weyl's extension of general relativity, especially the issue of "Parallelverschiebung," thus showing that, to Schrödinger's mind, there were close connections between Hamilton's analogy and another cluster of ideas to which we now turn.[42]

V. "ÜBER EINE BEMERKENSWERTE EIGENSCHAFT . . ."

We have argued that, among leading European physicists, Schrödinger was one of a small minority who would not have been prejudiced against de Broglie's ideas either on account of de Broglie's scientific reputation or on account of the type of solution of the problem of quanta for which de Broglie stood. There is, however, one further, possibly decisive, reason why it was precisely Schrödinger who applied himself seriously and successfully to the problem of developing an atom mechanics of de Broglie's phase waves.

In December 1926, Fritz London, who had himself been contrib-

[41] Schrödinger, three notebooks on "Tensoranalytische Mechanik," 118 pp., Sources for History of Quantum Physics microfilm nr. 39, section 3. The general treatment employed there, with the kinetic energy determining the metric of the configuration space and connections with Heinrich Hertz's mechanics, appears again in Schrödinger's second paper on wave mechanics, "Quantisierung als Eigenwertproblem (Zweite Mitteilung)," *Ann. d. Phys.*, 79 (1926), 489–527.
[42] In connection with the affinity Schrödinger perceived between general relativity and quantum theory it is perhaps pertinent to note that he published nothing on the former topic between 1918 and 1925—excepting the paper discussed in the following section—but that he returned to this field in the late spring of 1925: "Die Erfüllbarkeit der Relativitätsforderung in der klassischen Mechanik," *Ann. d. Phys.*, 77 (1925), 325–336, received 6 June 1925.

uting to the development and application of the recently discovered wave mechanics, wrote to Schrödinger from Stuttgart:

Dear Professor,

I must have a serious word with you today. Are you acquainted with a certain Mr. Schrödinger, who in the year 1922 (*Zeits. für Phys., 12*) described a "bemerkenswerte Eigenschaft der Quantenbahnen?" Are you acquainted with this man? What! You affirm that you know him very well, that you were even present when he did this work and that you were his accomplice in it? That is absolutely unheard of.... You even had the resonance character of the quantum postulate in your hands long before de Broglie.... Will you now immediately confess that, like a priest, you kept secret the truth which you held in your hands, and give notice to your contemporaries of all that you know! [43]

[43] F. London to Schrödinger, ca. 10 Dec. 1926 (not "[Spring 1927]" as catalogued by Sources for History of Quantum Physics), S.H.Q.P. microfilm nr. 41, section 9. The transcription which follows *is* accurate.

Stuttgart Hoferstr 5.

Hochverehrter Herr Professor.

Ich muss heute mit Ihnen ein ernstes Wort reden. Ist Ihnen ein gewisser Herr Schrödinger bekannt, der im Jahre 1922 (Zs. f. Ph. 12) eine „bemerkenswerte Eigenschaft der Quantenbahnen" beschrieben hat? Ist Ihnen dieser Mann bekannt? Was, Sie behaupten, Ihn ganz gut zu kennen, Sie seien damals, als er die Arbeit schrieb, gar bei ihm gewesen und seien an der Arbeit mitschuldig? Das ist ja unerhört. Sie haben also vor 4 Jahren schon gewusst, dass man in dem Kontinuumsgeschehen, in welches man die Atomvorgänge aufzulösen hat, keine Massstäbe und Uhren zur Definition eines Einstein-Riemannschen Masszusammenhanges hat und also zu sehen hat, ob vielleicht mit den allgemeineren Massprinzipien, wie sie durch Weyls Theorie der Streckenübertragung ausgedrückt ist, weiterhilft, und Sie haben vor 4 Jahren sehr wohl gemerkt, dass Sie sogar sehr gut weiterhilft. Während nämlich gewöhnlich Unsinn herauskommt bei Weyls Streckenübertragung (Einsteins Einwände, Weyls sehr faule Ausrede mit der „Einstellung") haben Sie gezeigt, dass auf den diskreten wirklichen Bahnen die Eicheinheit sich (bei $\gamma = 2\pi i/h$) reproduziert bei räumlich geschlossenem Wege; und zwar haben Sie dabei bemerkt dass auf der n-ten Bahn n-mal die Masseinheit anschwillt und zusammenschrumpft genau wie bei der stehenden Welle, welche den Ort der Ladung beschreibt. Sie haben also gezeigt dass die Weylsche Theorie erst dann vernünftig—*d.h. zu einer eindeutigen Massbestimmung führend*—wird, wenn man sie mit der Quantentheorie verknüpft und zwar bleibt Einem ja garnichts anderes übrig, wenn die ganze atomare Welt ein Kontinuums-[ge]schehen ohne alle identifizierbaren Fixpunkte darstellt. Das haben Sie gewusst und nichts davon gesagt und ausgesprochen. Das ist beispiellos. Ganz schüchtern schreiben Sie in der Arbeit (S. 14): Sie seien—um es gleich zu gestehen—bei der Diskussion der eventuellen Bedeutung der Tatsache nicht sehr weit gekommen. Und dabei haben Sie in dieser Arbeit nicht nur der heillosen Konfusion der Weylschen Theorie ein Ende gemacht, Sie haben sogar den Resonanzkarakter der Quantenforderung lange vor De Broglie in den Händen gehabt und überlegen sich, ob Sie $\gamma = h/2\pi i$ oder $\sim e^2/c$ nehmen sollen! (S. 23.)—Werden Sie nun schnellstens bekennen, dass Sie einem Pfaffen gleich die Wahrheit, die Sie in den Händen hielten, geheim gehalten, und der Mitwelt alles kund geben, was Sie wissen! Das entscheidenste ist ja noch zu machen, jene Bemerkung aus 1922 war ein Satz der alten Quantenmechanik. Man kann mit Sicherheit voraussahnen, dass er in den sinnvollen Zusammenhang der Undulationsmechanik gebracht (ich habe es bis jetzt noch nicht getan) erst seinen ganzen Charakter zeigen wird. Ich finde, dass es Ihre Pflicht ist, nachdem Sie die Welt so mystifiziert haben, jetzt alles aufzuklären.

Nun genug. Haben Sie vielen Dank, dass Sie auf meinen blöden Brief [of 1 Dec. 1926, to which Schrödinger replied on 7 Dec. 1926] so eingegangen sind. Ich habe die Sache für *vorläufig* liegen lassen. Ich finde, im ganzen bedeutet die Kaluza—Kleinsche Raumtheorie einen *Rückschritt*, nachdem es die schöne Weylsche Raum-Theorie gibt, und ich möchte das erst näher verfolgen. Ich habe verschiedenen Anhalt, dass es nicht schwer sein wird, Weyl und Kaluza unter einen Hut zu

The paper to which London refers, "Über eine bemerkenswerte Eigenschaft der Quantenbahnen eines einzelnen Elektrons," is dated "Arosa, 3. Oktober, 1922."[44] Schrödinger began by stating briefly the principal result of Hermann Weyl's extension of general relativity. (Weyl's theory had originally been published in 1918, but somewhat revised in the 4th edition of his *Raum, Zeit, Materie,* published in 1921, and Schrödinger cites this latter work.[45])

---

bringen (tragen Sie sich an jedem Welt-Punkt die *Eicheinheit* als 5. Dimension auf, so sieht man gleich allerlei schönes!). Ich bin nun sehr begierig auf Ihr Manuskript [on relativistic wave equations mentioned in Schrödinger's letter of 7 Dec. 1926] (bis jetzt ist es noch nicht da) besonders nach den Andeutungen von Fues. Wenn es auch nur auf einen Tag wäre, möchte ich es doch *sehr, sehr* gern sehen können.

Ubrigens das Rockefell ist bewilligt, gestern kam das Telegramm! Ich bin so froh, dass es nun gewiss ist, dass ich mit Ihnen werde arbeiten dürfen.

Ich wünsche Ihnen nun alles Gute für Ihre Reise, ich freue mich schon auf Ihre Rückkehr.

Mit herzlichem Grusse bin ich

Ihr ganz ergebener

Fritz London.

Schrödinger's reply to this letter—and it would certainly be of considerable moment for the thesis presented in this and the following section—is not known to us. (Upon the kind initiative of Mrs. Edith London, search was made by Dr. Mattie Russell, Curator of Manuscripts, Duke University Library, Durham, North Carolina, in Fritz London's correspondence, which Mrs. London deposited, and has made available for research, at that institution.) Inasmuch as London's attitude toward the "bemerkenswerte Eigenschaft" as expressed in papers he submitted for publication in the next few months did not differ fundamentally from that stated in this letter of ca. 10 December, we surmise that Schrödinger had not repudiated London's evaluation in the meantime. (Schrödinger left Zurich for Madison on 18 December, so that it is also quite possible that London got no reply at all.) London reported on his attempt to find this "sinnvollen Zusammenhang" between the "bemerkenswerte Eigenschaft" and the wave mechanics at the meeting of the Gauverein Württemberg of the Deutsche Physikalische Gesellschaft in Stuttgart on 18 December 1926, sent off a brief notice on 19 January 1927 ("Die Theorie von Weyl und die Quantenmechanik," *Naturwiss., 15* [25 Feb. 1927], 187), and a longer essay on 27 February ("Quantenmechanische Deutung der Theorie von Weyl," *Zeits. für Phys., 42* [1927], 375–389). By going to the five-dimensional relativistic theory of O. Klein, Fock, and Kudar, London believed he could identify the Weyl-Schrödinger "tract factor" (*Eicheinheit*) with the wave function, Ψ.

[44] Schrödinger, *Zeits. für Phys., 12* (9 December 1922), 13–23. M. Jammer, *C D Q M* (1966), p. 96 has cited this paper as one of a number of "attempts at reconciling the quantum conditions with the general theory of relativity." Wm. T. Scott, *Erwin Schrödinger* (1967), p. 47, has given a summary of this paper, suggesting that Schrödinger was "hinting at a sort of 'phase factor' considerably before the wave idea was applied to electrons." Still, Scott does not raise the question of the significance of this work for Schrödinger's subsequent response to, or work on, de Broglie's ideas.

[45] H. Weyl, "Gravitation und Elektrizität," *Preuss. Akad. d. Wiss., Sitzungsber.* (1918), trans., with additional notes, in H. A. Lorentz, *et al., The Principle of Relativity* (London, 1923; reprinted, New York, 1952), pp. 201–216. Weyl, *Raum, Zeit, Materie,* 4th ed. (Berlin, 1921); trans. by H. L. Brose as *Space-Time-Matter* (London, 1922; reprinted, New York, 1952). A clear account of this now defunct theory is given in the article on relativity which W. Pauli prepared in 1920–1921 for the *Encyklopädie der math. Wiss.,* trans. by G. Field as *Theory of Relativity* (New York, 1958), pp. 192–202, 223–224. The widespread interest in Weyl's theory is indicated by Whittaker, *op. cit.* (note 2), 188–191. The strong and permanent influence of Weyl's ideas upon Schrödinger is evident in the organization and content of Schrödinger's *Space-Time Structure* (Cambridge, England, 1950).

In Weyl's world-geometry, in addition to the well known quadratic form of the differentials, which determines the metric at the individual world points, there also arises a linear form

$$\phi_0 \, dx_0 + \phi_1 \, dx_1 + \phi_2 \, dx_2 + \phi_3 \, dx_3 = \phi_i \, dx_i$$

which determines the metric relationship between the individual world points.[46]

That is, at each world point $P \equiv (t, x_1, x_2, x_3)$ there is a metric tensor, $\{g_{ij}\}$, which determines the square of the length of any vector $dx \equiv (dt, dx_1, dx_2, dx_3)$ from P to an infinitesimally near point P':

$$ds^2 = |dx|^2 = \sum_{i,j=0}^{3} g_{ij} \, dx_i \, dx_j,$$

where, following Weyl and Schrödinger, we write x_0 for t. Moreover if $\xi \equiv (\xi_0, \ldots, \xi_3)$ is an arbitrary finite four-vector at the point P, the g_{ij} also determine its "tract" (*Strecke*), the square of its length:

$$l \equiv |\xi|^2 = \sum_{i,j=0}^{3} g_{ij}\xi_i\xi_j.$$

Now ξ may undergo a "parallel displacement" (*Parallelverschiebung*) to P'—which means, in rough analogy with the usual notion of displacing a vector parallel to itself, that in a suitably chosen coordinate system the components of ξ at P' are the same as at P.[47] Nonetheless, its "tract" will in general be altered:

$$l(P') = l(P) + dl.$$

Weyl set $dl = -l \cdot d\phi$, with $d\phi$ independent of l, and found that under reasonable restrictions $d\phi$ must be a linear differential form: $d\phi = \sum_{i=0}^{3} \phi_i \, dx_i$, or, adopting the convention that a summation is to be understood when indices are repeated, $d\phi = \phi_i \, dx_i$.[48] If then a vector with tract l_0 is "congruently transplanted" along a finite length of a world-line, i.e. undergoes a repeated infinitesimal parallel displacement subject to the condition $dl = -ld\phi = -l\phi_i dx_i$, its tract becomes:

$$l = l_0 e^{-\int d\phi} = l_0 e^{-\int \phi_i \, dx_i}.$$

The value of the integral, and the length of the vector, will be in-

[46] Schrödinger, *op. cit.* (note 44), 13.

[47] *Raum, Zeit, Materie* (1921), pp. 100 ff.; *Space-Time-Matter*, pp. 112 ff.

[48] *R Z M* (1921), pp. 109–110; *S T M*, pp. 122–123.

dependent of the path (world-line) only if $d\phi$ is a total differential. But exploration of the conditions to be imposed on the ϕ_i in order that $\phi_i\, dx_i$ be a total differential revealed that the ϕ_i satisfy the same equations as the electromagnetic potentials: [49]

$$\frac{\partial \phi_i}{\partial x_k} - \frac{\partial \phi_k}{\partial x_i} \equiv f_{ik}, \quad \frac{\partial f_{kl}}{\partial x_i} + \frac{\partial f_{li}}{\partial x_k} + \frac{\partial f_{ik}}{\partial x_l} = 0.$$

In the general theory of relativity the g_{ij} were interpreted as (proportional to) the potentials of the gravitational field; Weyl proposed to extend the physical content of the theory by interpreting the ϕ_i as (proportional to) the four potentials of the electromagnetic field, $V, -\frac{1}{c}A$.[50] Schrödinger wrote the proportionally constant as e/γ, so that the "tract factor" became

$$e-\frac{1}{\gamma}\int\left(eV\,dt - \frac{e}{c}A_x\,dx - \frac{e}{c}A_y\,dy - \frac{e}{c}A_z\,dz\right).$$

The great difficulty with Weyl's theory had been precisely that with the identification of the ϕ_i as the electromagnetic potentials the condition that $d\phi$ be a total differential,

$$\frac{\partial \phi_i}{\partial x_k} - \frac{\partial \phi_k}{\partial x_i} = 0,$$

also required the electromagnetic fields, *E, H,* to be zero. In the presence of an electromagnetic field a rod or a clock carried about a closed path returned with a different length or keeping different time.[51] This untoward behavior had been suppressed by Weyl in the discussions in *Space-Time-Matter,* but Schrödinger now took full advantage of it in order to reproduce the quantum conditions of the hydrogen atom.

"Now," Schrödinger explained, "the property of the quantum orbits which is announced in the title and which appears so remarkable to me is that the 'genuine' quantum conditions, i.e., those which suffice to determine the energy and thus also the spectrum, are also precisely sufficient to make the exponent of the tract factor an integral

[49] *R Z M* (1921), p. 111; *S T M,* p. 124.
[50] *R Z M* (1921), pp. 256 ff.; *S T M,* pp. 282 ff.
[51] As Einstein immediately pointed out in a "Nachtrag" to Weyl's original paper: *Preuss. Akad. d. Wiss., Sitzungsber.* (1918), 478. This item is not included in the most recent Einstein bibliography: Nell Boni, *et al., A Bibliographical Checklist and Index to the Published Writings of Albert Einstein* (Patterson, New Jersey, 1960).

multiple of h/γ (which is a pure number) for all approximate periods of the system." [52] Letting it be known that he didn't really understand why this relation obtained, Schrödinger proceeded immediately to demonstrate the fact for all the one-electron motions: unperturbed elliptical orbits, Zeeman effect, Stark effect, combined parallel electric and magnetic fields, relativistic precession. For example: the "genuine"—i.e., the Bohr—quantum condition for an unperturbed ellipse is $2\tau\overline{T} = nh$, where τ is the period of the orbit and $\overline{T}$ is the time average of the kinetic energy. Since for an inverse square force the time average of the kinetic energy is half that of the potential energy,

$$\frac{nh}{2\tau} = \overline{T} = \frac{1}{2}e\overline{V} = \frac{e}{2}\cdot\frac{1}{\tau}\int_0^\tau V\,dt,$$

or,

$$nh = \int_0^\tau eV\,dt.$$

Now since the field through which the electron moves is, in this approximation, purely electrostatic, $A_x = A_y = A_z = 0$. Hence the integral in the tract factor, $\int e\phi_i\,dx_i$, reduces to $\int eV\,dt$; but the quantum conditions make this integral, extended over the period of the motion, equal to nh. Thus if the electron carried a vector along with itself about this orbit, and in such a way that the vector was always displaced parallel to itself, then each time the electron returned to its original position the tract of the vector would be multiplied by an integral power of $e^{-h/\gamma}$.

"It would be difficult to believe," Schrödinger said, "that this result is solely an accidental mathematical consequence of the quantum conditions and without deeper physical significance." [53] Yet what that significance may be, Schrödinger did not know. He conceded that it "is more than doubtful that the electron really does carry around some

---

[52] Schrödinger, *op. cit.* (note 44), 14.

[53] *Ibid.*, 22. The few lines which Scott, *Erwin Schrödinger* (1967), p. 47, devotes to the *bemerkenswerte Eigenschaft*—perhaps in consequence of Scott's desire to make the result appear familiar and suggestive—are somewhat misleading. Scott writes $\Sigma_k p_k\,dq_k$ for Schrödinger's $\Sigma_i\phi_i dx_i$, thus implicitly identifying the ϕ_i with the p_k. Scott's p_k, however, are presumably the momenta conjugate to the coordinates q_k, and are thus the sums of the mechanical momenta and the electromagnetic momenta, $-(e/c)A_k$. But Schrödinger's $\phi_1 = -A_1/c$, $\phi_2 = -A_2/c$, $\phi_3 = -A_3/c$ represent only the electromagnetic momenta, and do not refer at all to the mechanical momenta. It would hardly have been a very remarkable property of the quantum orbits that, as Scott has it, the expression $\Sigma_k p_k dq_k$ integrated around a path determined by the conditions $\oint p_k dq_k = nh$ should also turn out to be an integral multiple of h. But this was not the expression Schrödinger integrated. Neither was Schrödinger using the quantum conditions in *that*, i.e., Sommerfeld's, form. Rather, Schrödinger followed "durchweg" Bohr's procedure in which the periods rather than the degrees of freedom are quantized. Scott, in effect, asserts that Schrödinger's *bemerkenswerte Eigenschaft* was

sort of tract in its motion," and thought it very possible that instead of undergoing a parallel displacement the electron "adjusts" (*einstellt*) to the conditions at each world point, as, in Weyl's simile, a magnetic needle adjusts to a magnetic field. The significance of the *bemerkenswerte Eigenschaft* might then be, Schrödinger supposed, "that not every tempo of alignment is equally possible for the electron." [54]

Despite all the uncertainty about its true physical interpretation, the mathematical expression of this *bemerkenswerte Eigenschaft* of the electronic orbits subsists, and Schrödinger concluded his paper with a discussion of the value to be assigned the universal constant γ. There are, he emphasized, many ways to form a constant with the dimensions of an action from elementary constants, but the obvious choice, and the one towards which he leaned, is to relate γ to h. And this choice "leads one to consider if it is not conceivable that γ has the pure imaginary value $\gamma = h/2\pi\sqrt{-1}$." [55] If we rewrite—although Schrödinger did not—the expression for the tract using this value of γ,

$$l = l_0 e^{-\frac{2\pi\sqrt{-1}}{h}\int e\phi_i \, dx_i},$$

we obtain an expression which bears a striking resemblance to the familiar expression for the amplitude of a wave. The fact, then, that the

---

physically and formally equivalent to de Broglie's phase wave as applied by de Broglie to atomic orbits. Had it been, the argument of this paper would be considerably more cogent; in fact it was not.

London, "Quantenmechanische Deutung ...," *Zeits. für Phys.*, 42 (1927), 382–383, was able to show that in a relativistic treatment the *bemerkenswerte Eigenschaft* could be derived for a limited, but important, class of potentials. Namely, the quantum condition $\sum_{i=1}^{3} \oint \frac{\partial W}{\partial x_i} \, dx_i = nh$, where W is the action function, reduces the relativistic relation $\int \left(\frac{\partial W}{\partial x_i} - \frac{e}{c}\phi_i \right) dx_i = -\int m_0 c^2 \sqrt{1 - (v/c)^2} \, dt$ to $\oint \frac{e}{c}\phi_i \, dx_i = nh$ if and only if the potential energy is a homogeneous function of the x_i of degree -1.

[54] Schrödinger, *op. cit.* (note 44), 23. Weyl, *R Z M* (1921), pp. 280–281, *S T M*, pp. 307–309, hoped to avoid the untoward consequences of his theory by arguing that we cannot be certain which physical quantities actually undergo a *kongruente Verpflanzung* along a world line, and which rather undergo *Einstellung* to unspecified obtaining conditions—thus exposing himself to the charge of resorting to a "sehr faule Ausrede" (London, note 43). However, in the 5th edition of *R Z M* (Berlin, 1923), p. 301, preface dated "Herbst 1922," Weyl claimed to have shown, "dass die Übertragung einer Strecke durch kongruente Verpflanzung zufolge der Wirkungsgesetze genau so vor sich geht wie durch Einstellung auf den Krümmungsradius."

[55] *Loc. cit.* In a footnote, on p. 14, Schrödinger added that Weyl had written him that the *bemerkenswerte Eigenschaft* for the case of unperturbed Keplerian motion had been found by A. D. Fokker in 1920, and that Fokker had also been led to consider purely imaginary values of ϕ_i. Fokker does not appear to have published anything on this. (Curiously, at the time Fokker made this discovery he also was in Arosa, convalescing. Interview by Sources for History of Quantum Physics, 1 April 1963.)

quantum conditions somehow just suffice to restore this amplitude to its original value when the wave is propagated around the orbit is, as London said, strongly reminiscent of de Broglie's "resonance" interpretation of the Bohr-Sommerfeld quantum conditions.[56]

VI. THE ROUTE TO THE WAVE MECHANICS

We have shown that Schrödinger's "bemerkenswerte Eigenschaft der Quantenbahnen," published two years before de Broglie's thesis, contained certain ideas which were both conceptually and formally similar to (but not identical with) de Broglie's notion that the quantum conditions expressed the requisite phase relations for a stationary undulatory regime within the atom. This circumstance thus constitutes a presumptive answer to the question "Why Schrödinger?" It is necessary to consider, however, whether or not there is any direct evidence that the fact of having previously found this "bemerkenswerte Eigenschaft" affected (i) Schrödinger's receptivity to de Broglie, and (ii) the direction which Schrödinger took in attempting to develop de Broglie's ideas.

Klein, in his essay on the route to the wave mechanics,[57] offered an interpretation which, implicitly, runs counter to ours on both these points. On the question of receptivity, Klein emphasized Schrödinger's admission of a debt to Einstein for the stimulus to his wave mechanics.[58] Although Schrödinger's public acknowledgement is ambiguous

---

[56] London, *op. cit.* (note 43), 381–382: "The quantum theory allows matter only a discrete sequence of states of motion. One speculates that these specially distinguished motions allow the standard length to be transported only in such a way that upon returning to the starting point the phase has gone through an integral number of cycles, so that, despite the nonintegrability of the displacement of a tract, the standard length is always realized in an unambiguous way at every place. In fact one is reminded of the resonance property of the de Broglie waves, the very property by which the old Sommerfeld-Epstein quantum conditions were first so successfully reinterpreted by de Broglie. . . . If one had added, axiomatically, the unambiguity of the concept of length to Weyl's theory as a generally recognized fact of experience, then one would have been led, necessarily, to the system of discrete states of motion of the classical quantum theory and to their de Broglie waves.

"I do not wish to lose this opportunity to point out that this resonance property of Weyl's measure of the tract, which we meet here as a characteristic theorem of the wave mechanics, had been conjectured by Schrödinger already in 1922 as a 'remarkable property of the quantum orbits,' and demonstrated for a number of examples, without its significance having been recognized at that time. . . . Thus already at that time Schrödinger had in his hands the characteristic wave-mechanical periodicities which he was obliged to encounter again later from viewpoints so entirely different."

[57] M. J. Klein, *op. cit.* (note 5).

[58] Schrödinger, "Über das Verhältnis der Heisenberg-Born-Jordanschen Quantenmechanik zu der meinen," *Ann. d. Phys.*, 79 (1926), 735. "Angeregt wurde meine Theorie durch L. de Broglie, Ann. de Physique (10) 3. S. 22. 1925 (Thèses, Paris 1924) und durch kurze aber unendlich weitblickende Bemerkungen A. Einsteins, Berl. Ber. 1925. S. 9 ff."

in that it cites de Broglie and Einstein together, Klein's point is strengthened by Schrödinger's private acknowledgement in a letter to Einstein (23 April 1926):

> Your and Planck's assent are more valuable to me than that of half the world. Moreover, the whole business would certainly not yet, and perhaps never, have been developed (I mean not by me) if the importance of de Broglie's ideas hadn't been put right under my nose by your second paper on gas degeneracy.[59]

In this paper,[60] published in February 1925, Einstein had drawn attention to de Broglie's association of a wave field with a material particle in order to suggest a physical foundation for the analogy between black body radiation and his own theory of the behavior of a gas at low temperatures. Inasmuch as Schrödinger began to work with de Broglie's ideas only after they had become widely known (and a full nine months after Einstein's paper had appeared) it is difficult to argue that the "bemerkenswerte Eigenschaft" had rendered Schrödinger especially receptive to those ideas. Rather we must concede that it was, as Klein argues, in large measure through Schrödinger's involvement in the fundamental problems of quantum statistical mechanics that he came to regard de Broglie's notion of a matter-wave as a serious physical possibility.[61]

[59] K. Przibram, ed., *Briefe zur Wellenmechanik* (Vienna, 1963), p. 24. A translation of this passage is given by Jammer *C D Q M*, p. 257; the entire volume has recently been translated by Klein as *Letters on Wave Mechanics* (New York, 1967).

[60] Einstein, "Quantentheorie des einatomigen idealen Gases. 2. Abhandlung," *Preuss. Akad. d. Wiss., Sitzungsber.* (9 Feb. 1925), 3–14. On page 9 of the text of this paper Einstein shows how de Broglie's $m_0c^2 = h\nu$ necessarily leads to a wave, exposes the relation between phase and group velocity, and the fact that the group velocity of the wave is the velocity of the particle. To the footnote citing de Broglie's dissertation, Einstein adds that, "Also to be found in this dissertation is a very noteworthy geometrical interpretation of the Bohr-Sommerfeld quantization rule."

[61] It is, moreover, quite conceivable that the *bemerkenswerte Eigenschaft*, which Schrödinger in all probability had already discarded as a false trail, actually served to inoculate Schrödinger against de Broglie's ideas. Thus one might construe Schrödinger's statement of his debt to Einstein as implying, indirectly, that Schrödinger had reason to expect himself to have taken up de Broglie's ideas without having his "Nase gestossen . . . auf" their importance by Einstein's theory of gas degeneracy. If so, then it was presumably on account of his own discovery of the *bemerkenswerte Eigenschaft*.

There is an anecdote told by E. Bauer (Jammer, *C D Q M*, p. 258) that de Broglie's thesis was given to Schrödinger by Victor Henri, but returned shortly with the remark, "That's rubbish." This view was reported to Langevin, who instructed Henri to insist that Schrödinger look at the thesis again. If we seek for the message in this story—and it is our view that such anecdotes are not to be regarded as historical accounts but as historical allegories, which are remembered and repeated precisely for the sake of the moral they are meant to convey—it would appear to be that even Schrödinger, the man who was in some way fated to develop de Broglie's ideas into a wave mechanics, did not recognize these ideas as promising when first exposed to them. In this form the anecdote is probably "true," but says, of course, little more than what Schrödinger conceded to Einstein.

The second issue—the starting point and initial direction of Schrödinger's efforts to develop de Broglie's ideas—is a more critical index
of the role of the "bemerkenswerte Eigenschaft" in preparing Schrödinger's perception of the possibilities inherent in de Broglie's thesis,
and thus of the relevance of the "bemerkenswerte Eigenschaft" to the
question "Why Schrödinger?" Here again Klein has suggested that
Schrödinger first considered, and worked with, de Broglie's phase
waves in connection with the problem of Bose-Einstein statistics and gas
degeneracy.[62] He is able to support this suggestion by citing Schrödinger's testimony towards the end of the first paper on wave mechanics: "the considerations concerning the atom reported above could
have been presented as a generalization of this work on the gas model."
The reference here is to a paper Schrödinger submitted to the *Physikalische Zeitschrift* some six weeks earlier (15 December 1925) which
"takes seriously the de Broglie-Einstein wave theory of a moving particle, according to which the particle is nothing more than a kind of
foamy crest on a wave radiation," in order to account, without a
"sacrificium intellectus," for the applicability of Bose statistics to a
gas.[63] And if it were indeed the case that this work, fully within the
quantum statistics tradition, was the point of Schrödinger's entrance
into de Broglie's ideas, then it would hardly be possible to attribute a
significant role to the "bemerkenswerte Eigenschaft."

Schrödinger's statement, however, is at the very least ambiguous.
The subjunctive "hätten sich . . . darstellen lassen" suggests rather that
while the wave mechanics of the atom *could* have been represented as
an outgrowth of the paper on quantum statistics, in fact it was not.
Be that as it may, the immediate issue is not what line of attack ultimately brought Schrödinger to the goal he sought, but rather the role
of the "bemerkenswerte Eigenschaft" in the definition of that goal and
in Schrödinger's initial steps toward it. On this issue Schrödinger's
testimony is far more explicit. The paragraph which concludes with
the bow to the gas theory opens with a declaration that "I owe the
stimulation to these considerations in the first place to the ingenious
thesis of Mr. Louis de Broglie and to reflection upon the spatial distribution of those 'phase waves' of which he has shown that, measured
along the orbit, there turns out always to be an integral number in
every period or quasiperiod of the electron." [64] That this connection

---

[62] Klein, *op. cit.* (note 5), 45–46; Schrödinger, *op. cit.* (note 4), 373.
[63] Schrödinger, "Zur Einsteinschen Gastheorie," *Physikalische Zeits.,* 27 (1 Mar. 1926), 95–101.
[64] *Op. cit.* (note 4), 372–373.

with the problem of the quantum conditions for a hydrogen atom was indeed Schrödinger's point of entrance into de Broglie's ideas is confirmed by remarks in a letter to Alfred Landé on 16 November 1925.

It pleases me greatly to hear that your paper[65] is intended to be a "return to the wave theory." I also am very much inclined to do so. Recently I have been deeply involved with Louis de Broglie's ingenious thesis. It's extraordinarily stimulating but nonetheless some of it is very hard to swallow. I have vainly attempted to make myself a picture of the phase wave of an electron in an elliptical orbit. The "rays" are almost certainly neighboring Kepler ellipses of equal energy. That, however, gives horrible "caustics" or the like as the wave front. At the same time, the length of the wave ought to be equal to [that of the orbit traced out by the electron in] one Zeeman or Stark cycle![66]

But Schrödinger's testimony does not merely support our general point that it was *qua* theoretical spectroscopist that he developed de Broglie's ideas into a wave mechanics. It also suggests that what Schrödinger saw in de Broglie's thesis was strongly influenced by the "bemerkenswerte Eigenschaft." There is an unmistakable affinity between Schrödinger's earlier concern with demonstrating the "Ganzzahligkeit des Streckenexponenten" for the several quasi-periods of the hydrogen atom in magnetic and electric fields, and his subsequent desire to visualize those same conditions in terms of de Broglie's phase waves.

VII. SUMMARY AND ACKNOWLEDGEMENTS

To recapitulate: We have raised, and have attempted to answer, the question why Schrödinger, of all theoretical physicists, took up the ideas put forward in de Broglie's thesis and developed them into a wave mechanics of the atom. Firstly, we have suggested that de Broglie's ventures into theoretical spectroscopy and the theory of the periodic

---

65 Evidently, A. Landé, "Lichtquanten und Kohärenz," *Zeits. für Phys.*, *33* (18 August 1925), 571–578.
66 Schrödinger to A. Landé, 16 Nov. 1925; Sources for History of Quantum Physics microfilm nr. 4. The concluding paragraph of the letter:

> Ganz besonders freut mich ihre Mitteilung, dass Ihre Arbeit ein "Zurück zur Wellentheorie" sein sollte. Auch ich neige sehr dazu. Ich habe mich dieser Tage stark mit Louis de Broglies geistvollen Theses beschäftigt. Ist ausserordentlich anregend, hat aber doch noch sehr grosse Härten. Ich habe vergebens versucht, mir von der Phasenwelle des Elektrons auf der Keplerbahn ein Bild zu machen. Als "Strahlen" kommen doch wohl benachbarte Keplerellipsen von gleicher Energie in Betracht. Das gibt aber greuliche "Brennlinien" oder dergl. für die Wellenfläche. Anderseits soll die Welle eine Längserstreckung gleich einem Zeemann- oder Starkzyklus haben!

system of elements had rendered him and his work highly suspect in the eyes of those physicists principally concerned with these problems. Schrödinger, however, because of his detachment and his marginal commitment to theoretical spectroscopy, would have been one of the Central European theoretical spectroscopists least likely to have been severely prejudiced against de Broglie on this account. Secondly, we have suggested that the pro- and anti-Copenhagen alignment usually associated with the interpretation of quantum mechanics had already formed by 1923 over the issue of the sort of quantum mechanics to be sought for. Here again both Schrödinger and de Broglie were among the relatively small minority of active quantum theorists who were allied with Einstein on this issue. Thus Schrödinger, unlike most of his colleagues, would not have been prejudiced against de Broglie's "manner of attacking the problem of quanta." Thirdly, we have noted a substantial congruence in the range of Schrödinger's and de Broglie's interests and the types of problems on which they had worked—quantum statistics and theoretical spectroscopy, relativistic mechanics and Hamilton's analogy. Lastly, we have pointed to Schrödinger's previous work "Über eine bemerkenswerte Eigenschaft der Quantenbahnen eines einzelnen Elektrons," and its striking similarity to de Broglie's explanation of the Bohr-Sommerfeld quantum conditions. We suggest that this prior work, if not a decisive factor in Schrödinger's initial assessment of de Broglie's ideas, nonetheless determined the starting point for Schrödinger's efforts to develop them into a wave mechanics.

ACKNOWLEDGEMENTS: We are very grateful to Martin J. Klein and especially to the editor of *Historical Studies in the Physical Sciences,* Russell McCormmach, for their numerous criticisms and valuable suggestions; to William F. Eberlein and Sanford L. Segal for clarifying discussions; to Mrs. Edith London for permission to publish the letter by her husband; and to Murphy Smith, manuscripts librarian, American Philosophical Society, for ready assistance.